Happy Father's Day

Greg

The Guide to Garden Shrubs and Trees

The Guide to Garden Shrubs and Trees

(INCLUDING WOODY VINES)

Their Identity and Culture

NORMAN TAYLOR

500 species most likely to be useful to
the home gardener,
with 323 species
illustrated in color
and 192 in
black and white

BY EDUARDO SALGADO

BONANZA BOOKS · NEW YORK

51711027X

Library of Congress Catalog Card Number: 64–10266

This edition is published by Bonanza Books,
a division of Crown Publishers, Inc.
by arrangement with Houghton Mifflin Company.
b c d e f g h
Manufactured in the United States of America.

To the memory
of
the late Alfred Rehder
of the
Arnold Arboretum
whose scholarly
Manual of Cultivated Trees and Shrubs
has been my unfailing consultant
in the preparation of this book

Preface

A GARDEN without shrubs and trees can scarcely be considered one. They give grace, dignity, and permanence to any garden whether a suburban lot or a country estate. But the number of different kinds is bewildering to the amateur and their identity and culture still more so.

This book was designed to fill that void. It presents, in simple form, the descriptions and culture of the 500 species most likely to interest the average home gardener, leaving out those difficult to find or of interest only to the specialist. It also suggests their chief use in the garden, which, in these days of high labor costs, make shrubs and trees of increasing interest—for their maintenance is much less costly than that of any flower bed or border.

In other words, the book answers three significant questions: "What is that shrub or tree?" "How do I grow it?" "What can I use it for?"

N. T.

Elmwood
Princess Anne, Maryland

Contents

Introduction

THERE are over 1200 species of shrubs and trees hardy within the United States and Canada. To keep the book within reasonable bounds this huge list has been cut to the 500 species most likely to be useful to the amateur. This drastic reduction has forced the omission of the following:

1. Species too rare, too difficult to grow, or unavailable from ordinary nurseries.
2. All tropical and subtropical species.
3. All cacti, palms, bamboos, and *Yucca*.
4. All fruit plants (cherry, apple, plum, fig, blackberry, etc.).

All species included here are grown for their beauty of foliage, flowers, fruit, or habit — in other words, all are what are usually classed as ornamentals. Many of them are superb garden plants.

As to their availability, all can be purchased from nurseries, but no one nursery is likely to stock all of them. Those who want a reasonably complete list of the leading nurseries in the United States and Canada should consult *Plant Buyers Guide,* 6th edition, issued by the Massachusetts Horticultural Society in 1958. Their address is Horticultural Hall, Boston 15, Mass. They list nearly all the commonly cultivated plants and the sources of them.

Shrubs, Trees, and Vines

Some confusion exists as to what is the difference between a shrub and a tree. Trees generally have a single woody trunk and are relatively tall (20–150 feet high). Shrubs usually have several woody stems and are lower (from 6–8 inches to 10 or 15 feet high). There are, however, borderline species (*i.e.,* shrublike trees and treelike shrubs) which may be difficult to assign to either category. In such cases, which are relatively rare, one's judgment is best based on a careful appraisal of the descriptions, particularly in dealing with immature trees.

Vines scarcely need a definition. They are really shrubs that are prostrate like the wintergreen or climb to great heights like wisteria. There are over 40 vines in the book.

How To Use the Book

Shrubs and trees belong to many plant families, which in technical books are generally arranged in a definite sequence. But the differences between plant families are often obscure and sometimes baffling to the amateur because they are based upon technical characters. Hence, the ideal botanical groupings and the sequence of plant families, obligatory in technical books, has usually been abandoned here for a simpler method.

The much easier scheme used here may well cause a bit of scorn by some of my professional colleagues, but its simplicity outweighs any objections to it, for it attempts to make a frankly difficult problem as simple as possible.

All the plants in the book are hence sorted into only five main categories, easily distinguished by anyone. They are listed, with some notes on them, at p. xxi at the general Key. The plants are mostly arranged without much regard to family relationships. But for those interested in knowing what family a plant may belong to, this is always mentioned. Also, in the Systematic Tabulation (p. 427) will be found all the plant families in the book, the genera that belong to each and a page reference to each genus. And the general index, of course, includes all Latin and English names (if any).

The Descriptions

In writing a book like this, one's technical vocabulary must be all but thrown out the window. But trees and shrubs have obvious (or sometimes rather obscure) features, which the seeker must notice if he is seriously interested in identifying the plant in hand. While technical terms have thus been banished, there are some nontechnical features of plants that may puzzle the seeker. What, for instance, is the difference between a simple and a compound leaf, an alternate or opposite one, or a regular or irregular flower? Because such features are diagnostic and imperative, a list of them will be found at the Picture Glossary, which is just before the Index.

The Latin and common names of the plants have been made to conform to the fourth edition of *Taylor's Encyclopedia of Gardening* (Houghton Mifflin, Boston, 1961).

In the descriptions of the individual species a definite order is maintained throughout the book. These, and some notes on them, are as follows:

1. An introductory sentence telling whether the plant is a shrub, tree, or vine, its origin, and its chief gardening features.
2. **Size.** The extremes are given, as 6–12 feet, which means that somewhere between these figures is the average height as usually cultivated. Immature specimens will be below that average, but many trees, especially in their natural habitat, will be far higher than they usually are in cultivation. The spread of the plant is also noted, but this varies much with crowding.
3. **Leaves.** Whether they are simple or compound, opposite or alternate, have marginal teeth or none, and their shape and average size are always mentioned, as well as their value for shade or beauty, especially in the evergreens.
4. **Flowers.** The size, color, and period of bloom is always noted, and whether they are borne in clusters or are solitary. Many amateurs mistake a flower cluster for a flower, which is really only an individual component of the cluster. Also some think that a *solitary flower,* such as in magnolia, for instance, means that there are perhaps no other flowers on the plant — a wholly inaccurate assumption. Actually a solitary flower is one borne on a single stalk, never in clusters, so that some plants that have solitary flowers may still be strikingly beautiful from their profusion of bloom. Those flowers favored by bees are always noted.

5. **Fruit.** All plants that flower are followed by a fruit, whether inedible, as are most of the fruits in this book, or edible, such as apple, pear, cherry, etc., which are omitted here. In the descriptions of the ornamental plants here included it is specified whether the fruits are dry or fleshy, their color, size, and whether they persist through the winter or not, as such features are often diagnostic as well as decorative. Their attractiveness for birds is also noted.
6. **Hardy.** The notes on hardiness follow those in the fourth edition of *Taylor's Encyclopedia of Gardening,* issued in 1961. See the map and its explanation on pp. xviii–xix. Unfortunately the *Plant Hardiness Zone Map* issued by the U.S. Department of Agriculture in 1960 was not available during the revision of the *Encyclopedia,* or it would have been used there and here. Failing that, it seems best to make the *Encyclopedia* and this book agree.
7. **Varieties.** This is used as generally understood, *i.e.,* both for named horticultural varieties, such as Blue Queen, and for actual botanical varieties, such as *Tsuga canadensis pendula* (Sargent's Weeping Hemlock). That such names as Blue Queen designate what pedants call *cultivars* is a truism, but *variety* as here used has wide currency, while *cultivars* is still a bit esoteric.

   Because *varieties* often differ in habit or color from the species from which they have (reputedly) been derived, these and other features of varieties are always mentioned. Often they are much finer garden subjects than the derivative species.
8. **Culture.** This is noted wherever the plant needs other than ordinary garden soil, especially if the plant has specific requirements as to acidity, moisture, shade, etc. Fortunately most of the plants in the book are grown in a great variety of garden soils, as evidenced by their extremely wide cultivation, and hence need few special directions. Where necessary these are always given, and the chief use of the plant in the garden is emphasized. For details of planting see pp. xv–xvii.

How To Plant Shrubs and Trees

THE OPERATION is basically simple if a few essential facts are kept in mind. These are:

Time To Plant

All trees and shrubs that lose their leaves in the fall (deciduous) should only be planted sometime *after* leaf fall in autumn and *before* leaf expansion in spring. It is generally a matter of convenience, largely dictated by your local climate, whether you choose fall, winter, or spring planting. Snow cover often makes winter planting impossible, but many warm regions permit easy winter planting. The only real exceptions are plane trees and magnolias which are better planted in early spring.

All coniferous evergreen shrubs and trees, and practically all broad-leaved evergreens (box, camellia, holly, etc.) can be planted in early spring, or in late summer and early fall. Late fall planting is more hazardous, for it does not give the newly planted stock time to get established before bitter winter winds practically stop all growth. None of these evergreens should ever be planted "bare root," and any good nursery will deliver them with a ball of earth containing most of their roots tightly wrapped in bagging, commonly called balled and burlapped and in most catalogs usually abbreviated B. & B. Insist that all such stock be B. & B.

Digging the Hole

When plants arrive from the nursery, measure the size of the ball in B. & B. stock and the spread of the roots in those deciduous shrubs and trees that always come bare root, *i.e.,* packed in a little moist moss or excelsior.

Having determined the root spread, dig the hole at least one foot deeper and wider than the actual measurement of root spread, or preferably a bit more. Separate carefully all the topsoil, making a separate pile of the subsoil. You will need the topsoil to cover the roots at planting time. If you have little or no topsoil it is better to get some elsewhere, as it invites failure to plant a shrub or tree in indifferent subsoil, rocks, cinders, ashes, etc. In most ordinary gardens there will be 6–8 inches of fairly good topsoil and if this is separated carefully you are ready for planting.

Planting

While digging holes see that the nursery stock is put in the shade and out of the wind. In all bare-root stock cut off with a sharp knife or pruning shears any broken or damaged roots. You are then ready for actual planting.

Place the spread-out roots on a layer of firmly tamped topsoil at the bottom of the hole. See that there is enough of this to put the shrub or tree, when planting is completed, at the same level as it was in the nursery, *i.e.,* do not bury it or leave it higher in the hole than necessary. It is usually easy to see its old ground level from the discoloration of its bark at that point.

Begin putting topsoil on the roots a shovelful at a time, and tamp it down with a stout stick or your feet. Gradually add more soil and repeat the tamping each time until the hole is about ⅔ to ¾ full. Then, according to the size of the hole, pour in one or two buckets of water. This will give the roots needed moisture and eliminate small air pockets that your tamping may not have corrected. Air pockets in the soil are useless to the roots.

When the water has completely settled, you can fill up the hole to within an inch or two of the general ground level, using topsoil if you have it or subsoil. There are practically no roots at or above this level and the quality of the soil is less important. The shallow "well" that remains after you have completed the job is useful for catching rainfall for the first year or two, but it can then be filled up to the general ground level.

Most shrubs need no staking, but all trees over 4 feet high should have a stout stake set in the hole before planting to which the tree must be fastened by raffia or strong tape (no wire). This keeps the trunk rigid, preventing the shifting of the roots before they get thoroughly anchored in the soil. Some, in order to avoid staking, use three guy wires put in a piece of old hose around the trunk and tightly fastened to three stakes driven in the ground. Such a tree cannot move no matter what the velocity or direction of the wind. This, too, can be removed in a year or two.

For all B. & B. stock the directions above should be followed. Do not remove the bagging from the ball of earth, but just before filling in the soil cut the string, or pull out the nails that sometimes are used to pin the bagging tighter around the ball of soil. Never break that ball, and just before filling the hole slit the bagging in three or four places. It will rot away and do no damage, and its function is to hold the ball of soil so tight that neither sun nor wind will strike the roots within it.

To Prune or Not

It is unnecessary, and often harmful, to prune coniferous evergreens or broad-leaved evergreens (box, camellia, holly, etc.). But it is essential to cut back the twigs or stems of all bare-root nursery stock at planting time. This should be done with pruning shears, reducing the upper branches and twigs from ¼ to ⅓ of their arrival height. This will lessen the number of leaves the next growing season and thus reduce by that much the water requirements of the shrub or tree until the shock of planting is over.

Final Caution

If, for any reason, planting must be delayed more than a day or two after the arrival of nursery stock, it is essential to "heel them in." This involves digging a trench about 18 inches deep, placing the nursery stock in it, tipped to about a 45° angle and covering *all* the roots with soil. Such trenched plants can safely be left for several days or even weeks, in the autumn, before planting. It is hazardous to leave heeled-in stock too long in the spring, as growth and leaf expansion may overtake you if a warm spell develops. This, of course, would defeat the object of always planting bare-root stock when it is completely dormant.

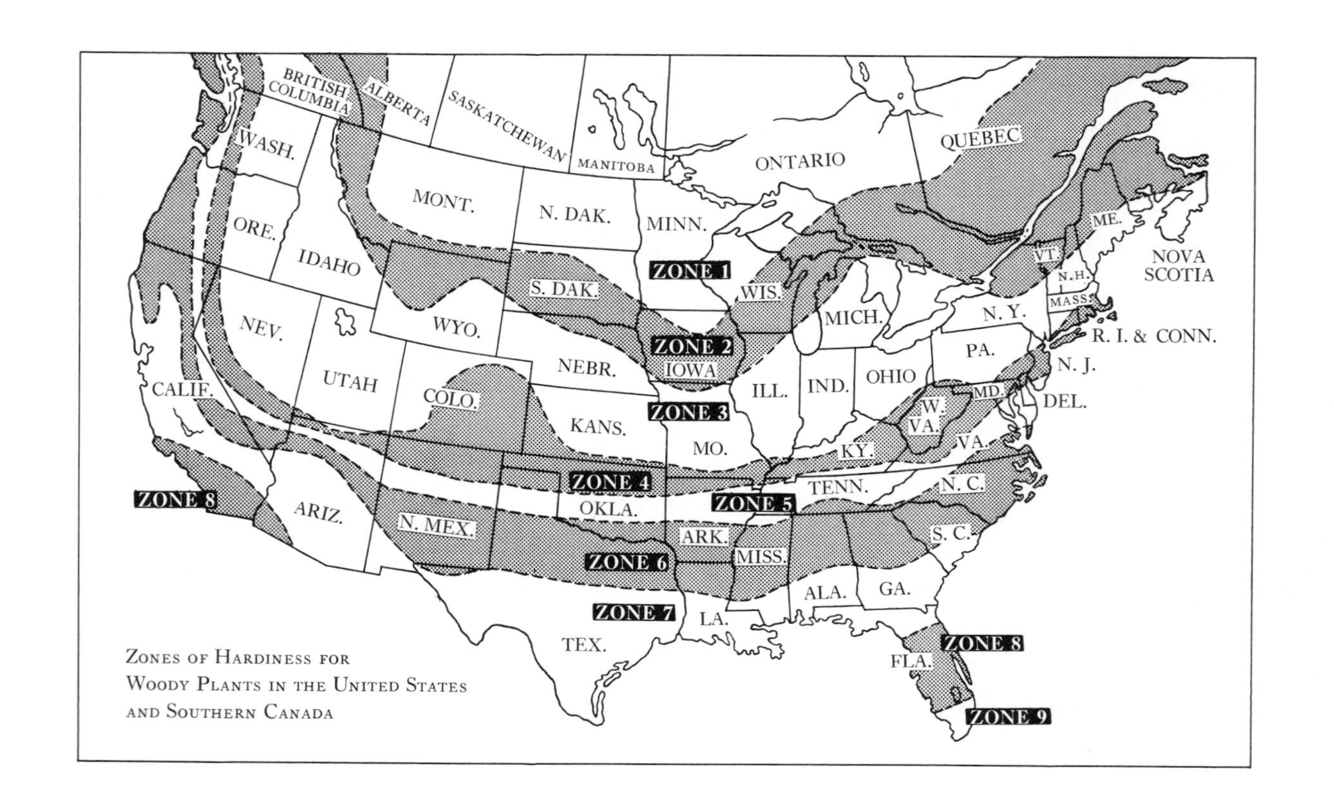

Zones of Hardiness for Woody Plants in the United States and Southern Canada

Hardiness Zone Map

THE MAP on the opposite page divides the United States and Canada into nine zones of hardiness. It is based on weather bureau figures kept for many years. The critical factor in the hardiness of all woody plants is the average minimum temperature of the coldest month. These are as follows:

Zone 1. Zero or below	Zone 4. 20°–25°
Zone 2. Zero to 10°	Zone 5. 25°–30°
Zone 3. 10°–20°	Zone 6. 30°–40°

Zones 7, 8, and 9 are omitted from this book because they contain mostly tropical or subtropical plants, the inclusion of which would have doubled the size of the book. It should be understood, however, that quite a few subtropical shrubs and trees creep northward into zone 6 or even protected parts of zone 5. All such are included in the book.

It should also be recognized that local factors may modify zone limits, especially proximity to the sea or elevations above 800–1000 feet. In other words the Hardiness Zone Map is a guide to the gardener rather than an expression of inflexible facts. Some cool-region shrubs and trees need a prolonged winter chilling in order to flourish. They do poorly or not at all in warm regions, their "hardiness" being dictated not by winter cold but by too much winter heat. All such are noted where they occur and their culture in warm regions is always hazardous.

Explanation of Key

A KEY is designed to lead the seeker to the identity of a plant, without, for the moment, concerning himself with any other item in the book.

In woody plants like shrubs or trees, the obvious characters used to separate whole groups are whether or not they are coniferous evergreens (*i.e.,* cone-bearing plants) or drop their leaves in the autumn. Hence our first subdivision is:

Coniferous evergreen shrubs and trees
as against
Trees and shrubs not coniferous evergreens.

In the second category one will be faced again with two alternatives:

Vines (*i.e., clambering or prostrate plants*)
as against
Erect trees or shrubs (*i.e., not clambering or prostrate*).

And so on to the end of the Key. Each time one is asked to choose between only two alternatives, and in all the larger groups these are defined at the pages cited. The smaller keys throughout the book, the descriptions, pictures, and index should make the final determination of the identity of the plant in hand as easy as it can be.

Key to Garden Shrubs and Trees

Note: *There are only five main categories, all of them numbered near the left-hand margin. Under each there are many subdivisions, unnumbered but indented, to show contrasting characters. At all the cited pages there will be found an elaboration of that section of the key, the exceptions to it and other data. These should always be consulted before going further in the main key.*

1. Coniferous evergreen trees and shrubs, mostly, but not always, cone-bearing and with needlelike or scale-like leaves — 1

   All the rest of the trees and shrubs bear no cones and usually have expanded leaves, most of which drop off in the fall. They are included in Sections 2–5.

2. Vines, trailing or erect — 48
   - Erect and often high-climbing — 55
   - Prostrate or trailing — 48
3. Erect trees or shrubs, the flowers of which have **no petals** — 78

   Here are many shade trees and shrubs grown for foliage and habit such as oak, elm, beech, birch, etc. They are subdivided as follows:
   - Flowers mostly in hanging catkins
     - Leaves compound (Walnut family) — 79
     - Leaves simple — 84
   - Flowers not in catkins (Elm family) — 116
4. Trees and shrubs the flowers of which have obvious, **separate petals** — 123

NOTE: *Most of the terms used in this key are defined and illustrated in the Picture Glossary, p. 426.*

The Guide to Garden Shrubs and Trees

NOTE

All main *plant names* are cross-referenced to the appropriate illustrations.

All *illustrations* are cross-referenced to the plants to which they belong.

I. Coniferous Evergreens

THERE could scarcely be a more misleading title, for some of the plants below are not evergreen, and some of them do not bear cones. But "coniferous evergreens" is such a widely used popular term, in spite of its inaccuracy, that it may be surprising to hear that it coincides precisely with the strictly botanical term *Gymnospermae, i.e.,* trees and shrubs which are usually evergreen and bear cones but no flowers in the garden sense of that term.

Everyone knows that a Christmas tree is an evergreen and that pine trees always bear the familiar pine cones. So do many other "coniferous evergreens," but the *Ginkgo* drops its leaves in the fall and has no cones, while the yew is truly evergreen and bears what looks like a red berry. These confusing exceptions to the concept of coniferous evergreens are so numerous that the identification of individual species is never easy and may be impossible without mature leaves (often small and scalelike) from a mature twig, and cones or other types of fruit if the tree does not bear cones. Trying to identify a tree from immature specimens and those snipped off a young twig leads only to frustration or failure.

To say that coniferous evergreens bear no flowers in the garden sense of *flower* is perfectly correct, but it is equally correct to say that technically they bear highly important flowers. This apparent paradox is explained by the botanical term *Gymnospermae, i.e.,* plants having a completely naked ovule. This is strictly true of all the coniferous evergreens, but difficult to see in so many of them that the details of ovules, stamens, etc., and how they function are best left to more technical books than this one.

Yet it is upon these technical, but popularly incomprehensible, characters that all coniferous evergreens are divided into 3 families (so far as the plants below are concerned). Even the distinctions among these 3 families present so much difficulty for the amateur that it seems best to separate 5 genera that, instead of being evergreen, drop their leaves in the fall (for which one must wait for certain identification). All of them bear cones except the ginkgo, which bears an edible seed in a fleshy, evil-smelling fruit. These 5 genera are easily distinguished thus:

Leaves broad and fanlike; fruit fleshy. Ginkgo
Leaves narrow, needlelike; fruit always a cone.
  Leaves clustered in small bunches.
    Leaves about 1/8 inch wide and 1½–3 inches long. Golden Larch
    Leaves narrower and shorter. Larch
  Leaves not clustered.
    Leaves about ½ inch long; cones nearly globular. Bald Cypress
    Leaves about ¾ inch long; cones not globular, about ¾ inch long. Metasequoia

GINKGO (*Ginkgo biloba*). Family Ginkgoaceae p. 22
An interesting survivor of an ancient group of trees, never found in its wild state. Probably from China, where it was perpetuated as a temple tree. It is rather gawky in youth, ultimately a fine round-headed shade tree for lawn or avenue planting. **Size:** 100–125 feet high, about 20–25 feet wide. **Leaves:** Fan-shaped, wedge-shaped at the base, 2½–3½ inches wide, turning a beautiful soft yellow before autumn leaf-fall. **Flowers:** Negligible. **Fruit:** Fleshy, yellow, and of such foul odor that only male (*i.e.,* non-fruiting) trees should be planted. **Hardy:** Throughout our range. **Varieties:** *aurea* has yellow leaves in youth; *fastigiata* has upward-pointing branches; *pendula* has hanging branches; *variegata* has yellow-blotched leaves. **Culture:** Easy in all garden soils, preferably in full sun.

GOLDEN LARCH (*Pseudolarix amabilis*)
Family Pinaceae p. 22
An ornamental "evergreen" that drops its leaves after they have turned a golden yellow in autumn; a fine lawn tree. **Size:** 60–100

feet high, 20–30 feet wide. **Leaves:** Linelike, very narrow, 1½–2½ inches long, borne in decided clusters. **Flowers:** Negligible. **Fruit:** An egg-shaped, reddish-brown cone, 2½–3½ inches long, its woody scales notched at the tip. **Hardy:** Throughout our range. **Varieties:** None. **Culture:** Easy in most garden soils, but avoid those in a limestone area. Plant in full sun and avoid windy sites.

LARCH (*Larix decidua*). Family Pinaceae p. 22
Valuable European "evergreen" tree that drops its clustered needles in the autumn; a fine lawn tree, but of irregular branching in age. **Size:** 90–120 feet high, about 35 foot spread. **Leaves:** Needlelike, about 1 inch long, borne in clusters, the midrib prominent on the under side, the foliage rather thin and sparse. **Flowers:** Negligible. **Fruit:** An oval cone about 1–1½ inches long, its 40–50 rounded scales downy on the back. **Hardy:** From Zone 2 southward. **Varieties:** None. **Culture:** Plant in a moist soil with good drainage, preferably in a mildly acid site.

Related plants (not illustrated):

American Larch (*Larix laricina*). 40–60 feet. Bog tree.

Japanese Larch (*Larix leptolepis*). 70–90 feet. Bark scaly and leaving handsome red scars.

Dahurian Larch (*Larix gmelini*). 70–90 feet, its branches wide-spreading.

BALD CYPRESS (*Taxodium distichum*)
Family Pinaceae p. 22
The southern cypress with very durable wood, inhabiting swamps in the southeastern states, but can be grown as a lawn specimen. **Size:** 90–150 feet high (less as cultivated) and 20–35 feet wide. **Leaves:** Foliage graceful and feathery, the needles about ½ inch long, becoming orange-yellow before autumn leaf-fall, not borne in clusters. **Flowers:** Negligible. **Fruit:** A small nearly globular cone, not over 1 inch long. **Hardy:** From Zone 5 (doubtfully in 4) southward. **Varieties:** None. **Culture:** Preferably in a moist or wet soil, in full sun, but many specimens in Del. and Md. are thriving in ordinary garden soil. Keep away from a limestone area.

METASEQUOIA (*Metasequoia glyptostroboides*)
Family Pinaceae p. 22

The fastest-growing "evergreen" tree in the eastern states, some specimens becoming 40 feet high in 12 years. Originally known only as a fossil, in 1946 living trees were discovered in China and all our specimens are derived from imported plants. **Size:** 100–150 feet high, spread unknown. **Leaves:** Narrow needles, arranged opposite each other and falling in the autumn, about ¾ inch long. **Flowers:** Negligible. **Fruit:** A small oval cone, about ¾ inch long. **Hardy:** From Zone 4 southward. **Varieties:** None. **Culture:** Thrives in a variety of soils. It should be planted in full sun. It has already acquired the doubtful vernaculars of "dawn redwood" (it is not a redwood) and "living fossil," which is nonsense. Actually it is the most interesting tree to be found living, when everyone supposed it to be only a fossil ancestor of *Sequoia*.

*

THE REST of the coniferous evergreens are truly evergreen, but some of them do not bear obvious cones. The cones of pines, spruces, and firs are readily recognized, but the word *cone* still requires a definition. In pine cones there are numerous woody scales, between which are the ripened, usually winged seeds, if the cone is mature.

Such cones are obvious enough, but some of the coniferous evergreens bear only slightly conelike fruit and others bear only dry or fleshy berrylike fruit. All of these non-cone-bearing shrubs and trees are here segregated from the true cone-bearing evergreens for convenience and ease of identification.

There are thus 3 different groups of evergreens, separated by the type of fruit they bear:

Fruit an obvious cone, usually with woody scales. Pine Family. p. 14
Fruit not an obvious cone.
- *Fruit dry and berrylike.* Junipers. p. 9
- *Fruit fleshy* . . . Yew Family comprising the 4 genera below.

YEW FAMILY
(*Taxaceae*)

VERY beautiful evergreen shrubs and trees containing not only the familiar English Yew and its relatives, but the less-known plum-yews, also *Torreya* and *Podocarpus.* Leaves alternate, needlelike or wider. Male and female flowers on separate plants, followed, in the female, by an apparently fleshy fruit, actually a dry seed covered by a bony shell, which is invested with a usually brightly colored covering (an aril). Of about 10 genera and many species only 4 are here considered:

Leaves pinkish when unfolding, ½ inch wide and 3–4 inches long. — *Podocarpus*
Leaves always green, narrower and shorter.
- *Fruit plum-like.*
  - *Leaves with raised midrib on upper side.* — Plum-Yew
  - *Leaves without an obvious midrib above.* — *Torreya*
- *Fruit berrylike, scarlet.* — Yews

PODOCARPUS (*Podocarpus macrophyllus*) p. 6
A fine evergreen in its Chinese home, but here of rather indifferent growth, except in Calif. and the Gulf Coast. Recommended only as a specimen plant in protected places northward. **Size:** 40–60 feet high (less as cultivated), about 15–20 feet wide. **Leaves:** Thick, leathery, evergreen, lance-shaped, 3–4 inches long, dark green above, paler beneath, pinkish as they unfold. **Flowers:** Negligible. **Fruit:** Fleshy, greenish purple, egg-shaped, about ½ inch long, its fleshy stalk purple. **Hardy:** Not reliable north of Zone 6. **Varieties:** *Maki,* the maki of the Japanese, is a shrubby form, less hardy than the tree type. **Culture:** Easily grown in any ordinary soil, but it will not stand intense or prolonged cold or bitter winds.

PLUM-YEW (*Cephalotaxus drupacea*) p. 6
A beautiful evergreen tree (usually shrubby as cultivated) with dark green foliage, and a native of Japan. Well suited to the

lawn as a specimen plant, but casting no shade. **Size:** 20–30 feet high in maturity, but usually a round-headed shrubby plant as cultivated and rarely over 4–6 feet high. **Leaves:** Needlelike, 1–2 inches long, very numerous, and with 2 bluish-gray bands on the under side and a prominent midrib above. **Flowers:** Negligible. **Fruit:** Stalked, plum-like, fleshy and green, about 1 inch long. **Hardy:** From Zone 4 southward, but unhappy in regions of intense summer heat and in dry regions, hence thriving best near the seacoast. **Varieties:** Many, the most useful being *pedunculata,* with longer leaves, which is sometimes offered as *Cephalotaxus harringtoniana;* and *fastigiata,* with upright columnar habit. **Culture:** Suited to a variety of soils and thriving in most ordinary gardens, but resenting hot, dry winds. Hence put in a sheltered place, protected from prevailing summer winds.

Related plant (not illustrated):

Cephalotaxus fortunei. 20–30 feet high, the leaves 2–4 inches long, the fruit purple.

Left: Torreya (*Torreya nucifera*), p. 7. Center: Plum-Yew (*Cephalotaxus drupacea*), p. 5. Right: Podocarpus (*Podocarpus macrophyllus*), p. 5.

TORREYA (*Torreya nucifera*) p. 6
A Japanese evergreen tree with fissured bark and graceful branches arranged in tiers, but in cultivation only a small tree or large shrub. **Size:** 40–75 feet high at maturity, much less as cultivated. **Leaves:** Lance-shaped, prickle-pointed and stiff, ¾–1¼ inches long, dark green and shining above, 2-banded beneath. **Flowers:** Negligible. **Fruit:** Plum-like, practically stalkless, oblongish, about 1¼ inches long, green and faintly streaked with purple, ripening in the second season. **Hardy:** From Zone 3 southward. **Varieties:** None. **Culture:** Easy in most garden soils, but preferring cool, moist sites to hot and dry ones. Not suited to the prairie states. Plant in full sun.

Related plant (not illustrated):

Stinking cedar (*Torreya taxifolia*). A Florida evergreen, reliably hardy only from Zone 6 southward. Fruit egg-shaped, purple, about 1¼ inches long.

YEW
(*Taxus*)

The most widely planted and deservedly popular evergreens in the family Taxaceae, all of them with a poisonous juice. They comprise magnificent evergreen trees, like the English Yew, and many varieties of other yews suitable for accent plants and for hedges. Bark reddish brown on mature trees. Leaves small, needlelike, with 2 bands on the under side, not resinous or aromatic as in so many evergreens. Male and female flowers on separate plants, both rather inconspicuous. Fruit berrylike, brilliantly scarlet. There are many horticultural forms of yew and their identification is not easy.

For the species below, the following key may help to distinguish the most used plants.

Needles gradually tapering to the tip. English Yew
Needles abruptly tapering to the tip. Japanese Yew

The only other widely cultivated yew is *Taxus media,* a hybrid between the English and Japanese yews, resembling each of the parents in some features, and of chief interest for its horticultural varieties.

ENGLISH YEW (*Taxus baccata*) p. 22
A magnificent evergreen tree in England, but difficult to grow in the eastern states as it does not like summer heat or drying winds. But there are stately specimens in some old gardens in Va., Md., and Del., mostly because they were planted in protected sites. **Size:** 40–60 feet high, nearly as wide in maturity (rarely in the U.S.). **Leaves:** Needlelike, 1–1¼ inches long, gradually tapering to the tip, so numerous that the tree casts a dense shade. **Flowers:** Negligible. **Fruit:** Berrylike, scarlet, about ½ inch long, the seed inside poisonous, but the red coat eaten by some. **Hardy:** From Zone 4 southward, but not happy in most sites, and unsuited to city or suburban gardens, *i.e.,* it needs freedom from dust, smoke, wind, and factory fumes. **Varieties:** Generally more useful than the typical form: *adpressa,* a low shrub or small tree, easily sheared and useful for hedges; *repandens,* a nearly prostrate shrub and the hardiest of all the varieties of English yew; *stricta,* the Irish Yew, a columnar, erect tree, densely branched, forming a bushy outline. Not quite so hardy as the typical form: *washingtoni,* a wide-spreading shrubby form with golden-yellow foliage. **Culture:** The English Yew should not be attempted unless the conditions are as outlined under Hardy (above), but the varieties can be grown in a variety of soils. Their need for coolness and moisture is far more important than the soil type in which they are growing. All of them are unsuited to the prairie states.

JAPANESE YEW (*Taxus cuspidata*) p. 9
One of the most valuable evergreens ever imported from Japan; the only yew fit for suburban gardens and it even tolerates some city sites. Many experts think it is only a Far Eastern form of the English Yew, from which it differs mostly in rather technical features, and in its far greater tolerance of unfavorable conditions. **Size:** 20–40 feet high, nearly as wide, but these dimensions are rare in cultivation as the plant is mostly shrubby. **Leaves:** Very numerous, needlelike, about 1 inch long, suddenly tapering to a short dark-green tip. **Flowers:** Negligible. **Fruit:** Berrylike, scarlet, about ½ inch long. **Hardy:** From Zone 3, and in protected

Japanese Yew
(*Taxus cuspidata*), p. 8.

places in Zone 2, southward. **Varieties:** *nana,* a fine shrubby form 4–6 feet high; *densa,* a compact form, rarely over 3 feet high. **Culture:** Much more easy than for the English Yew. This valuable evergreen grows in a variety of soils and tolerates conditions unfavorable for most evergreens, although it thrives best in cool, moist sites.

Related plants (not illustrated):

Taxus media, a hybrid between the English and Japanese yews and chiefly valuable for its varieties: *hatfieldi,* shrubby and conical; *hicksi,* an upright dense shrub, standing shearing and one of the best for evergreen hedges; *Kelsey,* with more spreading branches.

Ground Hemlock (*Taxus canadensis*). A low native evergreen, nearly prostrate and useful as a ground cover in cool northern areas.

JUNIPER
(*Juniperus*)
Family Pinaceae

A single genus, the junipers are distinguished from all other coniferous evergreens by having small, dry, berrylike fruits. They comprise the familiar Red Cedar and many valuable garden shrubs and trees, nearly always aromatic, one unpleasantly so. Leaves either needlelike and sharp, or scalelike and hugging the twigs. Sometimes both sorts of leaves are on the same tree; in other cases a single species will have all needlelike leaves on one specimen and all scalelike leaves on another. Hence leaf characters are difficult to appraise and may lead to much confusion. The leaves are either opposite (often in the scalelike ones) or in threes in the needlelike types. Without the dry, berrylike fruit these plants can easily be mistaken for several other genera, notably *Libocedrus, Cupressus, Chamaecyparis,* and young specimens of the Big-tree (*Sequoia gigantea*).

The junipers are about the easiest of the coniferous evergreens to cultivate, for many of them, particularly those from our Southwest, grow in dry, wind-swept places. Their cultivation in gardens presents almost no problem, but they prefer light sandy soils to heavy clay ones. Some, like our Red Cedar, are spontaneous along fence rows from the excrement of birds who eat the fruit.

The ones here treated may be distinguished thus:

Leaves all needlelike and spreading, not scalelike.
- *Erect trees or shrubs.* — Common Juniper
- *Procumbent shrub, the leaves bluish-green beneath.* — *Juniperus procumbens*

Leaves on mature plants scalelike, but sometimes needlelike on young twigs.
- *Scalelike leaves blunt; fruit brown.* — Chinese Juniper
- *Scalelike leaves, sharp-pointed.*
  - *Trees with pleasantly aromatic foliage.* — Red Cedar
  - *Shrub with unpleasantly scented foliage.* — Savin

COMMON JUNIPER (*Juniperus communis*) p. 22

A cosmopolitan evergreen tree or shrub, found in North America and in Eurasia, and widely cultivated for ornament and for its aromatic fruits, which are used to flavor gin (juniper berries). **Size:** Usually 6–12 feet high, but sometimes up to 40 feet. **Leaves:** Needlelike and spreading, about ⅝ inch long, with a broad white band above. **Flowers:** Negligible. **Fruit:** Dry, berrylike, pea-size, bluish black and with a bloom. **Hardy:** From Zone 1 southward. **Varieties:** Many, and among the best for the small garden are *depressa,* a low-growing form, with wide-spreading branches, rarely over 3–4 feet high, and valuable as it grows well in dry, rocky places; *hibernica,* the Irish Juniper, a narrow, upright, columnar form, with dark green leaves; *oblonga-pendula,* a columnar form, with the tips of the branches hanging; *suecia,* the Swedish Juniper, a columnar form with bluish-green leaves, the tips of the branches hanging. **Culture:** See JUNIPER above.

Juniperus procumbens, below.

JUNIPERUS PROCUMBENS above

A low, prostrate Japanese juniper often planted as a ground cover, the ascending branches rarely over 20–30 inches high. **Size:** Essentially prostrate, but spreading 6–8 feet wide. **Leaves:** In threes, not over ½ inch long, spiny-tipped, bluish green above, paler beneath and with 2 white spots near the base. **Flowers:** Negligible. **Fruit:** Nearly globular, about ¾ inch in diameter, bluish. **Hardy:** From Zone 2 southward. **Varieties.** None. **Culture:** See JUNIPER above.

Related plants (not illustrated):

Juniperus squamata. A low, reclining Chinese shrub with needlelike leaves and purple-black fruit.

Creeping Juniper (*Juniperus horizontalis*). A prostrate North American shrub, with both needlelike and scale-like leaves, valuable for dry, sandy or rocky places.

CHINESE JUNIPER (*Juniperus chinensis*) below
A variable Asiatic evergreen, sometimes treelike but more generally shrubby, especially in some of its horticultural forms which are in wide use in the garden. **Size:** As a tree, 40–60 feet (but rarely here), as a shrub, see varieties below. **Leaves:** Prevailingly scalelike and blunt-tipped, but needlelike, sharp-pointed leaves are also common and have two white lines above. **Flowers:** Negligible. **Fruit:** Pea-size and purplish brown. **Hardy:** From Zone 2 southward. **Varieties:** Many, of which the best are better than the typical tree form; among the most useful, all shrubby, are *variegata,* with twigs white-tipped; *aurea,* an upright form, 3–5 feet high, the twigs golden; *pyramidalis,* a narrow, upright but pyramidal form, the leaves bluish green; *pfitzeriana,* usually called Pfitzer's Juniper, a partly upright form, its spreading branches making a bird's-nest-like hollow in the center of the plant that may be 4–6 feet high and 5–8 feet wide. One of the most popular evergreens as it will stand city conditions. **Culture:** See JUNIPER above.

Pfitzer's Juniper (*Juniperus chinensis pfitzeriana*), above.

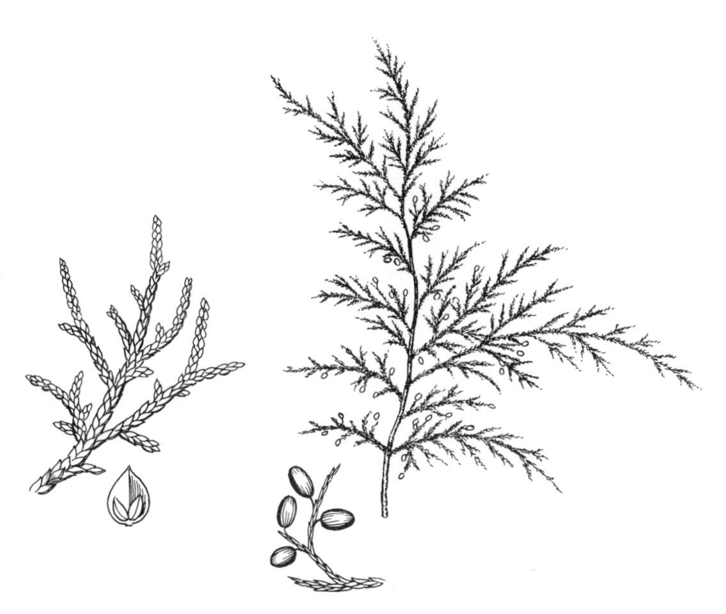

Red Cedar (*Juniperus virginiana*), below. At the left the mature form, at the right the juvenile form, which may be permanent.

RED CEDAR (*Juniperus virginiana*) above
This native evergreen tree was called "cedar" by the English colonists because they wrongly confused it with the true cedar (*Cedrus*). Its pleasantly aromatic wood (chests and pencils), fragrant foliage, and its almost perfect candle-flame outline make it an ever popular lawn or avenue tree (in the country). **Size:** 40–80 feet high, its red-brown bark peeling off in long strips. **Leaves:** Of two kinds, prevailingly scalelike and pointed, but needlelike leaves, especially on young twigs (sometimes on the whole plant), are common. They are sharp-pointed, and about ⅜ inch long. **Flowers:** Negligible. **Fruit:** Pea-size, bluish, and with a bloom. **Hardy:** Everywhere. **Varieties:** *elegantissima,* with tips of the twigs golden; *glauca,* a form with bluish-gray foliage; *keteleeri,* a compact upright form with only scalelike leaves; *tripartita,* a low shrubby form with spreading branches

and mostly needlelike leaves. **Culture:** Will grow almost everywhere, and on the author's farm frequently covered with brackish tidewater without injury.

Related plant (not illustrated):

Juniperus scopulorum. A western representative of our Red Cedar.

SAVIN (*Juniperus Sabina*) p. 17
A low, spreading Eurasian evergreen with unpleasantly scented foliage, its branches ascending. **Size:** Mostly 4–5 feet high (rarely 10 feet) and about as wide. **Leaves:** Of two kinds, needlelike, concave, and bluish green above, and scalelike on mature twigs. **Fruit:** About ¼ inch in diameter, brown, and covered with a bluish bloom. **Hardy:** From Zone 3 southward. **Varieties:** *cupressifolia,* a low, almost prostrate form with generally only scalelike leaves; *tamariscifolia,* a low form with generally needlelike leaves. **Culture:** It does not do well in acid soils, its wild home being in prevailingly limestone areas.

PINE FAMILY
(*Pinaceae*)

The most valuable evergreens in the garden world are found in the Pine Family, which comprises about 33 genera, over 250 species, and innumerable horticultural forms. Besides the true pines, the family includes all the firs and spruces, the true cedars, the hemlock, arborvitae, and others treated below. As garden subjects for the temperate zone, no trees and shrubs exceed the Pine Family, although their identification is difficult without cones and mature foliage. For those plants in this family that drop their leaves in fall see p. 2. For the junipers, which also belong to this family, but bear dry, berrylike fruits instead of cones, see p. 9. Leaves needlelike, or scalelike and hugging the twigs. Fruit a true cone, small and with few scales in some forms, but usually with many obvious woody scales, between which are the usually winged seeds (seen only in mature cones). For ease of identification the family may be divided thus:

Leaves needlelike, never hugging the twigs. See p. 20
Leaves scalelike, always hugging the twigs (an exception is found in young foliage of some horticultural forms of the arborvitae and of Chamaecyparis).
Cones more or less globular.
Cone scales with many seeds. Cypress
Cone scales with only 2 seeds. *Chamaecyparis*
Cones egg-shaped or oblongish.
Cone scales with 3–5 seeds. False Arborvitae
Cone scales with only 2 seeds.
Cone scales in 6–8 pairs. Arborvitae
Pairs of cone scales only 4. Incense Cedar

CYPRESS (*Cupressus sempervirens*) p. 22
A lofty evergreen tree, the cypress of history and the poets, especially in the variety noted below, which is the familiar Italian Cypress of the Mediterranean Coast, but unfortunately not reliably hardy in any of the area covered by this book. **Size:** 40–70 feet (but only in Calif. and on the Gulf Coast). **Leaves:** Scalelike, very small, hugging the twigs, which are so numerous as to make the tree densely foliaged. **Flowers:** Negligible. **Fruit:** A nearly globular cone, almost 1½ inches in diameter and composed of 6–12 rather sharp-pointed scales between which are many seeds. **Hardy:** Only precariously in the coastal part of Zone 6 and southward. **Varieties:** *horizontalis,* a form with branches that spread out in a flat plane; *stricta,* the familiar columnar, narrow cypress of Italian, French, and Spanish gardens, superb for avenue planting and in formal gardens, but really suited only to Calif. and the Gulf Coast. **Culture:** A sandy loam is the preferred soil for the cypress, but much more important are its climatic requirements noted above.

Related plants (not illustrated):

Arizona Cypress (*Cupressus arizonica*). 30–40 feet, foliage bluish green; stands wind and exposure better than the Italian Cypress, but not so fine a tree.

Shasta Cypress (*Cupressus macnabiana*). A bushy tree 10–35 feet high, with dark green foliage; the hardiest of all the cypresses.

Monterey Cypress (*Cupressus macrocarpa*). 50–75 feet high, the foliage green. Most suited to southern Calif.

CHAMAECYPARIS
(*Chamaecyparis*)

In nature tall, stately evergreen trees, not much grown, but as garden subjects the horticultural shrubby forms are widely used for lawn specimens and in foundation plantings. The leaves in mature trees are always scalelike and hug the twigs, but in most of the shrubby forms the leaves are needlelike but not sharp-pointed as in so many of the junipers, so that the foliage is soft rather than prickly. Flowers minute and of no garden interest. Cones more or less globular, but of slight decorative value.

These shrubs and trees have dense foliage and are among the most popular of all evergreens in spite of the fact that their cultural requirements are rather demanding. In addition to the hardiness notes below, their culture requires a somewhat acid soil, freedom from strong winds (summer or winter), and they do better near the coast than inland. They cannot be recommended for the prairie states. The plants are often called False Cypress and *Retinospora.* The two commonest species in cultivation are:

Leaves tightly hugging the twigs. Hinoki Cypress
Leaves more loosely hugging the twigs. Sawara Cypress

HINOKI CYPRESS (*Chamaecyparis obtusa*) p. 22
In Japan, but rarely here, this is a stately evergreen tree, but as a garden plant here, usually shrubby as outlined below. **Size:** 70–120 feet high in nature, much lower and shrubby as usually cultivated. **Leaves:** Scalelike, very small, and tightly hugging the twigs in mature trees, but softly needlelike in most garden forms. **Flowers:** Negligible. **Fruit:** Small and inconspicuous cones, with only 2 seeds at each scale. **Hardy:** From Zone 3 southward. **Varieties:** Innumerable, but cultivated mostly, and very widely, in the following forms, all shrubby: *aurea,* foliage golden yellow;

albo-spicata, young foliage white-tipped; *gracilis,* a compact and pyramidal form, also in a golden-foliaged form; *erecta,* a columnar, shrubby form; *crippsi,* a narrow-pyramidal form with golden foliage; *compacta,* a low, broadly conical form; *pygmaea,* a very dwarf form with almost prostrate branches. All of these varieties are best selected from nurseries where the labeling is reasonably correct as the confusion in naming varieties is considerable. **Culture:** See note above.

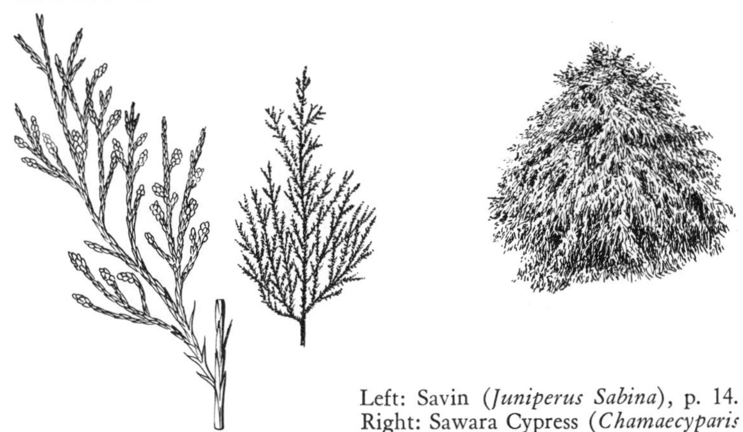

Left: Savin (*Juniperus Sabina*), p. 14.
Right: Sawara Cypress (*Chamaecyparis pisifera*), below.

SAWARA CYPRESS (*Chamaecyparis pisifera*) above

A Japanese tree, there up to 150 feet high, but much less as cultivated here, usually only in the shrubby forms. **Size:** 70–150 feet high in maturity, more often shrubby here. **Leaves:** Scalelike, whitish beneath, but not hugging the twigs so tightly as in Hinoki Cypress, and the twigs in a flattish, fanlike arrangement, the foliage hence feathery. **Flowers:** Negligible. **Fruit:** A small, inconspicuous cone of no decorative value. **Hardy:** From Zone 3 southward. **Varieties:** *plumosa,* a very popular form with dense conical outline, and feathery, frondlike foliage; *filifera,* a pyramidal form with the tips of the branches drooping; *squarrosa,* a dense bush or small tree, feathery, but the twigs not flattened or frondlike. **Culture:** See above.

Related species (not illustrated):

Port Orford Cedar (*Chamaecyparis lawsoniana*). A magnificent evergreen tree of Wash. and Ore., but not reliable in the East, except in coastal regions from Zone 4 southward, and precarious there.

Alaska Cedar (*Chamaecyparis nootkatensis*). Taller than the last and of similar climatic restrictions.

Southern White Cedar (*Chamaecyparis thyoides*). An evergreen bog tree of the Atlantic Coast, not much cultivated and less attractive than the Sawara or Hinoki cypresses.

FALSE ARBORVITAE (*Thujopsis dolabrata*) p. 23
A Japanese evergreen tree, often known as Hiba Arborvitae, but as grown here mostly shrubby. It differs from the true arborvitae only in the rather technical features noted in the key on p. 15. **Size:** 30–50 feet in Japan, here shrubby. **Leaves:** Scalelike, small, hugging the twigs, which are arranged in flattish, fanlike clusters. **Flowers:** Negligible. **Fruit:** An egg-shaped cone, hardly ½ inch long, the scales with 3–5 seeds each. **Hardy:** From Zone 4 southward, but only in regions of adequate summer rainfall; not for the prairie states. **Varieties:** Mostly cultivated in a dwarf, shrubby form known as *nana;* also *variegata* with white-tipped twigs. **Culture:** Same as for the true arborvitae; see below.

Arborvitae
(*Thuja*)

Evergreen trees of giant dimensions in the forest, but as cultivated here containing the most widely planted evergreen shrubs, in innumerable varieties. The leaves are small, scalelike, and very tightly hug the twigs, which are arranged in flat, fanlike sprays or fronds, the foliage hence dense and easily sheared. (In some horticultural or juvenile forms these are spreading, needlelike leaves.) Some of the varieties make fine hedge plants. Flowers and cones of little garden interest, as the great value of the group is its handsome evergreen foliage. It is distinguished from its close relatives only by the rather technical features in the key on p. 15.

Only 2 species are of outstanding garden importance, one from eastern Asia, the other American. There are perhaps 100 varieties of these and their identification is difficult and the names for them inclined to be ephemeral. The only safe way to select from the lists below is to visit a well-labeled nursery or arboretum.

The culture of arborvitae is relatively easy if it is restricted to moist or even swampy soils. It does not tolerate hot, drying summer winds, although winter cold is of little significance, as our American species grows wild as far north as Hudson's Bay, and the Asiatic one is equally hardy. Neither species is fit for the prairie states. The 2 chief cultivated species are:

Cone scales thin, the seeds winged; foliage fronds horizontal. American Arborvitae
Cone scales thick, the seeds without wings; foliage fronds vertical. Oriental Arborvitae

AMERICAN ARBORVITAE (*Thuja occidentalis*) p. 23
A medium-sized tree in nature, but as usually cultivated overwhelmingly shrubby, and in these forms perhaps the most widely grown evergreen in America. **Size:** 30–60 feet in the wild, much lower as a garden plant. **Leaves:** Small, scalelike, always hugging the twigs in mature specimens, but often needlelike and spreading in some horticultural forms. **Flowers:** Negligible. **Fruit:** A small cone, with 6–8 pairs of scales. **Hardy:** From Zone 1 southward, but see above for moisture requirements. **Varieties:** Over 65 and far too confusing to enumerate here. For the 15 best see *Taylor's Encyclopedia of Gardening,* 4th edition, p. 1208. There is every variety of outline and foliage in these, and all the columnar forms make very fine hedges as they stand shearing (once a year in July) very well. **Culture:** See above.

ORIENTAL ARBORVITAE (*Thuja orientalis*) p. 24
A rather low, bushy tree from eastern Asia, and often branching from the base. This form is not much grown, being replaced in popular favor by the varieties noted below. **Size:** 20–40 feet high (as a tree), more usually a shrub 4–10 feet high and as wide. **Leaves:** Scalelike, hugging the twigs, which form a frondlike,

vertical cluster of bright green foliage. **Flowers:** Negligible. **Fruit:** An egg-shaped cone with rather thick scales, the seeds within the scale wingless. **Hardy:** From Zone 4 southward, but precariously so if there is great summer heat and lack of moisture. **Varieties:** *atrovirens,* the foliage dark green; *aurea,* foliage yellow; *fastigiata,* a columnar form with erect branches; *pendula,* a form with drooping branches. These are all medium-sized bushes and stand shearing well. **Culture:** See above.

Related plants (not illustrated):

Western Red Cedar (*Thuja plicata*). A Pacific Coast tree 150–200 feet high, unsuited to regions of erratic summer rainfall.

Thuja standishi. A pyramidal Japanese evergreen, fine as a lawn specimen and hardy from Zone 4 southward.

INCENSE CEDAR (*Libocedrus decurrens*) p. 23

A beautiful Pacific Coast evergreen tree with reddish-brown, rather scaly bark, not so widely grown as it should be. **Size:** 50–100 feet high, about 30 feet wide in maturity. **Leaves:** Small, scale-like, tightly hugging the twigs except at the tip. **Flowers:** Negligible. **Fruit:** An oval or oblong cone ¾–1 inch long. **Hardy:** From Zone 4 southward. **Varieties:** None. **Culture:** A sandy loam in full sun suits it best, but it is nearly as tolerant of different soil types as our native Red Cedar, so that its culture is easy. A fine evergreen with a candle-flame-like outline, but not for the prairie states.

Leaves Needlelike, Never Hugging the Twigs

This group comprises all the rest of the Pine Family and cannot be mistaken for those in which the leaves are small, scalelike, and hug the twigs. All, of course, are true evergreens and all bear obvious cones, which are large in the pines but much smaller in the hemlock and a few other genera. In many cases identification is impossible without cones and mature foliage.

Leaves opposite or alternate, but never in clusters. See p. 32
Leaves in obvious clusters (except in a few scalelike leaves on shoots of Sciadopitys).
Leaves 15–35 in each cluster. Umbrella Pine
Leaves 7–12 in each cluster. True Cedars
Leaves 5 or less in each cluster. The Pines

UMBRELLA PINE (*Sciadopitys vertillicata*) p. 23
A magnificent Japanese evergreen, never so tall here as there, but a notable addition to any garden because of its unique dark green and lustrous foliage. **Size:** 80–120 feet (in Japan), less than one fourth this as cultivated here. **Leaves:** Of two kinds, a few inconspicuous ones scalelike and hugging the twigs. The others are linelike, in conspicuous clusters of 15–35, thick, furrowed both sides, nearly 6 inches long and soft-textured. The leaf clusters are very numerous so that the plant is profusely foliaged. **Flowers:** Negligible. **Fruit:** A stalkless, woody cone 3–5 inches long. **Hardy:** From Zone 4 southward. **Varieties:** None. **Culture:** A slow-growing tree best suited to a rich loam, in full sun, but it will not tolerate regions of high winds and poor rainfall, hence is unsuited to the prairie states.

TRUE CEDARS
(*Cedrus*)

Evergreen trees, all Eurasian, containing the famous Cedar-of-Lebanon, the timber of which was used to make the temple at Jerusalem. Only 3 species are in cultivation, but they are very popular for their fine foliage and spacious branching in favorable gardens. Leaves scattered, but most of them in clusters of 7–12. Flowers negligible. Fruit an erect cone, its scales tightly pressed, the seeds between them broadly winged.

The cedars are striking evergreens with usually a spread greater than their height, the lower branches sweeping the ground. One Cedar-of-Lebanon in the Bishop's garden in Tours, France, is 80 feet high and 125 feet wide. They thus need open places with plenty of room to spread. Any ordinary garden soil will do, but it should not be too moist. In England and Oregon the cedars

GINKGO FAMILY (*Ginkgoaceae*), 3
YEW FAMILY (*Taxaceae*), 5
PINE FAMILY (*Pinaceae*), 1, 2, 4, 6–9

1. **Golden Larch** (*Pseudolarix amabilis*) p. 2
   A beautiful "evergreen" that drops its golden-yellow needles in the fall; a fine lawn tree.

2. **Common Juniper** (*Juniperus communis*) p. 10
   A shrubby evergreen, usually not over 6–12 feet high, its needles rather prickly and white-banded.

3. **Ginkgo** (*Ginkgo biloba*) p. 2
   An "evergreen" tree that drops its fan-shaped leaves in the fall. The best evergreen for village streets.

4. **Bald Cypress** (*Taxodium distichum*) p. 3
   Another "evergreen" tree that drops its needles in the fall. A native of the southeastern states.

5. **English Yew** (*Taxus baccata*) p. 8
   The yew of legend and history, but not usually happy under cultivation in the eastern states.

6. **Larch** (*Larix decidua*) p. 3
   A fine European "evergreen" for the lawn; it drops its clustered needles in the fall. Hardy everywhere.

7. **Hinoki Cypress** (*Chamaecyparis obtusa*) p. 16
   As universally cultivated a fine shrubby evergreen with soft foliage; treelike in Japan.

8. **Cypress** (*Cupressus sempervirens stricta*) p. 15
   This magnificent columnar evergreen is better suited to California than to eastern gardens.

9. **Metasequoia** (*Metasequoia glyptostroboides*) p. 4
   A fossil cousin of the California Big-tree, discovered as a living tree in China and a fast grower.

1
2
3
4
5
6
7
8
9

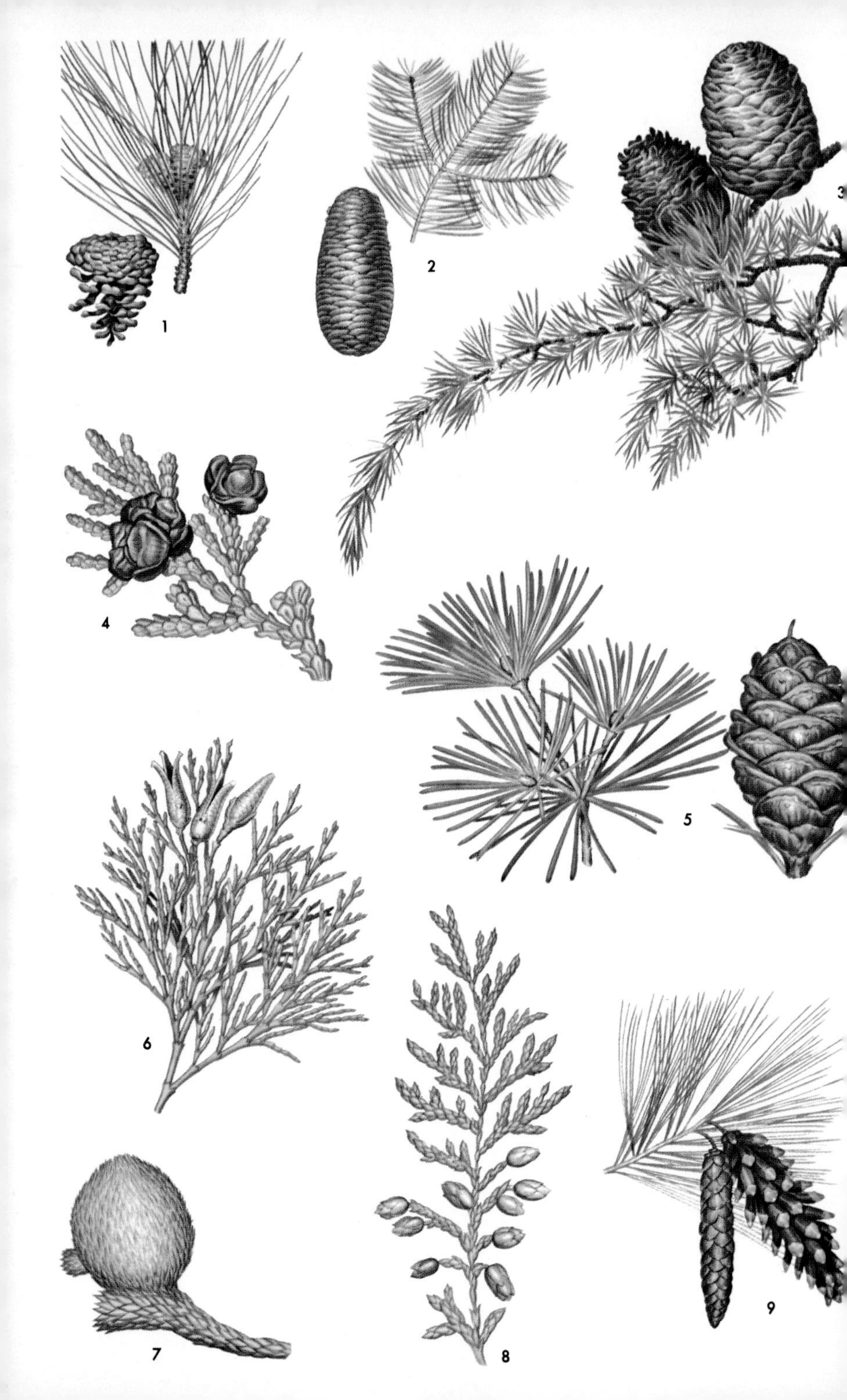

1
2
3
4
5
6
7
8
9

PINE FAMILY (*Pinaceae*), 1–9

1. **Austrian Pine** (*Pinus nigra*) p. 31
   Probably the most satisfactory of European pines for eastern gardens; even tolerating village streets.

2. **White Fir** (*Abies concolor*) p. 34
   A fine western evergreen tree, more tolerant of unfavorable sites than most firs; bluish-green foliage.

3. **Deodar** (*Cedrus Deodara*) p. 25
   A Himalayan evergreen with bluish-green needles; not hardy in the North, but thriving from Va. southward.

4. **False Arborvitae** (*Thujopsis dolobrata*) p. 18
   Shrubby as usually grown here but a tree in Japan. Foliage green, dense, in fanlike clusters.

5. **Umbrella Pine** (*Sciadopitys verticillata*) p. 21
   Lustrous dark-green foliage is a beautiful feature of this Japanese evergreen tree, which is hardy here.

6. **Incense Cedar** (*Libocedrus decurrens*) p. 20
   A tall evergreen from the Sierra Nevada, and quite hardy in most of the eastern states.

7. **Monkey-Puzzle** (*Araucaria araucana*) p. 32
   A grotesque Chilean evergreen thriving in the Pacific Northwest, but precarious in the East.

8. **American Arborvitae** (*Thuja occidentalis*) p. 19
   An eastern evergreen much better known in its many shrubby forms than as a tall tree.

9. **White Pine** (*Pinus Strobus*) p. 27
   The finest timber and ornamental pine of the eastern states; needles 5 and bluish-green.

TRUE CEDARS cont'd:
thrive better than in the eastern United States, but there are fine specimens in eastern gardens, especially along the coast. Not recommended for the prairie states.

Our three species are best distinguished thus:

Twigs upright or slightly nodding, not drooping at the tip.
Twigs densely short-hairy. Atlas Cedar
Twigs essentially smooth. Cedar-of-Lebanon
Tips of the twigs always drooping. Deodar

Left: Atlas Cedar (*Cedrus atlantica*), below. Center: Oriental Arborvitae (*Thuja orientalis*), p. 19. Right: Cedar-of-Lebanon (*Cedrus libani*), p. 25.

ATLAS CEDAR (*Cedrus atlantica*) above
A bluish-green evergreen tree from the Atlas Mountains in North Africa, widely spreading in age. **Size:** 40–100 feet high, nearly as wide or even wider, not usually attaining these dimensions in the East. **Leaves:** About 1 inch long, very narrow, stiff, and arranged in clusters. **Flowers:** Negligible. **Fruit:** An erect

brownish cone, 1–3 inches long, never produced in young specimens. **Hardy:** From Zone 5 southward, preferably along the coast, precariously hardy in protected places in Zone 4. **Varieties:** *glauca,* a beautiful form with grayish-blue foliage; *argentea,* foliage silvery, a very vigorous form. **Culture:** See above.

CEDAR-OF-LEBANON (*Cedrus libani*) p. 24
A tree famous since the time of Homer in Syria and Asia Minor and a magnificent evergreen often grown in the East, where it needs much space. **Size:** 60–100 feet high, as wide or wider in maturity. **Leaves:** About 1 inch long, dark green, arranged in clusters. **Flowers:** Negligible. **Fruit:** An erect brown cone, 2–4 inches long, only produced after 40–60 years. **Hardy:** From Zone 4 southward. **Varieties:** Comte de Dijon, a compact, dwarf form; *aurea,* a form with yellow foliage. **Culture:** See above, but the Cedar-of-Lebanon is more likely to do well in cultivation than the Atlas Cedar or the Deodar. Some authorities consider the two latter as mere forms of the Cedar-of-Lebanon, but from the garden standpoint the three species are quite distinct.

DEODAR (*Cedrus Deodara*) p. 23
A beautiful Himalayan evergreen, its leading shoot and most of its twigs drooping at the tip. **Size:** 100–150 feet high (in the Himalayas), nearly or quite as wide, its lower branches sweeping the ground. **Leaves:** Soft, nearly 2 inches long, bluish green. **Flowers:** Negligible. **Fruit:** A reddish-brown cone, 3–5 inches long, never produced on young specimens. **Hardy:** From Zone 6 southward, often doing well in protected places in Zone 5. **Varieties:** Several, but none of them are so desirable as the typical Deodar, which is an extremely graceful evergreen. **Culture:** See above.

The Pines
(*Pinus*)

Over 27 species of pines were admitted to the fourth edition of *Taylor's Encyclopedia of Gardening* because they are known to be in cultivation in the United States. Only 8 can be included here, and these comprise most of those likely to interest the average amateur. Eight others are given incidental mention.

Most pines are majestic trees, their branches usually arranged in tiers, but a few horticultural forms are shrubby, as in the Swiss Mountain Pine. They have needlelike leaves, found in clusters of 5, 3 or 2 (in those below), and at the base of the needles there is usually a small papery sheath which surrounds the leaf cluster. Male flowers produce much pollen (often a yellow nuisance near buildings), while the female flowers ultimately result in the familiar pine cone. This is composed of numerous woody scales, some prickle-tipped, and between the scales are the winged seeds, which are edible in the Swiss Stone Pine.

The cultivation of pines is relatively easy, for they are far more tolerant of heat, dryness, and poor soils than the spruces or firs. Some of them, like the Scotch Pine, will tolerate quite unfavorable sites. Generally a sandy loam suits them and almost no pines thrive if the soil is too wet. In regions of extreme summer heat and desiccating winds a north slope is better than a southern one.

The identification of the 8 species below is best understood by counting the number of needles in each cluster. The White Pine and its relatives have 5 needles in a cluster, the Longleaf Pine and its relatives have 3 needles, while the Austrian Pine has only 2 needles in each cluster. These statements do not apply to the many native pines which are of little or no garden significance, such as the Loblolly and the Pitch Pine. The only cultivated pines included here are shown in the key below:

Needles 5 in a cluster.

Young twigs not hairy, or only faintly so; needles bluish green, 3–5 inches long. White Pine

Young twigs decidedly hairy; needles dark green or blackish green. Swiss Stone Pine

Needles 3 in a cluster.

Needles slender, drooping, 8–18 inches long. Longleaf Pine

Needles stiff and straight, 5–11 inches long. Western Yellow Pine

Needles 2 in a cluster.
Spring shoots with several clusters of branchlets. Jack Pine
Spring shoots with only one cluster of branchlets.
A shrub, rarely a tree. Swiss Mountain Pine
Trees.
Needles bluish green. Scotch Pine
Needles not bluish green. Austrian Pine

WHITE PINE (*Pinus Strobus*) p. 23
A tall, stately evergreen of eastern North America and as a timber tree the initial source of the wealth of New England. As a cultivated tree few evergreens exceed it for the beauty of its foliage and its tierlike arrangement of branches. **Size:** 90–150 feet high, about a third as wide. **Leaves:** The bluish-green, soft needles are minutely rough to the touch, 4–5 inches long, and 5 in each cluster. **Flowers:** Negligible, but the pollen of its male flowers very copious. **Fruit:** The familiar pine cone which is cylindric, 4–7 inches long, the tips of its scales never prickly. **Hardy:** From Zone 2 southward. **Varieties:** Several, but the one of chief interest to those whose property is too small for the typical tree is the variety *fastigiata.* It is much narrower and does not spread so widely. **Culture:** See above.

SWISS STONE PINE (*Pinus Cembra*) p. 28
A rather slow-growing European tree, its twigs decidedly hairy, valuable for its dense, almost blackish-green foliage, and close pyramidal outline. **Size:** 40–75 feet high, nearly half as broad. **Leaves:** 5 in each sheath, 4–5 inches long, very dark green, minutely rough to the touch on the edges. **Flowers:** Negligible. **Fruit:** A more or less egg-shaped cone, about 3½ inches long, the scales never prickle-tipped, the rather large seeds edible and nutritious. **Hardy:** From Zone 3 southward. **Varieties:** None of any importance. **Culture:** See above.

Related species (not illustrated, all with 5 needles in each cluster):

Macedonian Pine (*Pinus Peuce*). A slow-growing, pyramidal evergreen, the needles minutely roughened to the touch. Cones yellowish, cylindric, 5–8 inches long. Balkans.

Himalayan Pine (*Pinus griffithi*). A tree 70–150 feet high, its needles minutely roughened to the touch, 5–8 inches long and drooping. Cones hanging, cylindric, resinous, 7–12 inches long.

Limber Pine (*Pinus flexilis*). A pine of the western U.S., not over 70 feet high, its needles stiff, dark green, 2–3 inches long. Cones egg-shaped, 4–6 inches long.

LONGLEAF PINE (*Pinus palustris*) below

A superb lawn tree, but only in southern regions, and also a valuable timber tree. **Size:** 90–120 feet high, the trunk with light orange-brown bark. **Leaves:** 3 needles to each cluster, which are

Left: Western Yellow Pine (*Pinus ponderosa*), p. 29.
Center: Swiss Stone Pine (*Pinus Cembra*), p. 27.
Right: Longleaf Pine (*Pinus palustris*), above.

showy, dark green, 15–25 inches long in young specimens, somewhat less in mature trees. **Flowers:** Negligible, except for the copious pollen from the male flowers. **Fruit:** A cylindric cone, 6–8 inches long, the scales with a recurved prickle at the tip. **Hardy:** Only from Zone 6 southward. **Varieties:** None. **Culture:** See above.

WESTERN YELLOW PINE (*Pinus ponderosa*) p. 28
A magnificent evergreen of western North America, with long, showy, dark green needles. **Size:** 80–150 feet high or even more, about one third as wide, the twigs fragrant when broken. **Leaves:** 3 needles to each cluster, 8–11 inches long, rather sharp. **Fruit:** An oblongish cone, nearly 6 inches long, the scales with a recurved prickle at the tip. **Hardy:** From Zone 4 southward, but more planted in the West than along the coast. **Varieties:** The Rocky Mountain Yellow Pine, the variety *scopulorum,* is a lower tree and hardy up to Zone 3. **Culture:** See above.

Related plants (not illustrated):

Jeffrey Pine (*Pinus jeffreyi*). A Pacific Coast evergreen, up to 190 feet high, its needles 3 in each cluster, 5–8 inches long. Cones 9–12 inches long. Hardy from Zone 4 southward.

JACK PINE (*Pinus banksiana*) p. 31
A shrubby, rather picturesque North American tree with straggling branches, most useful in dry, sandy places, especially the dunes along the Great Lakes (not on seacoast dunes). **Size:** Never over 60 feet high and inclined to low branches, hence almost shrublike. **Leaves:** 2 needles in each cluster, not over 2 inches long, stiff and twisted, the foliage rather sparse. **Fruit:** A cone, not over 2 inches long, oblongish, the scales minutely, or not at all, spiny-tipped. **Hardy:** From Zone 2 southward. **Varieties:** None. **Culture:** See above, but, with the Scotch Pine, the most useful of all the pines in unfavorable sites.

Related plants (not illustrated):

Aleppo Pine (*Pinus halepensis*). A European pine, not over 60 feet high, its light green needles 3–4 inches long, mostly 2, rarely 3 in each cluster. Cones egg-shaped or conical, about 3 inches long. Hardy only from Zone 6 southward.

SWISS MOUNTAIN PINE (*Pinus Mugo*) p. 31
An extremely popular European alpine pine as it furnishes varieties that are low, nearly prostrate shrubs, most suitable for massed plantings. **Size:** Usually shrubby, rarely a tree up to 25 feet high. **Leaves:** Very numerous, 2 in each cluster, bright green, 1½–2 inches long. **Fruit:** An egg-shaped cone, about 2 inches long, brownish yellow. **Hardy:** From Zone 2 southward. **Varieties:** Several, of which the two most important are *Mughus,* a prostrate shrub, widely and deservedly planted for a low, evergreen massed planting; and *compacta,* a low, shrubby, nearly globular shrub, useful as an accent (at the top or bottom of steps, etc.). **Culture:** See above.

SCOTCH PINE (*Pinus sylvestris*) p. 31
This Eurasian tree and the Jack Pine are the most useful of all the pines in wind-swept places with poor soil. Unfortunately it makes a rather straggly canopy that does not compare with better pines. **Size:** 40–75 feet high and about one third as wide. **Leaves:** 2 stiff, twisted, bluish-green needles in each cluster, about 2½ inches long. **Flowers:** Negligible. **Fruit:** An egg-shaped or conical cone, about 2 inches long. **Hardy:** From Zone 2 southward. **Varieties:** Several, but none of them better than the typical form. **Culture:** See above, but it will stand more neglect and exposure to wind than most pines.

Related plants (not illustrated):

Japanese Red Pine (*Pinus densiflora*). A round-headed evergreen up to 90 feet high. Needles 3–5 inches long, bluish green, 2 in each cluster. Cones oblongish, about 2 inches long. Hardy from Zone 3 southward.

Norway or Red Pine (*Pinus resinosa*). A North American quick-growing evergreen, 60–100 feet high. Needles bright green, very numerous, 4–6 inches long, 2 in each cluster. Cones egg-shaped or conical, about 2 inches long, the scales spineless.

AUSTRIAN PINE (*Pinus nigra*) p. 23
Perhaps the most widely planted of all the pines for its ease of culture, dense, dark green foliage and reasonably quick growth into a tall pyramidal tree. **Size:** 40–90 feet high, about one third as wide. **Leaves:** The deep green needles are 4–6½ inches long, 2 in each cluster. **Flowers:** Negligible. **Fruit:** An egg-shaped or conical cone, about 3 inches long, nearly or quite stalkless. **Hardy:** From Zone 3 southward. **Varieties:** There are several, but most of them are of little value compared to the typical form, the exception being the variety *calabrica,* the Corsican Pine, with a narrower crown than the typical species. **Culture:** See above, but the Austrian Pine will stand city conditions better than any other pine.

Related plant (not illustrated):

Japanese Black Pine (*Pinus thunbergi*). A close relative of the Austrian Pine, considered a mere form of it by some, but useful as it stands seacoast exposures very well.

Left: Jack Pine (*Pinus banksiana*), p. 29. Center: Swiss Mountain Pine (*Pinus Mugo*), p. 30. Right: Scotch Pine (*Pinus sylvestris*), p. 30.

Leaves Opposite or Alternate, Never in Clusters

Cones not erect; hanging or lateral. See p. 36

Cones erect

Leaves nearly ¾ inch wide, prickle-tipped. Monkey-Puzzle

Leaves narrower, often sharp-pointed but not prickle-tipped. Fir

MONKEY-PUZZLE (*Araucaria araucana*) p. 23
A grotesque Chilean evergreen tree, its main branches in tiers of 5, the branches horizontal, but tipping upward or downward at the tip and ultimately so inextricably mixed up that a monkey could scarcely climb the tree. **Size:** 40–100 feet high, about 30 feet wide, but much less as usually cultivated. **Leaves:** Leathery, stiff, prickle-pointed, about 2 inches long and ¾ inch wide, persisting for years, and greatly enhancing the grotesque aspect of the tree. **Flowers:** Negligible. **Fruit:** An egg-shaped cone, 6–8 inches in diameter. **Hardy:** Only along the coast from Zone 5 southward, and precariously so there. **Varieties:** None. **Culture:** Needs a rich loam, but far more important is its preference for cool, moist sites, *i.e.,* Wash. and Ore. An interesting curiosity in the East, but doubtfully hardy.

FIR
(*Abies*)

Impressive evergreen trees of which nearly 30 species are in cultivation in the United States. Of these the following are the most popular, together with a few others somewhat less known.

Firs usually have undivided trunks, a spirelike outline, smooth bark, and generally aromatic foliage. Leaves flattish, needlelike, usually with two whitish bands on the lower surface, not borne on a short spurlike cushion (as in the spruces) and hence leaving a smoothish twig at leaf-fall (rough in the spruce and hemlock). Cones erect, cylindrical, or more or less narrowly egg-shaped, the scales woody and sometimes colored. The firs are the best for Christmas trees as their needlelike leaves cling on even when withered, while in the spruces and hemlock the needles fall rapidly, while still green, if the room is warm.

As most firs are inhabitants of cool, mountain sites, they resent the lowlands and do poorly on the coastal plain. Their need for rich soil is negligible; a light, sandy loam is desirable and good drainage imperative. They will hence do poorly, or fail altogether, if there is standing water at their roots, and none of them will tolerate city conditions (except the White Fir) or the dry, windy prairie states. Two of the best for the somewhat unfavorable conditions of the coastal plain are the White Fir (*Abies concolor*) and the variety *glauca* of *Abies nobilis,* the typical form of which is unsuited to the East. Firs cast a dense shade and little or nothing will grow beneath them.

The firs here considered may be separated thus:

Leaves green above, paler beneath.
- *Buds distinctly resinous.*
  - *Foliage somewhat comblike.* — Giant Fir
  - *Foliage not comblike.* — *Abies veitchi*
- *Buds not resinous.* — Nordmann Fir

Leaves bluish green or grayish green both sides.
- *Bark gray; leaves 1½–2 inches long.* — White Fir
- *Bark reddish brown; leaves 1–1½ inches long.* — *Abies nobilis*

GIANT FIR (*Abies grandis*) p. 35

A mammoth Pacific Coast evergreen, scarcely suited to the East, but worth trying in New England and perhaps in the Maritime Provinces. **Size:** 175–300 feet high, and a third as broad, but never reaching these dimensions in cultivation. **Leaves:** Nearly 2½ inches long, rounded and notched at the tip. **Flowers:** Negligible. **Fruit:** An erect, greenish cone, 3–4 inches long. **Hardy:** See above. **Varieties:** None. **Culture:** See above.

Related plant (not illustrated):

Balsam Fir (*Abies balsamea*). A poor relation of most firs, scarcely 70 feet high, very difficult or impossible to cultivate, but furnishing the best of all Christmas trees in the Eastern States.

ABIES VEITCHI p. 35

A widely cultivated Japanese fir, without a common name, and very valuable as one of the easiest to grow. It is a medium-sized evergreen with smooth, grayish bark. **Size:** 40–70 feet high, about a third as wide, often less as cultivated. **Leaves:** About 1 inch long, blunt and notched at the tip, usually pointing forward or upward on the twig. **Flowers:** Negligible. **Fruit:** A bluish-purple erect cone, 1½–2½ inches long. **Hardy:** From Zone 2 southward, but not advised for the coastal plain. **Varieties:** None. **Culture:** See above, but it is one of the most satisfactory for the amateur.

NORDMANN FIR (*Abies nordmanniana*) p. 35

A fine evergreen from Asia Minor, the Caucasus, and Greece, with a densely branched canopy and dark green foliage. **Size:** 75–125 feet high, about one fourth as wide, but rarely reaching these dimensions as cultivated. **Leaves:** 1–1½ inches long, dark green, pointing upward or forward, rounded and notched at the tip. **Flowers:** Negligible. **Fruit:** An erect reddish-brown cone, 4–6 inches long. **Hardy:** From Zone 3 southward, but doing poorly on the coastal plain. **Varieties:** None. **Culture:** See above.

WHITE FIR (*Abies concolor*) p. 23

A western evergreen, often called the Colorado Fir, and the only one that stands much chance of growing in cities, if they are reasonably free of smoke and fumes. It has grayish bark, which is fissured and scaly when old. **Size:** 75–125 feet high, about one third as wide, usually less as cultivated. **Leaves:** Bluish green, nearly 2 inches long, not notched at the tip. **Flowers:** Negligible. **Fruit:** An erect greenish-purple cone, 3–5 inches long. **Hardy:** From Zone 3 southward. **Varieties:** One with bluish-gray foliage, another with still more bluish-gray foliage, neither of them superior to the typical White Fir. **Culture:** See above, and it is one of the best for the average grower.

Related plant (not illustrated):

Spanish Fir (*Abies Pinsapo*). From 40–75 feet high, the needles grayish green, about ¾ inch long, the erect cone brownish purple, about 4 inches long. Hardy from Zone 5 southward.

ABIES NOBILIS below

A majestic Pacific Coast evergreen, generally unsuited to the East, except for its variety *glauca,* which is widely and successfully grown there. **Size:** 150–250 feet high, with an immense spread, rarely or never reaching these dimensions as cultivated. **Leaves:** 1–1½ inches long, bluish green, scarcely or not at all notched at the tip, some of them upward-pointing. **Flowers:** Negligible. **Fruit:** A purplish-brown cone, 6–9 inches long, rarely produced

Left: *Abies nobilis,* above. Center: Giant Fir (*Abies grandis*), p. 33. Top right: Nordmann Fir (*Abies nordmanniana*), p. 34. Bottom right: *Abies veitchi,* p. 34.

in the East. **Hardy:** In its typical form generally unsuited to the East, but specimens are known up to Zone 3. **Varieties:** The variety *glauca* has grayish-blue leaves and is one of the most suitable firs for the East. **Culture:** See above. *Abies nobilis* is called by some *Abies procera,* but not here.

Related plant (not illustrated):

Abies magnifica, a Pacific Coast evergreen, 150–200 feet high, its leaves not notched at the tip, the cones cylindric, violet-purple, 6–9 inches long. Hardy from Zone 4, but not for the coastal plain.

Cones Not Erect; Hanging or Lateral

Leaves 1½–2½ inches long. China Fir

Leaves shorter.

Twigs smooth after leaf-fall. See p. 45

Twigs rough after leaf-fall (due to the leaves being inserted on a small, cushionlike base, which persists after leaf-fall).

Leaves flat, with 2 white lines on the under side. Hemlock

Leaves 4-sided, or at least not flat (but slightly flattened in the Serbian Spruce). Spruce

CHINA FIR (*Cunninghamia lanceolata*) p. 38

A medium-sized Chinese evergreen with reddish inner bark, the outer bark brownish, its branches spreading and drooping at the tip. **Size:** 40–75 feet high, about half as wide. **Leaves:** Crowded, narrow, about 2 inches long, broad at the base, but sharply pointed at the tip, green above and with 2 white bands beneath, minutely roughened on the edges. **Flowers:** Negligible. **Cones:** Nearly round, about 2 inches long, persisting after the shedding of its seeds. **Hardy:** From Zone 5 southward. **Varieties:** None. **Culture:** The soil matters less than a sheltered, wind-free site. If the plant is winter-killed, cut back to the stump and new shoots will make a bushy plant.

HEMLOCK
(*Tsuga*)

Evergreen trees, closely related to the firs and spruces, and often difficult to distinguish from either, especially some forms of the spruce. As in the spruces, the leaves are borne on a cushionlike base from which they are readily separable; thus the bare twigs are rough to the touch. Unlike the spruces the leaves are never sharp-pointed, and their soft tips are minutely notched in all those below except our common northern hemlock. All the leaves have two distinctly whitish bands on the lower surface. Cones hanging, about 1 inch long or less.

The hemlocks are at home in cool coniferous forests and, while tolerant of many soils, will not grow in regions of great summer heat, nor will they stand deficiency of rainfall. They are hence not suited to the prairie states and will not tolerate city smoke, dust, or drying winds.

Leaves not minutely notched at the tip. Common Hemlock
Leaves minutely notched at the tip.
  Leaves ⅓–½ inch long; twigs reddish brown. Japanese Hemlock
  Leaves longer; twigs yellowish brown. Carolina Hemlock

COMMON HEMLOCK (*Tsuga canadensis*) p. 38
A forest evergreen of eastern North America, its shade so dense that nothing will grow beneath it. The tree is quite unsuited to city conditions, does not thrive in hot, dry, windy places, but is a superb evergreen in its natural habitat. **Size:** 40–90 feet high, about half as wide, its branches gracefully drooping in age. **Leaves:** Dark green and shining above, about ⅝ inch long, usually without a notch at the blunt tip, almost microscopically saw-toothed on the edge (seen only with a lens). **Flowers:** Negligible. **Fruit:** A short-stalked, smooth cone, about ¾ inch long. **Hardy:** From Zone 2 southward, but often failing in cultivation because of unfavorable moisture conditions. **Varieties:** Sargent's weeping hemlock (variety *pendula*) is a compact bushy form, usually broader than high, its branches drooping at the tip; *com-*

PINE FAMILY (*Pinaceae*), 1–9

1. **Colorado Spruce** (*Picea pungens*) p. 43
   Magnificent western evergreen with bluish-green foliage and a typically conical outline.

2. **Douglas Fir** (*Pseudotsuga taxifolia*) p. 46
   A gigantic Pacific Coast evergreen, not well suited to eastern gardens, except in one form.

3. **Carolina Hemlock** (*Tsuga caroliniana*) p. 40
   A southern relative of our Common Hemlock, and well suited to regions too warm for the northern tree.

4. **Common Hemlock** (*Tsuga canadensis*) p. 37
   A densely foliaged evergreen growing in cool, reasonably moist places and casting a dense shade.

5. **Norway Spruce** (*Picea Abies*) p. 42
   The most widely planted spruce in this country but not long-lived and ultimately rather scraggly.

6. **Big-Tree** (*Sequoia gigantea*) p. 46
   The largest tree in the world, found in the Sierra Nevada, but quite unsuited to the East.

7. **China Fir** (*Cunninghamia lanceolata*) p. 36
   A fine Chinese evergreen tree, of medium height, with long, sharp, prickly needles.

8. **Japanese Hemlock** (*Tsuga diversifolia*) p. 40
   A splendid evergreen tree and the easiest of all the hemlocks to cultivate. Twigs reddish brown.

9. **Cryptomeria** (*Cryptomeria japonica*) p. 45
   A huge Japanese evergreen very popular here, where it may reach heights of 100 feet.

1
2
3
4
5
6
7
8
9

1
2
3
7
5
6
4
8
9

VINES

1. **Trumpet-Creeper** (*Campsis radicans*) p. 54
A rather rampant native vine with compound leaves and very showy, profuse orange-red flowers.

2. **Cowberry** (*Vaccinium Vitis-Idaea*) p. 53
A prostrate northern vine, difficult to grow, with evergreen foliage and red berries (the lingon).

3. ***Genista pilosa*** p. 50
A prostrate European vine with yellow pea-like flowers. Good ground cover in sandy places.

4. **English Ivy** (*Hedera Helix*) p. 50
An extremely variable Eurasian evergreen vine, useful as a shade ground cover or to climb on walls or trees.

5. **Common White Jasmine** (*Jasminum officinale*) p. 55
Its numerous white flowers are probably the most fragrant of any plant in this book. Not hardy northward.

6. ***Cytisus decumbens*** p. 49
A Canary Island prostrate vine, with yellow pea-like flowers. Its culture difficult and challenging.

7. **Wintergreen** (*Gaultheria procumbens*) p. 52
A highly aromatic prostrate native vine, suited to partially shady sites with sandy acid soils.

8. **Alpine Azalea** (*Loiseleuria procumbens*) p. 48
A northern prostrate evergreen vine, fine as a ground cover but only possible in cool, acid, gritty soils.

9. **Cross-Vine** (*Bignonia capreolata*) p. 54
A high-climbing native evergreen vine with only two leaflets and reddish-orange flowers in spring.

COMMON HEMLOCK, cont'd:
pacta is a dwarf, cone-shaped tree. There are several others, but they offer no advantages over the typical form. **Culture:** The most sensitive of all the hemlocks to dust, smoke, wind, and drought. It will make a magnificent evergreen hedge if sheared once a year from youth. Hedges 15 feet high and 6–8 feet thick are not uncommon in New England.

JAPANESE HEMLOCK (*Tsuga diversifolia*) p. 38
A pyramidal evergreen tree of slower growth than the Common Hemlock, but less demanding as to site. **Size:** 40–80 feet high, about half as wide, rarely reaching these dimensions as cultivated. **Leaves:** Less than ½ inch long, slightly notched at the tip, perfectly smooth on the margin. **Flowers:** Negligible. **Fruit:** A stalkless cone, more or less egg-shaped and about 1 inch long. **Hardy:** From Zone 3 southward and the easiest of the hemlocks to grow. **Varieties:** None. **Culture:** See above.

Related plant (not illustrated):

Tsuga sieboldi, a Japanese evergreen, 40–80 feet high, with shorter leaves and a little less hardy than *Tsuga diversifolia.*

CAROLINA HEMLOCK (*Tsuga caroliniana*) p. 38
A southern relative of the Common Hemlock, differing from it chiefly in having the leaves notched at the tip and in the more pronounced drooping of its terminal shoots. **Hardy:** From Zone 3 southward and often doing better in cultivation than the Common Hemlock. **Varieties:** None. **Culture:** See above, but it is easier to grow than the Common Hemlock and in Va. and Md. it makes splendid hedges if sheared.

SPRUCE
(*Picea*)

Next to the firs the spruces are among the finest of coniferous evergreens. Of the 40 known species, over 30 are cultivated in the United States, but only 6 can be considered here together with a few others that are less well known in cultivation. Typically the spruces are large evergreen trees with scaly bark, but many dwarf forms of the Norway Spruce are extensively used in foundation planting. Leaves usually pointed, more or less 4-sided or angled, but flattish in *Picea Omorika.* They are scattered on the twigs, each leaf inserted on a small cushionlike base from which it readily falls when dry or mature, leaving the naked twig rough to the touch (smooth twigs in the firs). Spruce, because of these readily shedding leaves, especially in warm rooms, is not good for Christmas trees. Cones usually small (except in the Norway Spruce), always hanging or lateral, never erect as in the firs.

All the spruces prefer well-drained, somewhat sandy loam, and tolerate water at their roots very little. Neither do they tolerate dry, windy sites, except the Norway and White Spruce, although they can stand any amount of winter cold. They thrive best in the northern part of the Atlantic seaboard and in cool mountain areas, especially in Oregon and Washington.

Our 6 spruces may be distinguished thus:

Leaves flattened, with white bands on upper side. — Serbian Spruce
Leaves 4-sided or angled, never flat.
  Cones 5–7 inches long. — Norway Spruce
Cones usually less than 3 inches long; rarely 4 inches long in the Colorado Spruce.
  Twigs not hairy.
    Leaves pointing upward or forward. — White Spruce
    Leaves spreading, not upward-pointing. — Colorado Spruce
  Twigs distinctly hairy.
    Twigs brown; leaves dark green. — *Picea orientalis*
    Twigs brownish yellow; leaves blue-green. — Engelmann's Spruce

SERBIAN SPRUCE (*Picea Omorika*) p. 43

A fine evergreen from Yugoslavia, more easy to grow than most spruces, its branches sweeping the ground, its general outline spirelike. **Size:** 50–100 feet high, about half as wide, but rarely reaching these dimensions in cultivation. **Leaves:** Flattened, keeled both sides, with 2 white bands on the upper side, scarcely ½ inch long. **Flowers:** Negligible. **Fruit:** A hanging, nearly stalkless, glossy brown cone, 2–2½ inches long. **Hardy:** From Zone 3 southward. **Varieties:** *pendula* has branches drooping at the tip. **Culture:** See above, but it is one of the most satisfactory of the spruces for the eastern states.

NORWAY SPRUCE (*Picea Abies*) p. 38
A ubiquitous and second-rate evergreen planted almost everywhere, but not long-lived and becoming straggly in age. **Size:** 60–150 feet high, about a third as wide, often less as cultivated. **Leaves:** More or less 4-sided, dark shining green, about ¾ inch long. **Flowers:** Negligible. **Fruit:** The cones are the largest of any of the spruces here treated, being smooth, cylindric, and 5–7 inches long. **Hardy:** From Zone 2 southward. **Varieties:** Over 30 and far too diversified to list here. They comprise low shrubs, prostrate forms, compact globelike forms, and foliage differences in color and texture. Many of these are very popular, and the only safe way to select them is to visit a well-labeled nursery or arboretum. **Culture:** By far the easiest of all spruces to grow in almost any soil, and some bedraggled specimens are even seen on the prairie.

Related plants (not illustrated):

Dragon Spruce (*Picea asperata*). A Chinese evergreen unique among spruces in being able to stand salt spray.

Tigertail Spruce (*Picea polita*). A Japanese evergreen with stiff, prickly leaves, so numerous that the twiggy branches of young specimens make an impenetrable hedge.

WHITE SPRUCE (*Picea glauca*) p. 43
A handsome evergreen from northern North America, widely grown, particularly in some of its varieties, with ascending branches but the tips of the twigs drooping. **Size:** 60–100 feet high, about a third as wide. **Leaves:** More or less 4-sided, pointing upward or forward, bluish green and about ¾ inch long, unpleasantly scented when bruised. **Flowers:** Negligible. **Fruit:**

A hanging, greenish-brown, shiny cone, 1½–2 inches long. **Hardy:** From Zone 1 southward, and especially suited to the prairie states. **Varieties:** The Alberta Spruce (variety *albertiana*) has shorter cones and is preferred in states like Minn. and Iowa and others near them; *caerulea* is a form with more bluish-green leaves; Black Hills Spruce (variety *densata*) is of compact habit and is more slow-growing. **Culture:** See above.

Left: White Spruce (*Picea glauca*), p. 42.
Right: Serbian Spruce (*Picea Omorika*), p. 41.

COLORADO SPRUCE (*Picea pungens*) p. 38
Often called Blue Spruce from its bluish-green foliage, this tree has perhaps the most classical outline of any evergreen. Typically broad at the base, with branches sweeping the ground, it is almost a perfect cone in outline, tapering to a spirelike tip. **Size:** 80–140 feet high, nearly half as wide, but less as cultivated. **Leaves:** More or less 4-sided, stiff, spreading, rigid, prickle-pointed, bluish green and about 1 inch long. **Fruit:** A hanging, brown, shiny, oblongish cone 3–4 inches long. **Hardy:** From Zone 1 southward. **Varieties:** Koster's Blue Spruce (variety *kosteriana*), a handsome form with bluish-gray foliage, is so much planted in suburban gardens that

it irks the sophisticated, but it is undeniably one of our finest evergreens, and is often called Colorado Blue Spruce; *compacta* is a low, dense form with twiggy branches; *moerheimi* is a form with silvery foliage, and is wider than Koster's Blue Spruce. **Culture:** See above.

PICEA ORIENTALIS below

A slow-growing but very beautiful evergreen from the Caucasus and Asia Minor, notable for its glossy green foliage. **Size:** 70–150 feet high, about a third as wide, but scarcely attaining these dimensions as cultivated. **Leaves:** More or less 4-sided, shining green, hardly ½ inch long, but very numerous. **Flowers:** Negligible. **Fruit:** A brownish-violet cone 2–3½ inches long. **Hardy:** From Zone 3 southward. **Varieties:** Two or three are offered, but they are no better than the typical form. **Culture:** See above.

Related plants (not illustrated):

Red Spruce (*Picea rubens*). A North American evergreen needing a cool, moist site. It is a source of paper pulp.

Black Spruce (*Picea mariana*). A low bog tree, also North American, sometimes cultivated for ornament.

Left: *Picea orientalis,* above. Right: Engelmann's Spruce (*Picea engelmanni*), p. 45.

ENGELMANN'S SPRUCE (*Picea engelmanni*) p. 44
A splendid evergreen from western North America, notable for its many tiers of branches and dense shade. **Size:** 80–150 feet high, about a third as wide, but not reaching such dimensions in cultivation. **Leaves:** Bluish green, a little flattened, about 1 inch long and slightly curved. **Flowers:** Negligible. **Fruit:** A cylindric, light brown cone, 2–3 inches long and essentially stalkless. **Hardy:** From Zone 1 southward. **Varieties:** *glauca,* a form with steel-blue leaves; *argentea,* a form with silvery foliage; *fendleri,* with drooping branchlets and slightly longer leaves. **Culture:** See above.

Twigs Smooth after Leaf-fall

THE 3 coniferous trees in this group differ from the related spruces and hemlocks in having smooth twigs after leaf-fall. They can be separated thus:

Bark reddish brown and shreddy; cones nearly globular. Cryptomeria
Bark reddish or reddish brown; cones more or less egg-shaped or oval.
  Cones 2–4 inches long. Douglas Fir
  Cones shorter. Big-Tree

CRYPTOMERIA (*Cryptomeria japonica*) p. 38
A gigantic Japanese evergreen, a specimen at the Shinto temple near Nara being 9 feet in diameter and reputedly 800 years old. Here it is considerably less, but there are many cultivated specimens in the U.S. that are well over 100 feet high, always with reddish-brown, shreddy bark. **Size:** As cultivated 40–125 feet high, about one third as broad. **Leaves:** Awl-shaped, the tips always curved toward the twigs, which are completely clothed with these small, keeled leaves, which turn bronzy in winter but become green the following spring. **Flowers:** Negligible. **Fruit:**

A nearly globular hanging cone about 1 inch in diameter. **Hardy:** Precariously so in exposed places in Zone 4, perfectly hardy from Zone 5 southward. **Varieties:** *lobbi,* the form most grown here, a little hardier than the typical tree, of more compact habit, and somewhat quicker growth. There are many other forms offered, some rather dwarf shrubs, but most of them are of interest only to specialists. **Culture:** Any good garden loam suits them, but far more important is a wind-free sheltered site, in full sunlight.

DOUGLAS FIR (*Pseudotsuga taxifolia*) p. 38
A huge evergreen timber tree of the Pacific Coast, unsuited to culture in the East, except for a Rocky Mountain form of it that is generally the only one grown here. **Size:** As cultivated 40–100 feet, about a third as wide, but reaching 300 feet in Ore. **Leaves:** Spirally arranged, linelike, scarcely curved, about 1 inch long, and with 2 pale bands beneath. **Flowers:** Negligible. **Fruits:** A hanging, egg-shaped cone, 2½–4½ inches long. **Hardy:** The Rocky Mountain form is hardy from Zone 4 southward. **Varieties:** Several, but of no advantage over the typical form. **Culture:** Plant in a good friable loam, well drained, but of more importance is freedom from dry winds and deficient rainfall, especially in the summer. Does poorly on the coastal plain.

BIG-TREE (*Sequoia gigantea*) p. 38
The largest living thing in the world and until recently supposed to be the oldest, this mammoth evergreen, isolated high in the Sierra Nevada in Calif., is only precariously hardy in the East, although there are scattered specimens along the Atlantic Coast. **Size:** As cultivated in the East, not over 100 feet high, but in Calif. up to 320 feet high and with a trunk 35 feet in diameter. **Leaves:** Narrow, scalelike, scarcely ½ inch long, more or less hugging the twigs. **Flowers:** Negligible. **Fruit:** A hanging, egg-shaped or ovalish cone, usually less than 2 inches long. **Hardy:** Precariously so in Zones 4 and 5, the summer heat and deficient rainfall appearing to be critical factors. **Varieties:** None worth growing in the East. **Culture:** Any good garden soil, but there must be no

water at its roots. Its great hazard in the East is our failure to provide the high, cool forests of the Sierra Nevada. For an interesting relative of the Big-Tree, recently found as a living tree but known previously only as a fossil, see METASEQUOIA, p. 4.

Related plant (not illustrated):

California Redwood (*Sequoia sempervirens*). A Pacific Coast evergreen, sometimes 340 feet high, and practically impossible to grow in the East.

2. Vines

VINES are merely shrubs that climb or sprawl. The high-climbing sorts like wisteria are known to practically everyone, but many erect vines never climb as high as wisteria, such as clematis for example, while still others sprawl over the ground like most forms of the English Ivy. As here defined, vines are woody plants that climb or sprawl, to be distinguished from herbaceous vines like the Morning-Glory, all of which are treated on pp. 1–8 of *The Guide to Garden Flowers,* a companion volume to this one.

The woody vines belong to many families, but as these are hard to distinguish, the identification of the different sorts is based on more easily detected characters. Of these the most obvious is whether the vines sprawl and are hence prostrate or trailing, or are erect and often high-climbing. The major division is thus:

Erect vines, often high-climbing. See p. 53

Prostrate or trailing vines, rarely over 10 inches high. (An exception is the high-climbing form of the English Ivy.)

Leaves alternate. See p. 49

Leaves opposite. *Alpine Azalea*

ALPINE AZALEA (*Loiseleuria procumbens*)

Family Ericaceae p. 39

A splendid evergreen, trailing ground cover, quite difficult to grow and impossible in warm regions as it grows naturally only in the coldest parts of the north temperate zone and even up to arctic regions. **Size:** 6–9 inches high, its trailing stems several times as long. **Leaves:** Opposite, oval or oblong, not over ¼ inch long, very numerous. **Flowers:** Bell-shaped, hardly ¼ inch long, borne in sparse terminal clusters in July, and of little garden interest, the plant being grown for its evergreen foliage. **Fruit:** A small dry capsule, not over 1/4 inch long. **Hardy:** From cooler parts of Zone 4 northward. **Varieties:** None. **Culture:** Quite difficult. It needs a gritty or sandy acid soil (pH 5), preferably in the rock garden and in full sunshine. If it needs watering, use only rain water. (For another opposite-leaved prostrate vine see PERIWINKLE, p. 360.)

Leaves Alternate

HERE belong several prostrate or trailing vines with alternate simple leaves. They are quite unrelated but have these features in common. They may be distinguished thus:

Flowers pea-like.
- *Pods merely hairy, 3–4-seeded.* — *Cytisus decumbens*
- *Pods silky, 5–8-seeded.* — *Genista pilosa*

Flowers not pea-like.
- *Flowers greenish yellow, rarely produced.* — English Ivy
- *Flowers white or pinkish, always obvious.*
  - *Petals 5, separate.* — *Cotoneaster*
  - *Petals united to form a bell-shaped corolla.*
    - *Flowers solitary.* — Wintergreen
    - *Flowers in clusters.* — Cowberry

CYTISUS DECUMBENS. Family Leguminosae p. 39
A sprawling, practically prostrate vine from the Canary Islands, suited to the rock garden and to be grown only in mild climates. **Size:** Not over 4–6 inches high, the trailing stems much longer. **Leaves:** Alternate, simple, oblongish, 1/4–1/2 inch long, usually hairy on the under side. **Flowers:** 1–3 in a sparse cluster, light yellow and pea-like, nearly 3/4 inch long, in May–June. **Fruit:** A small 3–4-seeded pod that is somewhat hairy. **Hardy:** Only from Zone 5 southward. **Varieties:** None. **Culture:** Best grown in the rock garden, in full sun, in a sandy or gritty soil, and as they transplant poorly they should be left alone when once established.

GENISTA PILOSA. Family Leguminosae p. 39
A low, prostrate European vine with warty stems, useful as a ground cover as it often roots at the joints and makes an inextricable mat. **Size:** Prostrate. **Leaves:** At least the upper ones alternate and simple, oblongish and scarcely ½ inch long, dark green and smooth above, usually hairy beneath. **Flowers:** Pea-like, yellow, 1–3 in a small cluster from the leaf joints, in May–July. **Fruit:** A small, silky, 5–8-seeded pod. **Hardy:** From Zone 4 southward. **Varieties:** None. **Culture:** Start only with pot-grown plants as it is difficult to transplant. It needs full sunlight, a sandy or gritty loam, and does well in open, rather dry sites.

Related plant (not illustrated):

Genista villarsi. A densely intertwined low shrub, with loose clusters of yellow flowers, hardy only from Zone 6 southward.

ENGLISH IVY (*Hedera Helix*). Family Araliaceae p. 39
Normally a protean, evergreen Eurasian vine, and one of the best of all ground covers, but if allowed to climb, frequently reaching 80–90 feet and completely covering walls in favorable places. **Size:** Usually prostrate, but high-climbing if permitted. **Leaves:** Very variable, in the typical form 3–5-lobed, 2–5 inches long, but in the flowering stage larger than this, only faintly lobed, or unlobed, and squarish. **Flowers:** Very rarely produced (never in the prostrate forms), small, greenish yellow, in globe-shaped clusters. **Fruit:** Rarely produced, round, black, nearly ½ inch in diameter. **Hardy:** Up to Zone 4, and even in Zone 3 if not planted in full sun. **Varieties:** Over 40, of which a selection might include:

arborescens. An upright form with nearly unlobed leaves, and perhaps originating from cuttings taken only from flowering branches.

argenteo-variegata. Vinelike, the leaves white-margined or white-variegated.

aureo-variegata. Yellow or yellow-variegated.

baltica. Baltic Ivy. A small-leaved form, very useful for ground covers and more hardy than the typical form.

conglomerata. A slow-growing, creeping form, with small, crowded, unlobed or only 3-lobed leaves, best suited to the rock garden.

gracilis. A small-leaved form, the foliage turning bronzy in winter; sometimes offered as Japanese Ivy.

minima. A very popular form with small, wavy-margined leaves.

hibernica. Irish Ivy. A good form for a quick ground cover.

Culture: Easy if it is remembered that the plant does best in cool, partially shaded places and is intolerant of hot, dry, and windy places. It will grow in a variety of soils, and roots very freely in sand, sandy soil, or in water. The best time to make cuttings for rooting is early September.

Related plants (not illustrated):

Algerian Ivy (*Hedera canariensis*). A stout, high-climbing vine, hardy only from Zone 7 southward.

Hedera colchica. A usually unlobed form with leaves 5–10 inches long; high-climbing. Not hardy above Zone 6.

COTONEASTER
(*Cotoneaster*)
Family Rosaceae

For a description of this important group of erect shrubs, see p. 285. The one below is a prostrate relative of the taller kinds, and has small but very profuse flowers with 5 separate petals, followed by a tiny, red, applelike fruit as showy as the flowers.

As the plants are impatient of moving, they are best started from potted specimens, preferably planted in spring, in open sunlight and in good garden loam. They do not do well in dry, windy places.

COTONEASTER ADPRESSA p. 52

A low, prostrate Chinese vine, its foliage not evergreen, grown mostly in the rock garden but useful as a ground cover in favorable places. **Size:** Prostrate. **Leaves:** About ½ inch long, wavy-

Cotoneaster adpressa, p. 51.

margined, simple, and alternate. **Flowers:** Small, pinkish, the 5 petals erect. **Fruit:** Apple-like, very small, red, effective in Oct. and Nov. **Hardy:** From Zone 4 southward. **Varieties:** *praecox,* a somewhat taller and more vigorous form with pink flowers. **Culture:** See above.

WINTERGREEN (*Gaultheria procumbens*)

Family Ericaceae p. 39

A prostrate, almost herblike North American evergreen woody vine, the juice of its foliage wintergreen-scented. It is scarcely a garden plant as it needs special conditions, preferably in the wild garden. **Size:** Prostrate and creeping, the tips of the shoots erect. **Leaves:** Simple, alternate, evergreen, ovalish, about 1¾ inches long, the marginal teeth bristly. **Flowers:** Urn-shaped, about ¼ inch long, solitary, nodding, white or pinkish; May–July. **Fruit:** Scarlet, pea-size, not persistent. **Hardy:** Everywhere. **Varieties:** None. **Culture:** Not easy. It demands the filtered light of pine woods or other thin-canopied trees, a light sandy, acid (pH 5) soil, and should only be started from potted plants as it transplants with difficulty. Grown chiefly for interest (the wintergreen flavor in its leaves) rather than for show.

COWBERRY (*Vaccinium Vitis-Idaea*)
Family Ericaceae p. 39

A prostrate evergreen vine from the cooler parts of the north temperate zone, suited only to such regions, but a beautiful groundcover in favorable places. **Size:** Prostrate, but the tips of the twigs erect and 4–8 inches high. **Leaves:** Simple, alternate, evergreen, shiny, about 1 inch long and ovalish. **Flowers:** Bell-shaped or urn-shaped, about ¼ inch long, pink, and in small nodding clusters. **Fruit:** A red berry (the lingon-berry), nearly round, about ½ inch in diameter, edible. **Hardy:** From the cooler parts of Zone 4 northward, nearly to the Arctic Circle. **Varieties:** None. **Culture:** Only to be started from potted plants in a sandy or gritty acid soil (pH 5), in full sun, set out in early spring, but only in a cool (or mountainous) region. Not easy to grow and impossible in dry hot regions.

Erect Vines, Often High-climbing

All the rest of the vines in this book are erect and often high-climbing. The group as a whole is most easily distinguished by those that have alternate or opposite leaves, thus:

Leaves alternate. See p. 66
Leaves opposite.
  Leaves simple. See p. 60
  Leaves compound.
    Flowers irregular and tubular.
      Plants with tendrils (see Picture Glossary). Cross-Vine
      Plants without tendrils. Trumpet-Creeper
    Flowers not irregular.
      Flowers small, white, very fragrant. Jasmine
      Flowers usually large, showy, with little or no fragrance, except in Clematis paniculata. *Clematis*

CROSS-VINE (*Bignonia capreolata*)

Family Bignoniaceae p. 39

A high-climbing woody vine of the southeastern U.S., grown for its handsome evergreen foliage and showy flowers. **Size:** Frequently climbing to 60 feet high. **Leaves:** Opposite, compound, with 2 stalked leaflets 4–6 inches long, without marginal teeth, replaced at the tip with a tendril. **Flowers:** Reddish orange, about 2 inches long, irregular, in small clusters from the leaf joints; May–June. **Fruit:** A long, narrow, slightly flattened pod, 5–8 inches long. **Hardy:** From Zone 6 southward. **Varieties:** None. **Culture:** Being a plant of moist, warm woods it needs conditions approximating this, and is more sensitive to cold than it is to the variety of soils in which it grows naturally.

TRUMPET-CREEPER (*Campsis radicans*)

Family Bignoniaceae p. 39

A moderately high-climbing vine that often sprawls over the ground if there is no support, not evergreen, and native from Pa. to Fla. and Tex. **Size:** Frequently climbing up to 30 feet high. **Leaves:** Opposite, compound, and composed of 9–11 ovalish leaflets, 1½–2½ inches long, sharply toothed on the margin, and without a terminal tendril. **Flowers:** Nearly stalkless, orange red, very showy, the tube 2–3 times the length of the expanded, partly irregular summit, the flowers in a terminal cluster; July–Aug. **Fruit:** A slender, ridged pod, 5–7 inches long, about pencil-thick. **Hardy:** From Zone 3 southward. **Varieties:** None. **Culture:** An invasive, rampant, very showy vine, difficult to eradicate when it becomes an almost weedy pest. It will stand smoke, dust, wind, and abuse, and is indifferent as to soils.

Related plants (not illustrated):

- Chinese Trumpet-Creeper (*Campsis grandiflora*). Climbs 10–20 feet high; flowers scarlet, nearly 3 inches wide.
- *Campsis tagliabuana.* A showy hybrid vine that climbs 10–20 feet high, its profuse flowers tawny orange; mostly offered under the varietal name of Mme. Galen.

Jasmine
(*Jasminum*)
Family Oleaceae

Most of the jasmines are erect shrubs, but the one below is a high-climbing vine with opposite, compound leaves and small, extremely fragrant flowers used for perfume since ancient times. For erect, shrubby jasmines see p. 325. The only vinelike one is:

COMMON WHITE JASMINE (*Jasminum officinale*) p. 39
A weak-stemmed, sprawling plant, usually high-climbing and vinelike, perhaps the most fragrant of any species in this book, and cultivated for centuries for perfume. It grows wild from Persia to China. **Size:** Reaching, with support, 20–40 feet high. **Leaves:** Opposite, compound, not evergreen, composed of 5–7 glossy leaflets, ½–2½ inches long, the terminal leaflet stalked and larger. **Flowers:** White, about 1 inch long, in a small cluster, its fragrance so enticing that it captivated Cleopatra as well as French perfumers. **Fruit:** A small black berry. **Hardy:** Only from the southern part of Zone 5 southward. **Varieties:** Several, but none of them superior to the typical form. **Culture:** Easy within the climatic zones noted above, requiring a sunny exposure and rich loamy soils.

Clematis
(*Clematis*)
Family Ranunculaceae

Very showy, climbing, woody vines, with fragile stems, commonly called Virgin's-Bower. Over 40 species are in cultivation in the United States, but the number of named horticultural varieties is over 60, as these hybrids are far more showy than the species from which they have been derived. The leaves are opposite and compound in our species, sometimes twice-compound, with 3–8 leaflets. Sometimes the terminal leaflet is replaced by a tendril. Flowers large and solitary, or in few-flowered clusters, or smaller and borne in profuse clusters. Petals none, but the sepals petaloid and showy, in a wide range of colors. Fruit dry, beautifully plumed in some species.

Clematis has so many fine hybrids that the size and color of the flowers is too variable to note in detail. The grower is urged to get descriptive catalogs from dealers before making a selection. Nine of the best hybrids might include: Ascotiensis (light blue), Crimson King (red), Gipsy Queen (purple), Henryi (white), Nelly Moser (pale mauve), Prince Hendrik (light blue), Ramona (blue), Ville de Lyon (red purple) and William Kennett (deep lavender).

Besides these hybrids there are 6 species of *Clematis* that are desirable garden vines and, like the hybrids, of relatively easy culture. All of them require a cool, moist site, and do better in a rich loam, preferably somewhat alkaline. The six species are divided thus:

Leaves twice-compound. Vine Bower
Leaves only once-compound.
  Flowers yellow. Golden Clematis
  Flowers not yellow.
    Flowers white, less than 1½ inches wide, in large showy clusters. Japanese Clematis
    Flowers mostly 2 inches wide or more, often much more; solitary or in few-flowered clusters.
      Flowers white. *Clematis montana*
      Flowers not white (except in some hybrids; see above).
        Flowers red; leaves tendril-bearing. Scarlet Clematis
        Flowers rosy purple or violet-purple, 4–7 inches wide. *Clematis jackmani*

VINE BOWER (*Clematis Viticella*) p. 57
A slender woody Eurasian vine, not rampant, useful for covering trellises. **Size:** Rarely growing above 12 feet high, and needing support. **Leaves:** Usually twice-compound, the leaflets ovalish,

¾–2 inches long, without marginal teeth, or sometimes 3-lobed. **Flowers:** Rose purple or violet, about 1¾ inches wide, in clusters of 1–3; June–Aug. **Fruit:** Dry, scarcely or not at all plumed. **Hardy:** From Zone 2 southward. **Varieties:** None. **Culture:** See above.

GOLDEN CLEMATIS (*Clematis tangutica*) p. 70
The only yellow-flowered clematis here included, a slender woody vine from northeastern Asia. **Size:** Rarely climbing over 6–9 feet high. **Leaves:** Once compound, the leaflets oblongish, 1½–2½ inches long, rather coarsely toothed, and sometimes 3-lobed. **Flowers:** Yellow, solitary, about 3 inches wide; June and often also in the fall. **Fruit:** Dry, its long silky plume quite showy. **Hardy:** Everywhere. **Varieties:** None. **Culture:** See above.

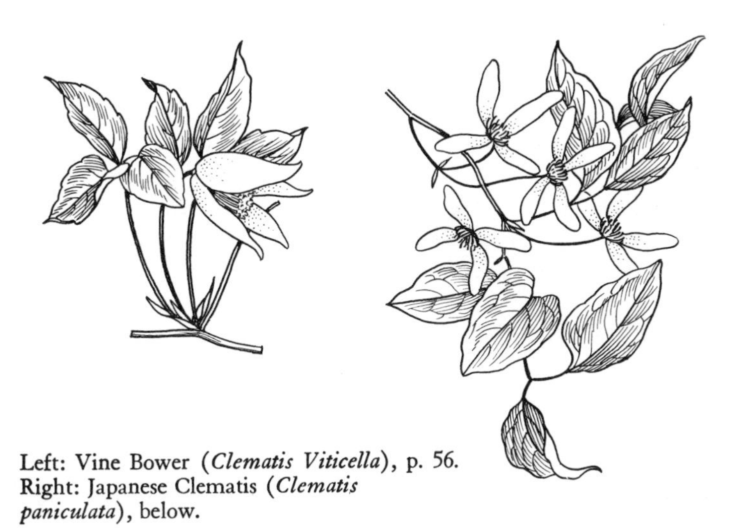

Left: Vine Bower (*Clematis Viticella*), p. 56.
Right: Japanese Clematis (*Clematis paniculata*), below.

JAPANESE CLEMATIS (*Clematis paniculata*) above
A stout woody climber with a profusion of small fragrant flowers, rather rampant and often growing high in the trees. **Size:** Often 20–30 feet high, or even more. **Leaves:** Compound, the 3–5 leaflets ovalish, 1½–4 inches long, without marginal teeth, but

sometimes lobed. **Flowers:** Scarcely 1½ inches wide, extremely profuse, in many-flowered clusters, very fragrant; Sept. and Oct. **Fruit:** Dry and plumed. **Hardy:** From Zone 2 southward. **Varieties:** None. **Culture:** See above.

Related plants (not illustrated):

Traveler's-Joy (*Clematis Vitalba*). European woody vine, up to 25 feet high. Leaves compound; leaflets mostly 5. Flowers white, profuse; July–Oct.

Woodbine (*Clematis virginiana*). A native vine, 12–18 feet high. Leaflets 3, long-stalked. Flowers white, profuse; July–Aug.

CLEMATIS MONTANA p. 59

A woody vine from China and the Himalayas, much grown on trellises for its handsome flowers. **Size:** Climbing up to 10–20 feet. **Leaves:** Compound, its 3 short-stalked leaflets ovalish, 1½–4½ inches long and sharply toothed (rarely with none). **Flowers:** White, slender-stalked, 1½–3½ inches wide, solitary or in clusters of 2–5; May. **Fruit:** Dry and plumed. **Hardy:** From Zone 4 southward, and with protection from Zone 3. **Varieties:** *rubens* has purplish young foliage and pink flowers; *wilsoni* has larger white flowers that bloom 1–2 months later than the typical form. **Culture:** See above.

SCARLET CLEMATIS (*Clematis texensis*) p. 59

A Texan clematis, only slightly woody, low-growing and grown chiefly for its showy flowers. **Size:** 4–6 feet high. **Leaves:** Compound, the 4–8 leaflets, bluish green, ovalish, 1¾–3½ inches long, the terminal one usually lacking and replaced by a tendril. **Flowers:** Solitary, stalked, urn-shaped, about 1½ inches long, scarlet or pink; July–Sept. **Fruit:** Dry, plumed. **Hardy:** From Zone 3 southward. **Varieties:** None. **Culture:** See above.

CLEMATIS JACKMANI p. 70

By far the most showy of all the garden clematis, itself a hybrid, and the source of many others still more showy. **Size:** Not usually over 8–10 feet high, its stems weak and needing support. **Leaves:**

Compound (rarely simple), the 3 coarsely toothed leaflets ovalish. **Flowers:** Solitary or in clusters of 2–3, very showy, typically violet-purple, 4–7 inches wide (more in some hybrids). **Fruit:** Dry, only slightly plumed. **Hardy:** From Zone 2 southward. **Varieties:** At least 60, some of them widely cultivated, and most of them more desirable than the typical form. For a selected list see the beginning of this treatment of *Clematis*.

Left: *Clematis montana,* p. 58.
Right: Scarlet Clematis (*Clematis texensis*), p. 58.

Erect, Often High-climbing Vines with Opposite Simple Leaves

Here belong a group of vines, often unrelated, that combine the features of having both opposite and simple leaves. The simplest clue to their identification is to separate them thus:

Leaves without marginal teeth. See p. 62
Leaves with obvious marginal teeth.
Leaves finely toothed; plant evergreen. *Euonymus fortunei*
Leaves coarsely toothed; plants not evergrèen.
August-blooming; sterile flower with 1 segment. *Schizophragma*
July-blooming; sterile flower with 3–4 segments. Climbing Hydrangea

Euonymus fortunei, below.

EUONYMUS FORTUNEI. Family Celastraceae above
A valuable evergreen Asiatic creeper, tightly clinging to brick or stone, less so to wood, because of its aerial roots at the joints which have minute sucking disks. **Size:** Not usually over 4–8 feet high. **Leaves:** Ovalish or broadly elliptic, opposite, simple, ¾–2¼ inches long, finely toothed on the margin, dark lustrous green. **Flowers:** Small, greenish white and of little interest. **Fruit:** Berrylike, small, nearly round, pale pink. **Hardy:** From Zone 3 southward. **Varieties:** *argenteo-marginatus,* with silvery-margined

leaves; *acutus,* leaves smaller than the typical form, and an especially good climber; *minimus,* a very small-leaved form useful in the rock garden, and often sold as variety *kewensis; vegetus,* a somewhat shrubby form. **Culture:** A superb evergreen vine for covering brick walls; needs little attention and can be planted in sun or shade and in a variety of soils as it thrives in many soil types. It is better started from potted plants in early spring. Long known as *Euonymus radicans.*

SCHIZOPHRAGMA HYDRANGEOIDES.

Family Saxifragaceae p. 70

A high-climbing Asiatic woody vine, with profuse clusters of white flowers and nearly round leaves. **Size:** Frequently reaching 30 feet high, and often clambering up trees. **Leaves:** Opposite, simple, nearly round, 3–4 inches wide, toothed on the margin and pale on the under side. **Flowers:** White and of two kinds, all in the same flat-topped cluster, the inner ones fertile and small, surrounded by a margin of sterile, more showy flowers which have only a single petal-like segment; Aug. **Fruit:** A small dry capsule. **Hardy:** From Zone 3 southward. **Varieties:** None. **Culture:** Plant in full sunlight in any ordinary garden soil. While it has aerial rootlets which cling to bark, it needs support or it will trail over the ground and do poorly.

CLIMBING HYDRANGEA (*Hydrangea petiolaris*)

Family Saxifragaceae. p. 70

A high-climbing Asiatic woody vine, superficially resembling the one above (*Schizophragma*) and just as showy. **Size:** Frequently reaching to 50–60 feet high. **Leaves:** Opposite, simple, broadly oval, 2–4 inches long, toothed on the margin, dark shining green above. **Flowers:** White, the flat-topped cluster 6–10 inches wide, composed of small, central, fertile flowers, and a marginal ring of larger, more showy, sterile flowers which have 3–4 petal-like segments (*Schizophragma* has only one); July. **Fruit:** A dry capsule. **Hardy:** From Zone 3 southward. **Varieties:** None. **Culture:** Plant in any ordinary garden soil, in full sun. It is a fine vine for clinging to masonry walls and the bark of trees.

Erect Vines with Opposite, Simple Leaves without Marginal Teeth

Leaves evergreen. See p. 65
Leaves not evergreen.
Fruit a berry; sap not milky. Honeysuckle
Fruit a long pod; sap milky. Silk Vine

HONEYSUCKLE
(*Lonicera*)
Family Caprifoliaceae

Here are included only the climbing honeysuckles, but many others are erect shrubs, for which see p. 383. Some of the vines are a pestiferous nuisance, such as the Japanese Honeysuckle, while others are handsome vines with beautifully colored and fragrant, often irregular flowers. Leaves opposite, simple, without marginal teeth, the upper pair sometimes joined at the base so that the stem passes through the united leaf-bases (perfoliate). Flowers nearly regular in some, but highly irregular in others, either borne in pairs or in terminal clusters of usually 6 flowers. Fruit a black or red berry.

The climbing honeysuckles are of the easiest culture in any good garden soil and often thrive on poor sites. Some of them are so rampant as to need control. Only four of the climbing honeysuckles are of much garden interest:

Flowers in pairs; fruit black. Japanese Honeysuckle
Flowers in terminal clusters of 5–6 flowers; fruit red or orange.
Flowers nearly regular, or only slightly 2-lipped. Trumpet Honeysuckle
Flowers irregular, obviously 2-lipped. Italian Honeysuckle.

JAPANESE HONEYSUCKLE (*Lonicera japonica*) below
A rampant curse in some regions, for it clambers over bushes or even trees if not kept under control, but also makes a tight ground cover if there is nothing upon which it can climb. **Size:** Often reaching 20–30 feet high, but prostrate without support. **Leaves:** Persistent or half evergreen, simple, opposite, without marginal teeth, oval to oblong, pointed, 1–3 inches long and usually downy beneath. **Flowers:** In pairs, white at first, fading to yellow, mostly purple-tinged, very fragrant, 1–1½ inches long; June–Aug. or later. **Fruit:** Small, black. **Hardy:** Everywhere. **Varieties:** *aureo-reticulata* has smaller leaves veined with yellow; *chinensis* has flowers nearly 2 inches long and reddish; *halliana,* Hall's Honeysuckle, has flowers without the purple tinge. **Culture:** Never start it without most rigid and continuous control, for it is undoubtedly the most pestiferous vine in this book. Will grow anywhere in any soil, and the chief problem is to check its growth and invasive spreading.

Related plant (not illustrated):

Lonicera henryi. A free-growing, half-evergreen Chinese vine, with purple-red flowers and black fruit.

Left: Japanese Honeysuckle (*Lonicera japonica*), above.
Right: Italian Honeysuckle (*Lonicera Caprifolium*), p. 64.

TRUMPET HONEYSUCKLE (*Lonicera sempervirens*) p. 70
A native American vine, nearly evergreen in the South, its foliage bluish green beneath. **Size:** May climb 40–50 feet high, usually much less. **Leaves:** Opposite, simple, without marginal teeth, oval to oblong, about 1⅓ inches long, the upper pairs united to form a disk. **Flowers:** In terminal clusters, the corolla about 2 inches long, red outside, yellow within, without any fragrance; May–Aug. **Fruit:** A small orange or scarlet berry. **Hardy:** From Zone 3 southward. **Varieties:** None of any importance. **Culture:** See above.

Related plant (not illustrated):

Lonicera tellemanniana. A showy hybrid honeysuckle, with dark yellow flowers in two distinct tiers of clusters. Hardy from Zone 3 southward.

ITALIAN HONEYSUCKLE (*Lonicera Caprifolium*) p. 63
A Eurasian vine of moderate stature, but valued for its showy clusters of flowers. **Size:** Rarely growing above 20 feet high. **Leaves:** Oval or elliptic, 2–4 inches long, bluish green beneath, the upper pairs joined to form a disk. **Flowers:** About 2 inches long, irregular, obviously 2-lipped, in clusters at the ends of the twigs; May–June. **Fruit:** A small orange berry. **Hardy:** From Zone 3 southward. **Varieties:** None. **Culture:** See above.

SILK VINE (*Periploca graeca*)
Family Asclepiadaceae p. 70
A rampant Eurasian woody vine, much given to twining about tree trunks, and a bit too exuberant for the small garden. **Size:** Climbing 25–40 feet high. **Leaves:** Opposite, simple, oblongish, without marginal teeth, 2–4½ inches long, the juice milky. **Flowers:** Not showy, about 1 inch long, greenish brown, borne in long-stalked, terminal clusters. **Fruit:** A collection of narrow, smooth pods, 3–5 inches long. **Hardy:** From Zone 5 southward. **Varieties:** None. **Culture:** Easy, if not too easy, in any garden soil, and as the plant is apt to be invasive it should only be planted where it can be controlled. Foliage very handsome.

Leaves Evergreen, Opposite, Simple, without Marginal Teeth

Flowers yellow.	Carolina Jasmine
Flowers white.	Confederate Jasmine

CAROLINA JASMINE (*Gelsemium sempervirens*)
Family Loganiaceae p. 70
A native American woody vine very popular in the South for its delicious odor and evergreen foliage. **Size:** Climbing 10–20 feet high. **Leaves:** Opposite, simple, without marginal teeth, oblongish, 2½–4 inches long, evergreen and with a dangerously poisonous juice (not to the touch). **Flowers:** About 1 inch long, extremely fragrant, yellow, borne in dense clusters from the leaf joints; Apr.–June. **Fruit:** A flattened short-beaked pod about ¾ inch long. **Hardy:** From Zone 6 southward. **Varieties:** None. **Culture:** This porch-climbing vine needs full sun and a rich garden loam, and plenty of water if there is a drought; otherwise easy to grow.

CONFEDERATE JASMINE (*Trachelospermum jasminoides*). Family Apocynaceae p. 70
A high-climbing Chinese vine with evergreen foliage and very fragrant flowers, much planted in the South. **Size:** Climbs 10–20 feet high, without tendrils and climbing by twining. **Leaves:** Opposite, simple, evergreen, without marginal teeth, oval-oblong, narrowed both ends, 2–3 inches long. **Flowers:** White, starlike, about ¾ inch wide, extremely fragrant. **Fruit:** A pair of long, slender pods. **Hardy:** From Zone 6 southward, and in sheltered places in Zone 5. **Varieties:** None. **Culture:** Needs shelter from drying winds, plenty of moisture in the growing season, full sunshine, and preferably a good garden loam. Even with these it is slow to get started. Use potted plants if possible.

High-climbing Vines with Alternate Leaves

ALL OF THE high-climbing vines so far described have opposite leaves, but all the rest of them have alternate leaves, though some of these have compound leaves while others always have simple ones. The simplest division of the remaining high-climbing vines is to divide them thus:

Leaves compound. See p. 72
Leaves simple.
Leaves with marginal teeth or even lobed. See p. 67
Leaves without marginal teeth.
Leaves kidney-shaped, 5–9 inches long; flowers U-shaped, yellow-green. Dutchman's-Pipe
Leaves ovalish, 2–3½ inches long; flowers white or pinkish white. Silver-Lace Vine

DUTCHMAN'S-PIPE (*Aristolochia durior*)
Family Aristolochiaceae p. 70
A native American, rather luxuriant woody vine, most useful as a screen on a north- or east-facing porch, but needing a trellis or lattice. **Size:** Frequently growing 20–40 feet high (or higher in the woods), but easily controlled by pruning unwanted shoots. **Leaves:** Numerous, casting a dense shade, roundish or kidney-shaped, long-stalked, 6–14 inches wide, and very handsome. **Flowers:** About 1½ inches long, bent like a U, yellowish brown and rather evil-smelling; June. **Fruit:** A dry capsule, about ⅝ inch long. **Hardy:** From Zone 3 southward. **Varieties:** None. **Culture:** Prefers a rich humus and at least partial shade; but will thrive in ordinary garden soil, especially on the north side of the house.

SILVER-LACE VINE (*Polygonum auberti*)
Family Polygonaceae p. 71
An Asiatic, very slender, high-climbing vine, one of the best for

city gardens and other unfavorable places. **Size:** Often 25–40 feet high, and twining freely over porches and fire escapes. **Leaves:** Alternate, simple, without marginal teeth, broadly lance-shaped, about 2½ inches long. **Flowers:** Small, greenish white, fragrant, borne in profuse drooping or erect clusters; Aug. **Fruit:** Small, dry, triangular, not showy. **Hardy:** From Zone 3 southward. **Varieties:** None. **Culture:** Will grow, often rampantly, in any sort of soil — even in city back-yards.

Related plant (not illustrated):

Polygonum baldschuanicum. Similar to the above, but with pink flowers.

Leaves with Marginal Teeth, or Even Lobed

Vines with tendrils. Grape Family. See p. 68

Vines without tendrils.

Pith white; leaf teeth not hairy. Bittersweet

Pith brown; leaf teeth hairy. Tara Vine

BITTERSWEET (*Celastrus scandens*)

Family Celastraceae p. 71

A rampant native American vine, not high-climbing, useful for covering wall-tops and rustic fences. **Size:** Usually not over 7–20 feet high, often 6–8 feet. **Leaves:** Alternate, simple, oblong-oval, 2½–5 inches long, tapering at the tip, the margins toothed. **Flowers:** Small, greenish yellow, mostly negligible. **Fruit:** Very handsome, because as its yellowish capsule splits, it reveals the brilliantly scarlet coating of the seeds. It is for this the plant is grown, as the scarlet seed coats persist until Nov. **Hardy:** From Zone 2 southward. **Varieties:** None. **Culture:** Simple in any ordinary garden soil, and will stand considerable exposure to wind and deficient moisture.

Related plant (not illustrated):

Celastrus orbiculatus. A close Chinese relative of the American bittersweet, with larger leaves that partly hide the showy fruits.

TARA VINE (*Actinidia arguta*). Family Dilleniaceae p. 71
A densely leafy Asiatic vine, valuable for its ability to cast complete shade if planted on trellises or arbors. **Size:** 40–60 feet high or long, if on an arbor, unless it is controlled by pruning. **Leaves:** Alternate, simple, broadly oval, about 5 inches long, the margins toothed. The leafstalk is about 2 inches long and reddish. **Flowers:** Half hidden by the leaves, brownish white, about ¾ inch wide, usually borne in clusters of 3, not showy; July. **Fruit:** A football-shaped yellowish berry about 1 inch long, edible and sweet. **Hardy:** From Zone 4, and in protected places in Zone 3, southward. **Varieties:** None. **Culture:** A quick-growing vine that thrives in any good garden soil, either in full sun or partial shade.

Related plants (not illustrated):

Yangtao (*Actinidia chinensis*). A shaggy-haired relative of the above, with larger white flowers and a gooseberry-like, edible fruit.

Kolomikta Vine (*Actinidia Kolomikta*). A lower-growing Asiatic vine, with often pink-blotched leaves, white flowers and fruit, both about 1 inch long.

GRAPE FAMILY
(*Vitaceae*)

Not many of the Grape Family have simple leaves, as the Virginia Creeper and its relatives have compound leaves and are treated at p. 72. One of those, however, the Boston Ivy, may have some simple leaves mixed with the compound ones and is hence wrongly assumed to be one of the vines noted below. Both those below have simple leaves, always toothed or even lobed on the margin. Their culture presents no difficulties in any ordinary garden soil. Only the ornamental plants are included here, as the edible grapes and other food plants belong to fruit culture. For the wild grapes, see the new Britton & Brown *Illustrated Flora,* by H. A. Gleason, issued by the New York Botanical Garden in 1952. The ornamental plants of the Grape Family, here treated, may be separated thus:

Bark shreddy; pith brown. Crimson Glory Vine
Bark not shreddy; pith white. *Ampelopsis brevipedunculata*

CRIMSON GLORY VINE (*Vitis coignetiae*) p. 71
A tendril-bearing ornamental Japanese relative of the edible grapes, grown mostly for its handsome foliage, which turns a bright crimson in the fall. **Size:** A very vigorous, quick-growing vine, which may be 60–90 feet high. **Leaves:** Alternate, simple, ovalish or roundish, 4½–9 inches wide, deeply heart-shaped at the base, unequally and shallowly toothed on the margin, usually grayish or rusty beneath. **Flowers:** Inconspicuous, greenish yellow, in small clusters. **Fruit:** Grape-like, black, and with a bloom, inedible. **Hardy:** From Zone 3 southward. **Varieties:** None. **Culture:** See above.

AMPELOPSIS BREVIPEDUNCULATA p. 71
A Chinese tendril-bearing vine, widely grown for its spectacular autumnal fruit. **Size:** 30–50 feet high, quick-growing, often as much as 15 feet in one season, when young. **Leaves:** Alternate, simple, more or less 3-lobed, the lobes coarsely toothed, about 5 inches wide. **Flowers:** Negligible. **Fruit:** A berry about ¼ inch in diameter, inedible, first green, then lilac, yellow, and finally blue, sometimes all these colors simultaneously in the same profuse, showy cluster. **Hardy:** From Zone 3 southward. **Varieties:** None readily available that are superior to the typical form. **Culture:** See above.

VINES

1. **Carolina Jasmine** (*Gelsemium sempervirens*) p. 65
   A native evergreen vine with dangerously poisonous juice and extremely fragrant yellow flowers. Not hardy northward.

2. ***Clematis jackmani*** p. 58
   Superb hybrid vines, now of many colors in dozens of varieties. They need support, for their stems are somewhat slender.

3. **Confederate Jasmine** (*Trachelospermum jasminoides*) p. 65
   An evergreen Chinese vine, with simple leaves and small white, starlike, very fragrant flowers. Not hardy northward.

4. **Climbing Hydrangea** (*Hydrangea petiolaris*) p. 61
   A high-climbing Asiatic vine with masses of flower clusters, the marginal ones white and with 3–4 segments.

5. **Golden Clematis** (*Clematis tanguitica*) p. 57
   An Asiatic high-climbing vine, superficially resembling No. 4, having yellow flowers.

6. ***Schizophragma hydrangeiodes*** p. 61
   An Asiatic high-climbing vine, superficially resembling No. 4, but the marginal flowers with only 1 segment.

7. **Dutchman's-Pipe** (*Aristolochia durior*) p. 66
   A fine shade-enduring vine, excellent for north trellises, its curious flowers curved like some pipes.

8. **Trumpet Honeysuckle** (*Lonicera sempervirens*) p. 64
   A high-climbing native vine, its foliage nearly evergreen, and its red flowers lacking any fragrance.

9. **Silk Vine** (*Periploca graeca*) p. 64
   An extremely rampant Eurasian vine, often invasive, with handsome foliage and small greenish-brown flowers.

1
2
3
4
5
6
7
8
9

1
2
3
4
5
6
7
8
9

VINES

1. **Japanese Wisteria** (*Wistaria floribunda*) p. 75
Not so high-climbing as the Chinese Wisteria, its flowers pea-like, usually violet-blue, but white in this variety.

2. **Silver-Lace Vine** (*Polygonum auberti*) p. 66
A splendid vine for city conditions, high-climbing and with a profusion of small white flowers in early fall.

3. **Boston Ivy** (*Parthenocissus tricuspidata*) p. 72
A better vine for city conditions than its close relative, the Virginia Creeper. It clings to stone or brick walls.

4. **Chinese Wisteria** (*Wistaria sinensis*) p. 76
The tallest-growing of the wisterias, with fewer leaflets than No. 1, its flowers bluish violet and fragrant.

5. **Tara Vine** (*Actinidia arguta*) p. 68
A midsummer-flowering Asiatic vine, more valuable for its dense foliage than for its brownish-white flowers.

6. **Bittersweet** (*Celastrus scandens*) p. 67
A native vine much more valued for its orange-red fruit than for its small greenish flowers.

7. ***Ampelopsis brevipedunculata*** p. 69
Grown almost entirely for its multicolored fruit, which may be of four different colors on the same plant.

8. ***Akebia quinata*** p. 73
A little-known Asiatic slender vine with rather inconspicuous but night-fragrant flowers in May.

9. **Crimson Glory Vine** (*Vitis coignetiae*) p. 69
A relative of the grape, this very vigorous Japanese vine is grown for its stunning crimson fall foliage.

High-climbing Vines with Alternate Compound Leaves

HERE belong only four commonly cultivated vines, if we exclude the climbing roses, all of which will be found at p. 301. The plants below can readily be separated thus:

Leaflets arranged feather-fashion; flowers pea-like. See p. 73
Leaflets arranged finger-fashion; flowers not pea-like.
Plants with tendrils. Virginia Creeper and Boston Ivy
Plants without tendrils. *Akebia*

VIRGINIA CREEPER (*Parthenocissus quinquefolia*)
Family Vitaceae p. 74
A rampant native American woody vine, valuable for covering walls and the bark of trees, as its tendrils are tipped with sucking disks that cling tightly. **Size:** 20–40 feet high, often more in favorable places. **Leaves:** Alternate, compound, the 5 leaflets arranged finger-fashion, the leaflets oblongish, pointed, 2–5 inches long, toothed on the margin, bright red in the fall. **Flowers:** Negligible. **Fruit:** A pea-size berry, bluish black and with a bloom, persisting into the winter. **Hardy:** Everywhere. **Varieties:** *Engelmann creeper* (the variety *engelmanni*) has smaller leaflets and is not quite so vigorous; *saintpauli,* smaller leaflets than the typical form and clings to stone or brick even better. **Culture:** Will grow in any ordinary garden soil, preferably in a reasonably moist site, and is not as well suited to city conditions as the next.

BOSTON IVY (*Parthenocissus tricuspidata*)
Family Vitaceae p. 71
An Asiatic representative of the Virginia Creeper and a far better vine for city conditions. Its tendency to produce both compound and simple leaves makes it confusing as it may be mistaken for *Ampelopsis brevipedunculata,* which, however, has smaller leaves and much more colorful fruit. **Size:** Frequently climbing by its sucking disks 40–60 feet high, densely covering masonry walls.

Leaves: Of two kinds, simple and 3-lobed, from 6–10 inches long, or on young shoots (sometimes on the whole plant) with 3 leaflets, all coarsely toothed. **Flowers:** Very small, almost unnoticed, but attracting hundreds of bees in early August. **Fruit:** Pea-size, bluish black, with a bloom. **Hardy:** From Zone 3 southward. **Varieties:** *lowi,* leaves much smaller, and the plant of more restrained growth; *purpurea,* a form with purplish leaves; *veitchi,* a form with smaller leaves that are purplish only when young. **Culture:** This is almost *the* vine for city conditions, as it will thrive in any kind of soil and stand smoke, dust, wind and the fumes of motors.

AKEBIA QUINATA. Family Lardizabalaceae p. 71
A slender Asiatic vine, without tendrils and climbing by twining, grown more for its foliage than for its inconspicuous, night-fragrant flowers. **Size:** 10–20 feet high. **Leaves:** Alternate, compound, its 5 leaflets oblongish, 1½–4 inches long, without marginal teeth, but notched at the tip, arranged finger-fashion. **Flowers:** Small and inconspicuous, night-fragrant; May. **Fruit:** A fleshy, black-seeded pod, 2½–5 inches long, purple-violet, but rarely produced in cultivation, the seeds edible (in Japan). **Hardy:** From Zone 4 southward. **Varieties:** None. **Culture:** Grows rapidly and will sprawl if there is no support, thus often smothering lower plants. It needs good garden soil, full sunshine. If it winter-kills it can be cut to the ground and a new leader will sprout up in the spring.

Leaflets Arranged Feather-fashion; Flowers Pea-like

Leaflets 3.	Kudzu-Vine
Leaflets 9–15.	*Wistaria*

KUDZU-VINE (*Pueraria thunbergiana*)
Family Leguminosae p. 74
The quickest-growing vine ever to come from eastern Asia, and

only to be planted if its invasive growth can be controlled, otherwise it can become a rampant nuisance. **Size:** 80–100 feet either growing up a support, if there is one, or sprawling over the ground to the exclusion of all else. It has been known to grow 60 feet in a single season. **Leaves:** Alternate, compound, its 3 coarse leaflets bean-like, so densely produced that it will shade out everything beneath it. **Flowers:** Pea-like, purple, about ⅝ inch wide, in dense, erect clusters, but these half hidden by the

Left: Virginia Creeper (*Parthenocissus quinquefolia*), p. 72.
Right: Kudzu-Vine (*Pueraria thunbergiana*), p. 73.

profuse foliage; July. **Fruit:** A hairy, flat, oblongish pod, 1¾-4 inches long. **Hardy:** Not certainly above Zone 5, but often killed to the ground in Zone 4 and springing up the next season. Not hardy above Zone 4. **Varieties:** None. **Culture:** Fatally easy and the only cultural direction needed is extreme caution in introducing it where it cannot be controlled. It is a weedy pest in southern gardens.

WISTERIA
(*Wistaria*)
Family Leguminosae

Perhaps our most beautiful and showy vines, perfectly at home in the country and often climbing to the tops of houses in the city. The finest ones are Asiatic, our native *Wistaria macrostachys* of garden interest only because it blooms later than the other two. They are stout, high-climbing, woody vines with alternate, compound leaves, the leaflets arranged feather-fashion, and most profuse. Flowers pea-like, borne in a long, drooping cluster and very showy. Fruit a hanging pod. Start with pot-grown plants put in a mixture of ⅔ garden loam and ⅓ old cow manure. Choose a sunny place and see that there is something for it to climb up. The plant bears no tendrils and must be tied to its support for the first year or two. After that it will do its own climbing, often very rapidly.

Many amateurs complain that the wisteria is a shy or negligent bloomer in youth. One way to correct this is to cut off in July and again in September the long straggling growth that all wisterias make in the summer. Cut back these shoots one half to one third of their length. This will induce the development of flowering spurs on the older wood. One such trimmed plant in the author's garden is so floriferous that it is next to impossible to see the twigs under the profusion of flowers.

The only two worth growing are:

Leaflets 13–19.	Japanese Wisteria
Leaflets 7–13.	Chinese Wisteria

JAPANESE WISTERIA (*Wistaria floribunda*) p. 71

A high-climbing vine, but not so vigorous as the next, rarely reaching housetops. **Size:** 15–28 feet high. **Leaves:** Alternate, compound, the 13–19 leaflets arranged feather-fashion, with an odd one at the end, the leaflets ovalish or oblong, 1¾–3½ inches long. **Flowers:** Violet-blue or violet, pea-like, arranged in a long drooping cluster, 12–18 inches long, always blooming before the leaves unfold, in early May. **Fruit:** A velvety hanging pod, 4½–7 inches long. **Hardy:** From Zone 3 southward. **Varieties:** Several, some with white or pink flowers, but the most important is variety *macrobotrys,* where the hanging flower cluster may be 3 feet long. **Culture:** See above.

Related plant (not illustrated):

Wistaria macrostachya. An American vine, with the flowers half hidden by the foliage, but it flowers later than the Asiatic species.

CHINESE WISTERIA (*Wistaria sinensis*) p. 71

Not much different from the Japanese Wisteria, but taller-growing, not quite so hardy, and with fewer leaflets. **Size:** 30–60 feet. **Leaves:** Alternate, compound, the leaflets 7–13 and arranged feather-fashion, with an odd one at the end. **Flowers:** Bluish violet, fragrant, the clusters about 12 inches long and drooping, always blooming before the leaves unfold in May. **Fruit:** A velvety hanging pod, 5–7 inches long. **Hardy:** From Zone 3 southward. **Varieties:** *alba* has white flowers, and there is a pink, double-flowered form. **Culture:** See above.

3. Erect Trees or Shrubs, Never Climbing or Prostrate Vines

EXCEPT the coniferous evergreens and the vines, which have already been treated, all the plants in this book are erect trees or shrubs, comprising several hundred species. Their identification without flowers or fruit is apt to be difficult, although many of the features used to distinguish the different categories are plain enough — such as simple or compound leaves, leaves that have teeth on the margin or none, leaves that are opposite or alternate in their insertion on the twigs. Other obvious characters, such as whether the plant is a shrub or tree, bears fleshy or dry fruits, and the color of the flowers, are easily seen.

While many amateurs think that some trees bear no flowers, this is never true, and all the rest of the trees and shrubs bear true flowers and fruit, however inconspicuous some of them may be. Botanically, and here, it is the *kind* of flower that divides great groups of plants and these main divisions of the plant world are best outlined at the beginning.

Some flowers bear no petals, others have separate petals, and in some the petals are united to form a cuplike, bell-like or funnel-shaped corolla. These distinctions are so fundamental that a simple outline of them is best understood at the outset. Upon the scheme below all the rest of the trees and shrubs in the book are divided:

Flowers without petals. See below
Flowers with petals. See p. 121

(Nature makes it difficult to see the petals in some trees and shrubs, for the petals are small or even lacking in some specimens. Such may

be mistaken for the plants below, to which they are not related and are hence found elsewhere in this book. Such apparent exceptions occur in the ash, plane tree, maple, sweet gum, the box, Elaeagnus, and in the katsura tree. If in doubt, see these entries and their pictures to settle the identity of these aberrant forms.)

Trees and Shrubs Whose Flowers Have No Petals

HERE belong many of our favorite trees and shrubs, among them the walnut, mulberry, bayberry, willow, poplar, oak, beech, birch, and the elms, as well as others noted below. None of these has petals in the flowers, many of which are small, borne in hanging or erect catkins, in catkin-like clusters, or, as in the elms, in small clusters.

These trees and shrubs that have no petals we grow for their foliage, for the beauty of their outline, their shade, and in a few cases, notably the beech, for the beauty of their bark. The group includes most of our shade trees. There are other trees, not treated here because their flowers have obvious petals. For these see the following in the index: *Paulownia,* maple, catalpa, horse-chestnut, locust, Honey Locust, Kentucky Coffee-Tree, linden, magnolia, mountain-ash, Tulip-Tree, Golden-Rain Tree, *Sophora,* hawthorn, *Malus, Prunus,* dogwood, *Halesia, Albizzia,* and a few others that are not so well known.

To distinguish between the trees whose flowers have no petals it is necessary to divide them thus:

Flowers in small clusters, never in catkins. — Elm Family See p. 116

Flowers in catkins, hanging or nearly erect.

Leaves simple. — See p. 84

Leaves compound. — Walnut Family

WALNUT FAMILY
(*Juglandaceae*)

TALL trees, the foliage often aromatic, the leaves always compound, and always with a terminal leaflet (there is no terminal leaflet in some compound-leaved trees, such as the Honey-Locust). The leaflets are usually short-stalked or stalkless. Male and female flowers separate, on the same tree, the male flowers in drooping catkins, the female small, often clustered, sometimes solitary. Fruit a winged nutlet in *Pterocarya,* and the familiar walnut or hickory in the other two genera.

The three groups may be divided thus:

Fruit a winged nut. Wing-nuts
Fruit a walnut or hickory.
Pith in plates; nut sculptured, its husk not splitting. Walnut
Pith solid; nut smooth, its husk splitting. Hickory

WING-NUT (*Pterocarya fraxinifolia*) p. 86
A Persian tree grown mostly for its hanging clusters of green winged nuts that are somewhat showy in midsummer. **Size:** 40–90 feet, and often dividing near the base so there are 2–4 trunks. **Leaves:** Alternate, compound, 9–24 inches long, composed of 11–21 oblongish, sharply toothed leaflets, 4–6 inches long, the foliage handsome. **Flowers:** In hanging clusters, 12–20 inches long, not conspicuous, greenish, small; Apr. **Fruit:** A hanging cluster, 12–20 inches long, of small, green winged nuts, quite showy in midsummer. **Hardy:** From Zone 4 southward. **Varieties:** None. **Culture:** Beyond its preference for moist sites there is no difficulty in growing it.

WALNUT
(*Juglans*)

Tall trees with aromatic foliage (when crushed), the leaves compound, with 7–21 leaflets, always with an odd leaflet at the tip, the other leaflets opposite. Bark usually scaly and furrowed.

Male flowers in hanging catkins, the female ones in terminal clusters, both sorts separate, but on the same tree. Fruit fleshy, and inside it a sculptured nut (the walnut or butternut), the husk of the fruit either very sticky-hairy or nearly smooth and not sticky.

The walnuts do well on deep, loamy soils, and as all of them produce a long taproot they are useless on shallow soils. Neither will they thrive if the soil is wet at the root depth. The Butternut is less demanding as to soils, but the English Walnut rarely does well in the East, except in a few varieties, thriving best in California. The species are divided thus:

Leaflets 5–9, the margins without teeth. English Walnut
Leaflets 11–23, the margins toothed.
Fruit smooth or slightly hairy, not sticky. Black Walnut
Fruit decidedly sticky-hairy. Butternut

ENGLISH WALNUT (*Juglans regia*) p. 86
A precariously hardy tree within the range of this book, but often tried by those who admire its fine canopy and delicious nuts. A native of Persia and unknown as a wild tree in England, its culture for nuts is practically confined to Calif. **Size:** 70–100 feet, the bark silvery gray. **Leaves:** Alternate, compound, the 5–9 leaflets oblongish, without marginal teeth, the foliage coarse and dense. **Flowers:** The male in hanging catkins, the female in erect clusters, both greenish, small, and inconspicuous. **Fruit:** Nearly round, green, about 2 inches thick, the enclosed nut much sculptured. **Hardy:** Only in some of its forms, from Zone 4 southward. **Varieties:** Many in Calif. for nut production. As an ornamental, and for nuts in the East, Carpathian is a hardier form; variety *laciniata* is a form grown mostly for its finely dissected leaflets which are very handsome. **Culture:** Precarious in the East; see above. The cabinet wood known as Circassian walnut is yielded by this tree, mostly in Europe.

BLACK WALNUT (*Juglans nigra*) p. 81
A beautiful native American tree valued for shade and for its hard, delicious nuts. **Size:** 70–150 feet high, never so high if the

ground water is too near the surface. **Leaves:** Alternate, compound, composed of 15–23 ovalish or narrower leaflets, 2½–5 inches long, irregularly toothed on the margin, minutely hairy but not sticky. **Flowers:** Inconspicuous. **Fruit:** Nearly round, about 2 inches thick, hairy, the enclosed nut irregularly sculptured. **Hardy:** From Zone 3 southward. **Varieties:** Thin-shelled varieties are found in Thomas, Thomas Improved, and Ten Eyck. **Culture:** There is no difficulty, except that a normally long-lived tree dies out young if the ground water is too near the surface.

Left: Black Walnut (*Juglans nigra*), p. 80.
Right: Butternut (*Juglans cinerea*), below.

BUTTERNUT (*Juglans cinerea*) above

A very hardy native American tree grown mostly for its nuts, as its canopy is generally too thin to cast much shade. **Size:** 40–90 feet. **Leaves:** Alternate, compound, composed of 11–19 sticky-hairy, oblongish leaflets, 2½–5 inches long. **Flowers:** Inconspicuous. **Fruit:** Egg-shaped or football-shaped, about 3½ inches long, densely sticky-hairy, the enclosed nut ridged, and with smaller ridges or furrows between the main ridges. **Hardy:** From Zone 2 southward. **Varieties:** None. **Culture:** See above.

Related plant (not illustrated):

Juglans sieboldiana. Hardier than the English Walnut, this Japanese tree has a variety *cordiformis,* which is the Heartnut, valued for its thin shell out of which the nut slips easily.

Hickory and Pecan
(*Carya*)

Tall trees, often resembling the walnut, but the branchlets have a solid pith (in plates in the walnut), and it bears a smooth nut (sculptured in the walnut). Leaves compound, with 5–17 opposite leaflets, always with an odd one at the tip, all of them toothed on the margin. Male and female flowers separate, but on the same tree, the female stalkless, the male in hanging catkins. Fruit with a splitting husk, the nut not sculptured and, except in the pecan, of much inferior flavor to the walnut and butternut.

There are many wild species of hickory, and some varieties of it, produced in the hope that the nuts would be easier to crack than at present. All the trees below have a deep taproot, are not easy to move, and when planted the taproot needs a deep root-run in reasonably good soil. Only the pecan is worth growing for its nuts, but its successful culture is mostly in the South. The hickories are scarcely garden trees, being most useful in places too hilly or rough for garden purposes. The cultivated species can be divided thus:

Leaflets 11–17. Pecan
Leaflets 5–7, very rarely 9. Shellbark Hickory

PECAN (*Carya Pecan*) p. 86

A brittle-wooded tree of the central and southern U.S., its branches and twigs often severely wind-pruned in a gale. **Size:** 60–150 feet, the bark furrowed. **Leaves:** Alternate, compound, composed of 11–17 oblongish leaflets that are 5–7 inches long and slightly sickle-shaped, the midrib not in the center of the leaflet. **Flowers:** Inconspicuous. **Fruit:** Fleshy at first, ultimately woody and splitting into 4 woody valves which enclose the cylindrical, smooth-shelled nut, 1½–2½ inches long, its shell (in the best varieties) thin, easily splitting, the kernel delicious. **Hardy:** As a desirable nut tree from Zone 5 southward. The tree is hardy up to Zone 4, but nuts will scarcely or never mature in that zone. **Varieties:** Over a hundred (mostly in the South). A selection for Zones 5 and 6 might include: Bradley, Curtis, Moneymaker, President, and Stuart. It is safer to select these only after tasting their nuts in a nut orchard, as, unless grafted trees are obtained, there is much uncertainty as to correct naming. **Culture:** See above, but it is imperative to have a deep hole to take care of the long taproot. The varieties above, plus the Burlington and Major, make good shade trees, except for the shedding of twigs and branches in a gale. The Pecan is called by some *Carya illinoensis.*

Shellbark Hickory
(*Carya laciniosa*), p. 84.

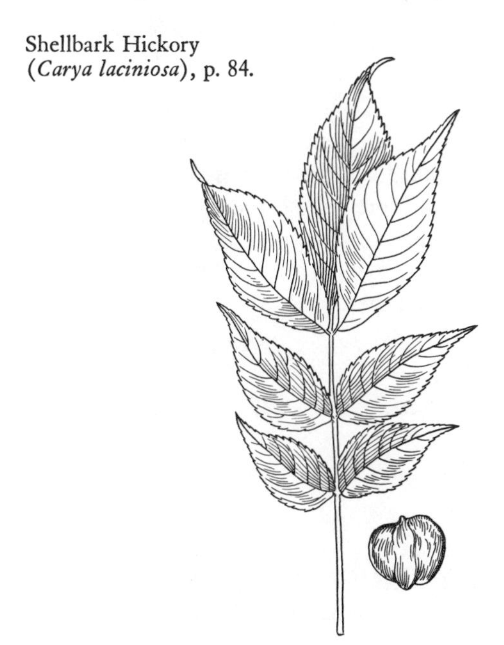

SHELLBARK HICKORY (*Carya laciniosa*) p. 83

A tall forest tree of the eastern U.S., its light gray bark shaggy and splitting off in long plates. **Size:** 80–120 feet high. **Leaves:** Alternate, compound, composed of 5–7 (rarely 9) oblongish leaflets 6–9 inches long, tapering at the tip and hairy beneath. **Flowers:** Inconspicuous. **Fruit:** The nut is nearly round, but obscurely 4-angled, pointed at the ends, its shell cracking with difficulty, the kernel sweet but meager. **Hardy:** From Zone 3 southward. **Varieties:** Several, among them Hales, Kentucky, Kirtland, Vest, and Weiper. These should be tasted at a nut orchard before planting as their identity is often obscure. Plant only grafted trees as seedlings may be erratic. **Culture:** See above.

Related plant (not illustrated):

Shagbark Hickory (*Carya ovata*). Taller than the Shellbark Hickory, with more shaggy bark and mostly with 5 (rarely 7) leaflets, their margins fringed with hairs.

Leaves Simple; Never Compound

Here belong a large group of shrubs and trees that have their flowers (or some of them) in catkins, and the flowers without petals. But unlike the walnuts and hickories the leaves are always simple, never compound. Here belong the bayberry, and on later pages the willow, oak, beech, chestnut, and the birch and its allies. Such a diversified group demands the following simple key to discriminate between them:

Juice or sap milky. Mulberry Family. p. 113

Juice or sap not milky.

Foliage resinous-dotted, pleasantly aromatic when crushed. Bayberry Family

Foliage not resinous-dotted, of indifferent or no odor when crushed. See p. 88

BAYBERRY FAMILY
(*Myricaceae*)

The bayberry and its relatives are among the most aromatic

plants, as their foliage has resinous dots on the upper or lower side, the dots filled with an aromatic oil, which may not be noticed until the leaf is crushed. Leaves alternate, simple, either toothed or the margins cut into narrow lobes, or sometimes with no teeth. Flowers without petals or sepals, mostly in small catkins. Fruit berrylike, often coated with wax or resinous dots.

The bayberries and the Sweet-Fern are of the easiest culture in sandy or gritty soil, but large specimens are hard to transplant. Both are natives and can often be dug from the woods, preferably when quite young.

Leaves without teeth or small ones. Bayberry
Leaf margins deeply cut into oblique lobes. Sweet-Fern

Bayberry
(*Myrica*)

Medium-sized shrubs, rarely treelike, with highly aromatic foliage when crushed. Only the three below are likely to be cultivated, and two of them can be dug from the wild. As the male and female flowers are borne on separate plants, only the latter bear the aromatic fruits, which are small and berrylike. The two below may be distinguished thus:

Leaves evergreen, less than ¾ inch wide. Wax Myrtle
Leaves not evergreen, ¾–1½ inches *wide.* Bayberry

WAX MYRTLE (*Myrica cerifera*) p. 91
An aromatic evergreen shrub or, rarely, a small tree, native in the southeastern U.S., grown chiefly for its aromatic evergreen foliage. Often called Bayberry or Tallow-Shrub. **Size:** 6–10 feet as a shrub, rarely a tree up to 35 feet. **Leaves:** Simple, alternate, more or less lance-shaped, 1–3 inches long, mostly evergreen, but sometimes merely persistent northward, resinous. **Flowers:** Inconspicuous. **Fruit:** Gray, wax-coated, about pea-size, borne in small clusters along the twigs. **Hardy:** From Zone 4 southward. **Varieties:** None. **Culture:** See above.

TREES AND SHRUBS, THE FLOWERS OF WHICH HAVE NO PETALS

1. **Wing-Nut** (*Pterocarya fraxinifolia*) p. 79
   Grown mostly for its green, winged nuts, this Persian tree often has 2–4 trunks.

2. **Weeping Willow** (*Salix babylonica*) p. 94
   Grown almost entirely for its handsome foliage on long pendulous branches that may reach the ground.

3. **Bayberry** (*Myrica pensylvanica*) p. 88
   An aromatic, nearly evergreen shrub, bearing the highly spicy bayberries used to scent candles.

4. **White Oak** (*Quercus alba*) p. 103
   The most majestic and long-lived of all the garden oaks, but difficult to start except from small specimens.

5. **English Walnut** (*Juglans regia*) p. 80
   A Persian tree, not usually thriving in the East. The picture shows young fruit, but not the nuts which are sculptured at maturity.

6. **White Poplar** (*Populus alba*) p. 90
   An almost weedy, quick-growing tree, useless for permanent effects, the underside of the leaves white.

7. **Pecan** (*Carya Pecan*) p. 82
   Native in the southern states but few wild trees bear such fine nuts as the hybrids. See text.

8. **Sweet-Fern** (*Comptonia peregrina*) p. 88
   A low, highly aromatic native shrub, useful in open, sandy places or in partial shade.

9. **Pin Oak** (*Quercus palustris*) p. 100
   A native, medium-sized, reasonably quick-growing tree, one of the best for making garden pictures.

1
2
3
4
5
6
7
8
9

1
2
3
4
5
6
7
8
9

TREES AND SHRUBS, THE FLOWERS OF WHICH HAVE NO PETALS

1. **Spanish Chestnut** (*Castanea sativa*) p. 106
   Grown almost entirely for its fine nuts, but not certainly hardy over much of the eastern seaboard.

2. **Osage Orange** (*Maclura pomifera*) p. 115
   A useful native tree for windy sites with poor soil. Its fruit large, green, and puckered.

3. **Black Alder** (*Alnus glutinosa*) p. 111
   Useful for low, wet places this Eurasian tree has little else to commend it. It grows 50–75 feet high.

4. **Hop-Hornbeam** (*Ostrya virginiana*) p. 112
   A medium-sized native tree best grown in cool, moist sites; rather attractive in fruit.

5. **Paper Birch** (*Betula papyrifera*) p. 108
   A beautifully white-barked native tree, more suited to the North than to warm, dryish sites.

6. **Mulberry** (*Morus alba*) p. 114
   A quick-growing tree with a profusion of blackberry-like, whitish fruit that is insipid and litters the lawn.

7. **Hornbeam** (*Carpinus Betulus*) p. 113
   An extremely hard-wooded, slow-growing, medium-sized tree, useful as sheared to make a dense hedge.

8. **European Hazel** (*Corylus Avellan*) p. 112
   Several varieties of this European shrub are far more decorative than the typical form. See text.

9. **European Beech** (*Fagus sylvatica*) p. 105
   A magnificent lawn tree, especially in some of its forms, such as the Copper Beech and the Weeping Beech.

BAYBERRY (*Myrica pensylvanica*) p. 86
An aromatic shrub, the foliage not evergreen but often hanging on until early winter. **Size:** 3–8 feet high, nearly as wide. **Leaves:** Resinous-dotted, very aromatic, more or less elliptic, or broadest toward the tip, 3–4 inches long. **Flowers:** Inconspicuous. **Fruit:** Conspicuously grayish-waxy, about pea-size, used in making bayberry candles. **Hardy:** Throughout its wild range from Newfoundland to Fla., mostly in sandy or rocky soils near the coast. **Varieties:** None. **Culture:** See above, but it is safer to dig wild plants while still young.

Related plant (not illustrated):

Sweet-Gale (*Myrica Gale*). A bog shrub, 2–4 feet high, found throughout the north temperate zone, useful only in very acid sites in the bog garden.

SWEET-FERN (*Comptonia peregrina*) p. 86
A very pleasantly scented, aromatic low shrub from eastern North America found growing in sandy or rocky soils and suited only to such sites. **Size:** 2–5 feet high, often less. **Leaves:** Alternate, stalked, rather narrow, 4–5 inches long, the margins obliquely cut into rounded lobes. **Flowers:** Inconspicuous. **Fruit:** A small nutlet, surrounded by tiny leaflike bracts, the fruit thus bur-like. **Hardy:** From Zone 2 southward. **Varieties:** None. **Culture:** An excellent low shrub for dry-sandy or peaty-sandy soils. It will not tolerate moist or heavy loamy soils.

Foliage Not Pleasantly Aromatic and Not Resinous-dotted

Here are grouped shrubs and trees with simple leaves, some of their flowers in catkins, always without petals. Many of them are among our most valuable cultivated trees. The group is difficult to separate without the use of technical characters, but the following scheme, plus the pictures, should make identifications relatively simple.

Fruit a small pod that splits. Willow Family
Fruit not a pod.
Fruit an acorn, or a nut partly enclosed by a woody husk. Beech Family. p. 97
Fruit not an acorn; either a nut enclosed by a leafy husk, or the fruit like a small woody cone. Birch Family
See p. 107

WILLOW FAMILY
(*Salicaceae*)

A VERY large family of shrubs and trees, including the willows and poplars, each with many species, and, in the willows, a bewildering number of natural and induced hybrids. The willows, particularly, are difficult or impossible to tell apart with only the foliage in hand. Leaves simple, alternate, never divided into lobes, usually toothed on the margin. Flowers always in catkins, which hang in the poplars but are erect in the willows. Male and female flowers always on separate plants. Fruit a capsule which splits lengthwise, revealing the seeds which have silky hairs at their base. The two genera can be distinguished thus:

Catkins hanging, the scales of the catkins minutely fringed. Poplar
Catkins erect, the scales of the catkin not fringed. Willow

POPLAR
(*Populus*)

Over 35 different species of poplar are known to be in cultivation in the U.S., not counting many others known only as wild trees. Of these only the four below can be included here, together with three others that may interest the amateur. All are trees of rapid growth, soft wood, brittle branches, and all of them are unsuited for permanent plantings as they are relatively short-lived. As temporary planting for quick growth they have many uses as screens and for pulpwood.

Leaves alternate, relatively long-stalked, the stalks in some flattened (the quaking aspens). Buds always resinous. Flowers in hanging catkins, always in bloom before the leaves unfold, the sexes on different trees. Fruit a splitting pod, which is ripe before the leaves are fully expanded. Seeds plentiful and brown, each with copious silky hairs at the base.

Poplars are of the easiest culture and some of them, due to plentiful seeding, can become a weedy nuisance, especially along roadsides and waste places. Their ease of propagation by cuttings is notorious, as any twig in reasonably moist soil will quickly strike root. Some of the wild species are known as cottonwood, others as aspen. The four here included may be distinguished thus:

Leaves white-felty beneath. White Poplar
Leaves not white-felty beneath; smooth or somewhat hairy, or pale beneath.
Leafstalks round, the leaves not quaking in the wind. Balm-of-Gilead
Leafstalks flattened, the leaves quaking in the wind.
Leaf margins minutely hairy. Carolina Poplar
Leaf margins without hairs. Black and Lombardy Poplar

WHITE POPLAR (*Populus alba*) p. 86
A decidedly weedy Eurasian tree, often escaping to roadsides and thickets and sometimes a nuisance, but quick-growing and its foliage striking. **Size:** 30–70 feet high, about half as wide. **Leaves:** Not quaking in the wind, 3–5-lobed, the lobes coarsely toothed, 3–5 inches long, the under side conspicuously white-felty. **Flowers:** Small, in hanging catkins, which bloom before the leaves unfold. **Fruit:** A small pod. **Hardy:** Everywhere. **Varieties:** *nivea,* a form with the under side of the leaves still whiter than the typical tree; *pyramidalis,* a narrow, columnar, upright form useful as an accent and sometimes offered as *Populus bolleana.* Many think it superior to the Lombardy Poplar, when used for the same purpose. *See* below.

Related plant (not illustrated):

European Aspen (*Populus tremula*). A larger, round-headed tree, the quaking leaves pale beneath, but not white-felty.

Left: Wax Myrtle (*Myrica cerifera*), p. 85. Center: Carolina Poplar (*Populus canadensis*), p. 92. Right: Balm-of-Gilead (*Populus candicans*), below.

BALM-OF-GILEAD (*Populus candicans*) above

A poplar of unknown origin, often called Balsam Poplar, of which only female trees are known. **Size:** 60–90 feet high, half as wide or more. **Leaves:** Not quaking, broadly oval or triangular, heart-shaped at the base, coarsely toothed, 4–6½ inches long, whitish and hairy beneath. **Flowers:** In hanging catkins. **Fruit:** A small pod. **Hardy:** Everywhere. **Varieties:** None. **Culture:** See above. Its wide-spreading habit makes it useless for confined spaces.

Related plant (not illustrated):

Populus simoni. A narrow-headed Chinese tree, 20–36 feet high, the ovalish leaves not quaking, 3–5 inches long and pale beneath.

CAROLINA POPLAR (*Populus canadensis*) p. 91
A hybrid poplar, not native in the Carolinas, of which only male trees are known, and often undesirable near drains or water pipes because its roots often clog them. **Size:** 50–90 feet high, about a third as wide. **Leaves:** Quaking, the leafstalks flattened, more or less triangular in outline, coarsely toothed, the margins minutely hairy, whitish and slightly hairy beneath. **Flowers:** In hanging catkins. **Fruit:** (?). **Hardy:** Everywhere. **Varieties:** *eugenei,* by some considered the true Carolina Poplar, is a form with a narrowly pyramidal habit. **Culture:** See above, but not recommended for planting on streets or lawns as it litters both with twigs and catkins.

BLACK and LOMBARDY POPLAR (*Populus nigra*) below
A hardy Eurasian tree, 60–90 feet high, with quaking leaves, much less grown than its variety *italica,* the Lombardy Poplar, which is the most planted of all poplars in the U.S. in spite of its short life, brittle wood, and invasive roots. The Lombardy Poplar is 30–50 feet high, its narrow columnar outline very effective as an accent or to line a drive (only temporarily). **Leaves:** Quaking, triangular or wedge-shaped, 3–4 inches long, the margins finely toothed but not hairy along the edge. **Flowers:** In hanging catkins. **Fruit:** A small pod. **Hardy:** Everywhere. **Varieties:** The

Lombardy Poplar
(*Populus nigra*), above.

Lombardy is a variety of the Black Poplar, and is perhaps overplanted in the U.S. because of its striking columnar habit, but it cannot be recommended. Many think the columnar form of the White Poplar (*Populus alba pyramidalis*) is superior to that of the Lombardy.

WILLOW
(*Salix*)

Shrubs or trees, the commonest being the Pussy Willow and the Weeping Willow. But the 63 species in cultivation in the U.S., and twice that number of natural or induced hybrids, are little known to the amateur and quite difficult to identify. Twenty species were listed in the fourth edition of *Taylor's Encyclopedia of Gardening,* but of these only 6 can be included here, together with a few other forms.

Leaves alternate (except in *S. purpurea*), generally narrowly lance-shaped. Flowers without petals or sepals, always in erect catkins (the Pussy Willow) and the sexes on different plants. They bloom before the leaves unfold, and the scales of the catkins are smooth on the margin (hairy in the poplars). Fruit a small, splitting capsule (a pod).

The cultivation of willows is practically foolproof. They generally tolerate moist or wet places and any twig or branch will root with ease in moist soil. By far the most desirable as landscape subjects are the weeping willows — most of them of uncertain hybrid origin. The six here considered are best divided only on habit and leaf characters as the technical differences are rather forbidding.

Leaves green on both surfaces. Black Willow
Leaves distinctly paler beneath, sometimes bluish or grayish green.
  Trees with pendulous, "weeping" branches. Weeping Willow
  Trees or shrubs with erect branches.
    Leaves nearly opposite. Purple Osier
    Leaves alternate.
    Leaves slightly wrinkled above. Goat Willow

Leaves not wrinkled above.
Leafstalk about 1 inch long. Pussy Willow
Leafstalk less than 1 inch long. *Salix Matsudana*

BLACK WILLOW (*Salix nigra*) p. 95
A native American tree, valued chiefly for its nearly black bark which makes dramatic contrasts with white-flowered shrubs or trees planted with it. **Size:** 20–35 feet high, less than a third as wide due to its erect branches. **Leaves:** Narrowly lance-shaped, 3–5 inches long, finely toothed. **Flowers:** In erect catkins. **Fruit:** A small pod. **Hardy:** From Zone 2 southward. **Varieties:** None. **Culture:** See above.

Related plant (not illustrated):

Laurel Willow (*Salix pentandra*). A European tree, 40–60 feet high, with shining green leaves 3–5 inches long, and relatively showy, golden-yellow catkins.

WEEPING WILLOW (*Salix babylonica*) p. 86
The most popular of all the willows because of its slender drooping branches, which may reach the ground. **Size:** 30–40 feet high, half as wide, its branches yellowish brown when young. **Leaves:** Very narrow, 5–6 inches long, finely toothed and one of the earliest to unfold in the spring. **Flowers:** In erect catkins. **Fruit:** A small pod. **Hardy:** Everywhere. **Varieties:** *aurea,* a form with yellowish foliage; *crispa,* a form with folded and slightly twisted leaves. **Culture:** See above.

Related weeping willows (not illustrated):

Wisconsin Weeping Willow (*Salix blanda*). A round-headed hybrid tree with weeping branches.

Thurlow's Weeping Willow (*Salix elegantissima*). A medium-sized tree with long, pendulous branches, probably of hybrid origin.

Niobe. A trade name for a weeping willow, probably of hybrid origin and related to the Wisconsin Weeping Willow.

Note: The identification of these 4 weeping willows is extremely difficult as their differences are slight, and based mostly on features not easily appraised.

PURPLE OSIER (*Salix purpurea*) below
A shrub, not over 8 feet high, a native of Eurasia, not much grown except in the variety *nana* (a trade name only) — a dwarf form widely grown for its purplish twigs. **Size:** 1–3 feet high, the young twigs purplish, later grayish. **Leaves:** Nearly opposite, lance-shaped or broader, 2–4 inches long, toothed toward the tip, pale beneath. **Flowers:** In erect catkins. **Fruit:** A small pod. **Hardy:** From Zone 2 southward. **Varieties:** The plant is a variety of *Salix purpurea.* **Culture:** See above.

Related plants (not illustrated):

Rosemary Willow (*Salix Elaeagnos*). A shrub or small tree, usually trimmed and then boxlike. Hardy from Zone 3 southward and Eurasian in origin.

Salix gracistyla. A Japanese shrub, 4–6 feet high, the twigs grayish-hairy. Leaves bluish green, and hairy beneath. Hardy from Zone 4 southward.

Left: Goat Willow (*Salix Caprea*), below. Center: Purple Osier (*Salix purpurea*), above. Right: Black Willow (*Salix nigra*), p. 94.

GOAT WILLOW (*Salix Caprea*) above
A Eurasian shrub or, more rarely, a small bushy tree grown mostly for its bright yellow catkins. **Size:** 10–25 feet high, usually much less as cultivated. **Leaves:** Oblongish or broader, 3–4 inches long, faintly toothed, slightly wrinkled above. **Flowers:**

Bright yellow and in fairly showy catkins. **Fruit:** A small pod. **Hardy:** From Zone 3 southward. **Varieties:** None that are superior to the typical form. **Culture:** See above.

PUSSY WILLOW (*Salix discolor*) below
The familiar Pussy Willow of the florists and easily forced into bloom by bringing cut twigs into the house any time after January 15th. **Size:** A shrub or small tree 10–18 feet high, about half as wide. **Leaves:** Elliptic or oblongish, 3–4 inches long, finely wavy-toothed or with none, bluish green beneath. **Flowers:** The female flowering catkins are the familiar Pussy Willow. **Fruit:** A small pod. **Hardy:** Everywhere. **Varieties:** None. **Culture:** See above.

Left: Corkscrew Willow (*Salix Matsudana tortuosa*), below.
Right: Pussy Willow (*Salix discolor*), above.

SALIX MATSUDANA above
An Asiatic tree scarcely known in gardens except for the variety *tortuosa,* the Corkscrew Willow, which is an interesting variation. **Size:** Usually a small tree or large shrub, 6–15 feet high, its twigs twisted and contorted. **Leaves:** Narrowly lance-shaped, 2–4

inches long, spirally twisted or half coiled. **Flowers:** In erect catkins. **Fruit:** A small pod. **Hardy:** From Zone 3 southward. **Varieties:** The Corkscrew Willow is itself a variety of *Salix Matsudana.* **Culture:** See above. When grown in any garden it is immediately a conversation piece because of its fantastic tortuous habit.

Fruit an Acorn, or a Nut Partly Enclosed by a Woody Husk

BEECH FAMILY
(*Fagaceae*)

STATELY, long-lasting trees are grouped in this family, which besides the beech includes the oak and chestnut. Leaves alternate, mostly toothed or lobed, but without either in some oaks. Flowers without petals, mostly in hanging catkins, the male and female separate, but on the same tree. Fruit an acorn in the oaks, but a nut partly or completely covered by a somewhat woody and often prickly husk in the chestnut, a less prickly one in the beech. The three genera are easily separated:

Fruit an acorn. Oak
Fruit not an acorn.
  Nuts obviously triangular; bark smooth and gray. Beech
  Nuts round, with one flattened side; bark not smooth and gray Chestnut

OAK
(*Quercus*)

Majestic trees, rarely shrubs, the wood hard and lasting. The oaks comprise a group of nearly 300 species of which about 60 are in cultivation in the U.S. Of these perhaps 20 are fairly well known as garden specimens, but here only the 8 most likely to interest the amateur are considered, plus a few more that are less known. The many native oaks, unless they are in common cultivation, cannot be included here. For them one should turn to the new *Illustrated Flora* of Britton and Brown, by H. A. Gleason, issued by the New York Botanical Garden in 1952.

Leaves simple, alternate, dropping in the fall or by late winter in all those below except the Live Oak. The leaves are variously toothed or lobed in all but the Live Oak and the Willow Oak, which have no lobes and few, if any, teeth. In those that have marginal teeth, the white oak group has no bristles at the ends of the lobes or teeth, while the red oak group always has such bristles at the tip of their teeth or lobes. Flowers without petals, the sexes separate on the same tree, the male flowers in drooping catkins, blooming before or with the unfolding of the leaves. Fruit the familiar acorn, set in a woody cup. The acorns in the white oaks mature in one season, while the red oak group matures its acorns at the end of the second season. The oaks are much given to hybridization and their identity is thus often puzzling. Without acorns it is hazardous to try to identify most of them.

The oaks are among the finest deciduous trees for the garden, but they need room. None should be planted nearer than 60 feet to any other tree, and as oaks are mostly slow-growing and attain great age and spread, this spacing is a minimum. The White Oak needs even more space and is difficult to transplant without a tight ball of soil. Most other oaks, particularly the Red and Pin Oaks, are easier to grow and tolerate any ordinary garden soil so long as it is not wet. The oaks below may be divided thus:

Leaves evergreen; a southern tree. Live oak
Leaves not evergreen.
  Leaves without marginal teeth or lobes. Willow Oak
  Leaves with marginal teeth or lobes.
    Lobes of the leaf or its teeth ending in bristles; acorns maturing in 2 years (the red oaks).
      Main branches in wide-spreading, nearly horizontal tiers. Pin Oak
      Main branches more or less erect.
        Leaves hairy beneath. Black Oak

Leaves smooth beneath, except for some tufts of hair near the midrib.
Leaves cut almost to the center. Scarlet Oak
Leaves less deeply cut. Red Oak
Lobes of the leaf or its teeth never ending in a bristle; acorns maturing in 1 year (the white oaks).
Leafstalk less than ⅜ inch long; leaves slightly eared at the base. English Oak
Leafstalks longer; leaves mostly wedge-shaped at the base. White Oak

LIVE OAK (*Quercus virginiana*) p. 100
The only evergreen oak native in the eastern states (many evergreen species in Calif. and China), but unfortunately not hardy in the North. It is a magnificent wide-spreading tree, needing much space in maturity. **Size:** 50–70 feet, nearly as wide. **Leaves:** Leathery, evergreen, elliptic or oblong, 3–5 inches long, without lobes, very rarely with a few teeth, but no bristles, green above, white-felty beneath. **Flowers:** In hanging catkins. **Fruit:** An egg-shaped acorn, about 1 inch long, its turban-shaped cup covering about a fourth of the acorn. **Hardy:** From Zone 6 southward, but worth trying in the southern part of Zone 5. **Varieties:** None. **Culture:** See above.

Related plants (not illustrated):

Holm Oak (*Quercus Ilex*). An evergreen tree from southern Europe, with prickly, holly-like leaves. Hardy only from Zone 6 southward.

Water Oak (*Quercus nigra*). Native from Del. to Fla., and nearly evergreen in the South, its bluish-green leaves unlobed, or with 3 lobes at the tip. Hardy from Zone 4 southward.

WILLOW OAK (*Quercus Phellos*) p. 100
A medium-sized tree with rather thin canopy and hence not casting a dense shade (lawns will grow under it), native from Long Island, N.Y. to Fla. **Size:** 40–60 feet high, about half as

Left: Live Oak (*Quercus virginiana*), p. 99.
Right: Willow Oak (*Quercus Phellos*), p. 99.

wide. **Leaves:** Superficially resembling willow leaves, without teeth, lobes, or bristles, 4–5 inches long, persisting into the winter, but not evergreen. **Flowers:** In hanging catkins. **Fruit:** A nearly round acorn, about ½ inch long, enclosed only at the base by the saucer-shaped cup. **Hardy:** From the southern edge of Zone 4 southward. **Varieties:** None. **Culture:** See above. Its rather shallow root system makes it easy to transplant, but violent winds in exposed places make it vulnerable to uprooting.

Related plant (not illustrated):

Laurel Oak, Darlington Oak (*Quercus laurifolia*). A native tree, with half-evergreen leaves, 4–6 inches long, without teeth or bristles, but sometimes faintly lobed. Acorn about ⅝ inch long, its cup one fourth the length of the nut. Hardy from Zone 4 southward.

PIN OAK (*Quercus palustris*) p. 86

A native tree with distinctive arrangement of its branches in flattish tiers, and one of the best and quickest-growing of the oaks for lawn or street planting. **Size:** 80–90 feet high, often less as cultivated, about a third as wide. **Leaves:** Elliptic, 4–5 inches long, shining green, sharply and deeply 5–9-lobed, the lobes bristle-tipped; scarlet in the fall. **Flowers:** In hanging catkins. **Fruit:** A flattened egg-shaped acorn, about ½ inch long, enclosed only a fourth or a third its length by the cup. **Hardy:** From Zone 3 or 4 southward. **Varieties:** None. **Culture:** See above, but it will thrive in moister sites than most oaks.

BLACK OAK (*Quercus velutina*) p. 102

A native tree useful as a tall, stately lawn specimen, its inner bark conspicuously yellow-orange. **Size:** 100–125 feet high, nearly half as wide. **Leaves:** Large, handsome, shining, dark green (red in autumn), 7–9 inches long, hairy beneath, 7–9-lobed, the lobes sharp-pointed and bristle-tipped. **Flowers:** In hanging catkins. **Fruit:** An elliptical acorn, about ¾ inch long, covered nearly half its length by the cup. **Hardy:** From Zone 3 southward. **Varieties:** None that are any better than the typical form. **Culture:** See above, but its deep taproot makes transplanting difficult.

SCARLET OAK (*Quercus coccinea*) p. 102

A native tree, its brilliant scarlet autumnal foliage superb. Most desirable as a street or lawn tree, but a little difficult to get established. **Size:** 50–80 feet high, more or less cylindric in outline, the foliage rather open. **Leaves:** Oblongish, smooth beneath (except for tufts or hair near the midrib), shining green above, oblongish, 4–6 inches long, cut almost to the center by the 7–9 sharp, bristle-tipped lobes. **Flowers:** In hanging catkins. **Fruit:** An egg-shaped acorn about ¾ inch long, its cup covering nearly one half of the nut. **Hardy:** From Zone 2 southward. **Varieties:** None superior to the typical form. **Culture:** See above, but the tree has a deep taproot and if this is not given sufficient root-run the tree will not thrive. Better to start with young specimens (2–4 feet).

RED OAK (*Quercus rubra*) p. 102

A native tree, one of the quickest-growing of all the oaks, its foliage turning a dull red in autumn. It is one of the best of the oaks for landscape or street planting, partly because of its ease in transplanting. **Size:** 60–80 feet high, round-topped in age, and needing plenty of space. **Leaves:** Oblong, 6–9 inches long, smooth beneath (except for tufts of hair near the midrib), cut nearly half its width into 5–7 (rarely 3–11) sharp-pointed, bristle-tipped lobes. **Flowers:** In hanging catkins. **Fruit:** An egg-shaped acorn nearly 1 inch long, its cup about a third the length of the nut. **Varieties:** None. **Culture:** See above, but it is one of the easiest of all the oaks to get established. Some prefer to call this tree *Quercus borealis maxima.*

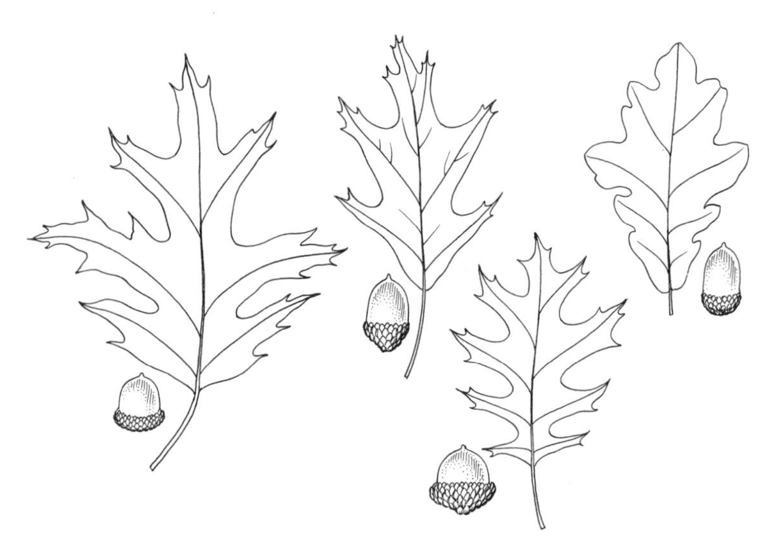

From Left to Right: Red Oak (*Quercus rubra*), p. 101; Black Oak (*Quercus velutina*), p. 101; Scarlet Oak (*Quercus coccinea*), p. 101; English Oak (*Quercus Robur*), below.

ENGLISH OAK (*Quercus Robur*) above

A Eurasian and North African representative of our native White Oak, not so large or lasting, but nearly as broad, its foliage still green when falling. **Size:** 60–100 feet, and half as wide or even more. **Leaves:** 3–5 inches long, broadest toward the tip, slightly eared at the base, with 6–14 shallow, rounded lobes which are not bristle-tipped, the leafstalk less than ⅜ inch long. **Flowers:** In hanging catkins. **Fruit:** An egg-shaped or oblongish acorn, about ¾ inch long, the cup enclosing about a third of the nut. **Hardy:** From Zone 3 southward, but often dying long before maturity in the northern part of its hardiness range. **Varieties:** Several, the best of them being *fastigiata,* a form with narrow columnar outline not unlike the Lombardy Poplar, but much more lasting; *atropurpurea,* leaves dark purple when young. **Culture:** See above.

WHITE OAK (*Quercus alba*) p. 86

The aristocrat of all the native oaks, of great size and spread, very hard wood, and living to a great age (500–800 years). **Size:** 60–100 feet high, nearly as wide or even more, its lateral branches huge. **Leaves:** Broadest toward the tip, narrowed at the wedge-shaped base, oblongish, 5–8 inches long, with 5–9 blunt lobes that are not bristle-tipped, green above, paler beneath, the leaf-stalk about ¾ inch long; autumnal color dull reddish, but the withered leaves persistent into the winter. **Flowers:** In hanging catkins. **Fruit:** Egg-shaped or oblongish, about ¾ inch long, the cup enclosing only about a fourth of the nut. **Hardy:** From Zone 2 southward. **Varieties:** None. **Culture:** Difficult. Only young specimens should be used, and the ball of earth surrounding the roots must be tightly roped up in burlap. Digging plants from the forest (a common practice) is courting failure.

Related plants (not illustrated):

Swamp White Oak (*Quercus bicolor*). A native tree 40–70 feet high, the lobes of the leaf cut nearly to the middle of the blade, not bristle-tipped. Hardy from Zone 2 southward.

Chestnut Oak (*Quercus Prinus*). A native tree 60–100 feet high, its leaves suggesting the chestnut, the shallow, blunt lobes not bristle-tipped. Hardy from Zone 2 southward.

Bur Oak (*Quercus macrocarpa*). A native tree 60–100 feet high, unlike all the other oaks here treated in having the cup of the acorn conspicuously fringed. Hardy from Zone 2 southward.

BEECH
(*Fagus*)

Beautiful stately trees with grayish, close bark and conspicuously golden buds. Leaves alternate, never lobed or cut (except in a variety of European Beech), but the margins with small or large teeth. Flowers without petals, inconspicuous, appearing with the unfolding of the leaves, the sexes separate but on the same tree. Fruit a small, obviously triangular nut, invested by a woody 4-hinged husk which is somewhat prickly on the outside. Only the native and European beech are listed below.

A full-grown beech is a stately lawn specimen and needs plenty of space. Its cultural requirements are few, but it must not go in wet land. The only real difficulty in getting it established is that it has a deep root and this should be preserved. Young root-pruned specimens with a good ball of earth usually do well, but the beech is not easily moved. It needs little subsequent attention.

Our two species are:

Leaves rather coarsely toothed; with 9–14 pairs of veins. American Beech
Leaves rather minutely toothed; with 5–9 pairs of veins. European Beech

AMERICAN BEECH (*Fagus grandifolia*) p. 105
A splendid lawn specimen, with beautifully gray, close and smooth bark, but not easy to transplant. **Size:** 90–100 feet high, usually columnar in outline, but at least 40 feet wide at maturity. **Leaves:** Simple, alternate, beautifully silky when unfolding, oblong-oval, coarsely toothed, 4–7 inches long, with 9–14 pairs of veins. The leaves are so numerous that the beech casts a dense shade. **Flowers:** Inconspicuous. **Fruit:** About 1 inch long. **Hardy:** From Zone 2 southward. **Varieties:** None. **Culture:** See above. The American Beech is not suited to city conditions, in which the next species is to be preferred.

Left: American Beech (*Fagus grandifolia*), p. 104.
Right: Chinese Chestnut (*Castanea mollissima*), p. 107.

EUROPEAN BEECH (*Fagus sylvatica*) p. 87
The finest beech for general planting and well suited for suburban and even city conditions if smoke, dust, and motor fumes are not too severe. Its chief difference from the American Beech is that the leaves are more finely toothed and have only 5–9 pairs of veins. Other differences are minor and technical and can be ignored by most gardeners. Its great value is its dense shade (nothing will grow under it) and majestic stature. **Hardy:** From Zone 3 southward. **Varieties:** At least 20, those noted below being deservedly the most widely planted of our ornamental trees: *atropunica* has purplish leaves; *cuprea,* the Copper Beech, has coppery or rosy-colored leaves; *pendula,* the so-called Weeping Beech, has many hanging branches so numerous that they may hide the trunk. It is a magnificent, gracefully drooping form which also comes in a purple-leaved type; *laciniata,* the Fern-Leaf Beech, has sharply cut, fern-like leaves which cover the tree with a dense, shimmering mantle of green; *fastigiata,* a type with erect branches, forming a narrow spirelike tree. **Culture:** See above.

CHESTNUT
(*Castanea*)

Medium-sized trees or shrubs with furrowed (never gray) bark and rather coarse wood. Leaves alternate, more or less toothed, and with many parallel veins. Flowers without petals, in more or less erect catkins, blooming as the leaves unfold, the sexes separate but on the same tree. Fruit nearly round, with one side flattened (the familiar chestnut), usually 1–3 of them surrounded by a prickly, bur-like husk which splits at maturity. The native and much-prized American Chestnut has been all but exterminated by the blight and is no longer cultivated. Hybrids with it and foreign species are the only trees worth cultivating, except the Spanish Chestnut, which so far seems to be relatively immune to the blight but is not hardy in severe climates.

Chestnuts are more tolerant of poor soil and dry weather than most trees, so their culture is no problem. They should not, however, be put in wet land. Only two species are here considered:

Leaves more or less felty beneath. Spanish Chestnut
Leaves not felty beneath. Chinese Chestnut

SPANISH CHESTNUT (*Castanea sativa*) p. 87
A round-headed tree grown mostly for its delicious nuts that are much larger than those of the defunct American Chestnut or the newer hybrids derived mostly from crossing the Chinese Chestnut with other types. **Size:** 60–90 feet high, nearly half as wide. **Leaves:** Oblongish, 7–12 inches long, coarsely toothed, more or less felty beneath. **Flowers:** Inconspicuous. **Fruit:** A prickly bur, 2–4 inches in diameter, containing 1–3 nuts that are nearly twice the size of those of the American Chestnut. **Hardy:** From Zone 5 southward, or even in protected sites in Zone 4 (hazardous there, however). **Varieties:** None of any importance to the gardener. **Culture:** See above, but the tree is not very well suited to most of our area, although there are fine fruiting specimens in Md. and Va.

CHINESE CHESTNUT (*Castanea mollissima*) p. 105
A relatively small tree, sometimes inclined to be shrubby, of chief interest as one of the parents of the many hybrid chestnuts being developed to replace the American Chestnut. **Size:** 30–50 feet high, about half as wide. **Leaves:** Oblong or elliptic, 4–7 inches long, coarsely toothed and often white-hairy on the veins beneath, but not felty. **Fruit:** A prickly bur enclosing 2–3 nuts that are nearly 1 inch wide and pointed. **Hardy:** From Zone 4 southward. **Varieties:** Many, as it has been much crossed with other species in developing a substitute for the American Chestnut, a program still continuing. **Culture:** See above.

Related plant (not illustrated):

Japanese Chestnut (*Castanea crenata*). A shrubby tree, 10–30 feet high, the leaves 4–7 inches long; of chief interest as used in hybridizing to create a blight-free substitute for the American Chestnut.

BIRCH FAMILY
(*Betulaceae*)

SHRUBS or trees with alternate, always stalked, and usually straight-veined leaves, which are not resinous-dotted (as in the bayberry) and have no odor when crushed, but the broken twigs of some species have a strong wintergreen odor. Flowers without petals, small, and in catkins, the sexes in separate flowers on the same tree. Fruit a nut, winged in the birches but surrounded by a leafy husk in the hazel and some other trees, or a persistent, woody conelike structure in the alders. None of the family is particularly ornamental except some of the birches. Our 5 groups can be divided thus:

Fruit a small winged nut, borne in a tight leafy cluster.
- *Fruit quickly falling; mostly tall trees.* Birch
- *Fruiting clusters woody and persistent; mostly shrubs.* Black Alder

Fruit surrounded by a leafy husk.
- *Nut large, edible; shrubs.* Hazel

Nuts small, not edible; trees.
Each nut in a closed bladdery husk, the cluster of them not unlike that of the hop. Hop-Hornbeam
Each nut not in a hop-like cluster. Hornbeam

BIRCH
(*Betula*)

The birches, except for the white-barked forms, are not particularly handsome garden trees, but some of the cut-leaved varieties are widely planted for ornament. Some species have a strong wintergreen odor and flavor in the bark of their twigs (the familiar "birch bark"), while in others there is no odor. Leaves alternate, stalked, always with marginal teeth or lobes, the veins straight. Flowers without petals, mostly in catkins, the male and female separate, but on the same tree. Fruit a 1-seeded small nut, often winged, borne in a small, tight, leafy cluster known as a strobile and often missed because it quickly falls when mature.

The birches are relatively quick-growing and not very lasting trees, suited to most garden soils, but tolerating moist or even wettish places better than some more desirable trees. The white-barked sorts make beautiful winter contrasts with evergreens. Only 4 of the birches are of much garden interest.

Bark white or grayish white, not wintergreen-flavored when stripped.
Young twigs hairy, sometimes a little sticky. Paper Birch
Young twigs smooth, nearly always sticky. European White Birch
Bark not white or grayish.
Twigs wintergreen-flavored when stripped. Black Birch
Twigs not wintergreen-flavored when stripped. River Birch

PAPER BIRCH (*Betula papyrifera*) p. 87
A beautiful, white-barked native American tree, often called Canoe Birch and White Birch, and often incorrectly listed as

Betula alba. **Size:** 30–50 feet, about a third as wide, its foliage rather thin. **Leaves:** More or less oval, but narrowed or wedge-shaped at the base, 3–5 inches long, double-toothed, with 3–7 pairs of veins. **Flowers:** Inconspicuous, in hanging catkins. **Fruit:** A small nut within a small leafy cluster (strobile). **Hardy:** From Zone 4 to the Arctic and not suited to the hot coastal plain. **Varieties:** None. **Culture:** Easy in the North, but to be avoided in the South. Its brilliant white bark makes a dramatic contrast to evergreens planted behind it.

Related plants (not illustrated):

Gray Birch (*Betula populifolia*). A weedy poor relation of the Paper Birch, its bark dirty white. Its great and only merit is that it will stand city dust, smoke, wind, and the cindery wastes of factory yards.

EUROPEAN WHITE BIRCH (*Betula pendula*) p. 110

A tree inferior to our native Paper Birch, but more tolerant of heat and hence useful in regions too hot for the American tree. Bark white, but not so brilliantly so as that of the Paper Birch, and obvious only in larger specimens. **Size:** 20–40 feet, about one third as wide, its foliage thin, and its twigs smooth but nearly always sticky. **Leaves:** Ovalish, more or less wedge-shaped at the base, double-toothed, 3–5 inches long, with 3–7 pairs of veins. **Flowers:** Inconspicuous, in hanging catkins. **Fruit:** A small nut within a small leafy cluster (strobile). **Hardy:** From Zone 2 southward, but not long-lived. **Varieties:** Several, of which the best are *youngi,* with gracefully hanging branches, a good form for small gardens; *fastigiata,* a columnar, narrow form; *laciniata* (sometimes offered as *gracilis*), a variety with much-cut, finely dissected leaves. **Culture:** See above. The plant is often incorrectly offered as *Betula alba*.

BLACK BIRCH (*Betula lenta*) p. 110

A tall native American tree, its bark dark reddish brown, its twigs wintergreen-flavored when stripped of bark. **Size:** 50–75 feet high, about a third as wide, the foliage rather dense. **Leaves:** Oblong-oval, 3–5 inches long, heart-shaped at the base, with 7 or more pairs of veins and double-toothed. **Flowers:** Inconspicuous. **Fruit:** A small nut within a small leafy cluster (strobile). **Hardy:** From Zone 3 southward. **Varieties:** None. **Culture:** See above.

Related plant (not illustrated):

Yellow Birch (*Betula lutea*). A native American tree, 80–90 feet high, its yellowish bark peeling off like shavings. Hardy from Zone 4 northward, and unsuited to the coastal plain.

Left: Black Birch (*Betula lenta*), p. 109.
Center: European White Birch (*Betula pendula*), p. 109.
Right: River Birch (*Betula nigra*), below.

RIVER BIRCH (*Betula nigra*) above

A native American tree preferring moist sites and useful in the garden only in such places. **Size:** 60–80 feet high, about a third as wide, its bark reddish brown and the tree is hence often called Red Birch. **Leaves:** Ovalish, 1½–3½ inches long, double-toothed, whitish below when young, with 7 or more pairs of veins. **Flowers:** Inconspicuous, in hanging catkins. **Fruit:** A small nut within a leafy cluster (strobile). **Varieties:** None. **Culture:** See above.

BLACK ALDER (*Alnus glutinosa*) p. 87

A Eurasian tree with sticky twigs, useful for places too moist for most other trees. **Size:** 50–75 feet, about a fourth as wide and of not much decorative value. **Leaves:** Nearly round or ovalish, about 4 inches long, coarsely double-toothed. **Flowers:** Inconspicuous, in hanging catkins. **Fruit:** A small, woody, conelike structure, within the scales of which are the small, narrowly winged nutlets. **Hardy:** Everywhere. **Varieties:** Many in Europe, and several are reputed to be here. The only readily available ones are *laciniata,* with deeply cut leaves; *imperialis,* with leaves still more deeply cut, the lobes narrow and without teeth. **Culture:** Its only requirement is a moist or wet place, and it has little to recommend it except for such sites.

Related plants (not illustrated):

Italian Alder (*Alnus cordata*). A tree from southern Europe with heart-shaped, finely toothed leaves. Hardy from Zone 4 southward, and suitable for wet places.

Speckled Alder (*Alnus incana*). A shrub or medium-sized tree, the twigs hairy. Leaves broadly elliptic, about 3½ inches long. Suitable only for wet places.

Hazel
(*Corylus*)

Shrubs of little garden interest, except for the production of edible nuts. Leaves alternate, mostly double-toothed on the margins. Male and female flowers separate but on the same shrub, the male flowers in catkins. The female flower is followed by the familiar hazelnut enclosed in a leafy, often fringed husk. The culture of these shrubs presents no difficulty in any ordinary garden soil. The Filbert, a close relative, does better in Washington and Oregon and is not here treated.

Husk about as long as the nut.	European Hazel
Husk nearly twice as long as the nut.	American Hazel

EUROPEAN HAZEL (*Corylus Avellana*) p. 87
A bushy shrub, of little decorative value in the typical form and grown chiefly in its horticultural varieties. **Size:** 6–10 feet. **Leaves:** Roundish or broadest toward the tip, heart-shaped at the base, 3–4 inches long. **Flowers:** Inconspicuous, in hanging catkins. **Fruit:** The familiar hazelnut, about ¾ inch long, enclosed in the leafy husk, which is about the length of the nut. **Hardy:** From Zone 2 southward. **Varieties:** Several, all more decorative than the typical form, among them being *aurea,* with golden-yellow leaves; *contorta,* with twisted and curled twigs; *fusco-rubra* (often offered as *atropurpurea*), a form with purplish or dull red foliage. **Culture:** See above.

AMERICAN HAZEL (*Corylus americana*) below
A native shrub, 5–8 feet high, which differs from the European Hazel chiefly in having the husk of the fruit 2–3 times the length of the nut, which is of little value. **Hardy:** From Zone 2 southward. **Varieties:** None. **Culture:** See above.

American Hazel
(*Corylus americana*), above.

HOP-HORNBEAM (*Ostrya virginiana*) p. 87
A moderately ornamental native American tree, more attractive

in fruit than in flower. **Size:** 15–20 feet or more in the wild, the wood very hard. **Leaves:** Alternate, simple, ovalish, 3–5 inches long, sharply double-toothed. **Flowers:** Small, greenish and inconspicuous. **Fruit:** A small inedible nut, enclosed in a closed, bladdery husk, the cluster of them not unlike the Hop, and the chief attraction of the tree. **Hardy:** From Zone 2 southward. **Varieties:** None. **Culture:** Prefers cool, moist, partially shady sites in deep, rich soil.

HORNBEAM (*Carpinus Betulus*) p. 87

A hard-wooded, slow-growing, medium-sized Eurasian tree, valued for its successful shearing and clipping into hedges or other forms. **Size:** 20–50 feet high, but usually much less as cultivated, and more often shrubby. **Leaves:** Alternate, simple, birch-like, ovalish or oblong, 3–4 inches long, sharply toothed, with 7–24 pairs of straight veins. **Flowers:** Small, greenish, inconspicuous, in hanging catkins. **Fruit:** A small winged nut, not enclosed within a hop-like cluster, as in the Hop-Hornbeam. **Hardy:** From Zone 2 southward. **Varieties:** *columnaris,* a form with a columnlike habit, very twiggy and most suitable for a hedge as it stands shearing; *fastigiata,* a type with upright branches forming a pyramidal tree 15–20 feet high; *pendula,* a form with drooping branches; *globosa,* a globe-shaped form, looking as if sheared. There are several others available in Europe but little known here. **Culture:** Needs full sun, a rich soil, and does better in cool regions than on the hot coastal plain.

Related plant (not illustrated):

American Hornbeam (*Carpinus caroliniana*). A close relative of the above, needing partial shade, a cool site and a rich soil. Hardy from Zone 3 southward.

MULBERRY FAMILY
(*Moraceae*)

Tall trees, always with a milky juice (easily seen by slitting a twig) with alternate, often lobed leaves. In some there may be lobed and unlobed leaves on the same tree. Flowers without petals, in catkins in some, but in close clusters in others, the sexes always separate, either on the same tree or in different ones. Fruits blackberry-like in the mulberries, but a small or large fleshy receptacle in the Osage Orange and Paper Mulberry. If commercial fruits were included, the Fig would belong here.

Fruit blackberry-like. — Mulberry
Fruit not blackberry-like.
  Trees thorny; fruit orange-yellow, 3–5 inches thick. — Osage Orange
  Trees not thorny; fruit about 1½ inches thick. — Paper Mulberry

MULBERRY (*Morus alba*) p. 87

A Chinese tree, cultivated there and in Japan for centuries as the food of silkworms and here as a doubtful ornamental. Its rather insipid fruit is favored chiefly by chickens, hogs, and children. **Size:** 30–50 feet high, round-headed and with dense foliage. **Leaves:** Alternate, simple, broadly oval, usually lobed and coarsely toothed, 3–5 inches long, smooth both sides. **Flowers:** Small, greenish, inconspicuous, in hanging catkins. **Fruit:** blackberry-like, typically white but sometimes pinkish violet, insipid and so plentiful it litters lawns or pavements. **Hardy:** From Zone 1 southward. **Varieties:** *tartarica,* the Russian Mulberry, is even more hardy and has red fruit; *pendula,* the Weeping Mulberry, has drooping branches and is much planted; a fruitless form is also offered. **Culture:** Almost too easy; the tree is overplanted and much inferior to many other shade trees, and the litter of its copious fruit is often a nuisance.

Related plant (not illustrated):

Red Mulberry (*Morus rubra*). A native American tree, 40–60 feet high, much resembling the above, but the leaves hairy below and rough above. Fruit red or purplish red, also insipid. Hardy from Zone 3 southward. Rather common in old gardens but not a recommended ornamental.

OSAGE ORANGE (*Maclura pomifera*) p. 87
A thorny, wind-resistant tree from the central U.S., often grown there as a windbreak and elsewhere for its interesting fruit and its ability to be pruned into a close-set thorny hedge. **Size:** 30–50 feet high and a third as wide, but often pruned into a tall hedge. **Leaves:** Alternate, simple, ovalish or oblong, pointed at the tip, 1½–4 inches long, without teeth or lobes. **Flowers:** Male and female on separate trees, greenish and small, in hanging clusters in the males, but in small, dense clusters in the females. **Fruit:** Spherical, 2–5 inches thick, green, its skin suggesting a puckered orange, and persisting late in the fall. **Hardy:** From Zone 3 southward. **Varieties**: None. **Culture:** Very easy in any ordinary garden soil, and a good tree in windy, exposed places, even if the soil is poor.

PAPER MULBERRY (*Broussonettia papyrifera*) p. 134
An Asiatic thornless tree to be planted with caution as its propensity to spread may often make it a weedy nuisance, as in some parts of the eastern states where it has escaped from cultivation. **Size:** 25–40 feet high, nearly half as wide, its shade dense. **Leaves:** Very rough above, hairy beneath, oval, 6–8 inches long, sometimes unlobed, but often deeply and roundly lobed, often both kinds on the same tree, the margins toothed. **Flowers:** Small, greenish, the male and female on separate trees, the males in hanging catkins, the females in small, compact clusters. **Fruit:** Round, about ¾ inch thick, densely hairy, orange-red, rather showy. **Hardy**: From Zone 4 southward. **Varieties:** None available here, several in Europe. **Culture:** So easy as to be dangerous, as its ability to become a pest is notorious. It will stand any amount of abuse, wind, dust, smoke, or fumes of motors and if cut down will not only sprout from the stump but many suckers will spring up from the roots.

ELM FAMILY
(*Ulmaceae*)

USUALLY magnificent forest trees, especially among the elms, but generally more moderate-sized in the Hackberry and in *Zelkova.* Leaves alternate, usually oblique at the base and inequilateral (*i.e.,* more of the blade on one side of the midrib than on the other). Margins always toothed, sometimes doubly so (*i.e.,* the teeth are themselves minutely toothed). Juice and sap watery, not milky. Flowers without petals, in short clusters (not in catkins), the sexes not separate, followed by a fruit which in the elms is small and winged, but in the two other genera, bony and nearly nutlike. Our three genera are easily separated.

Leaves with at least 7 pairs of nearly parallel veins, none of the prominent ones arising at the base.
- *Fruit flat, winged; leaves usually doubly toothed.* Elm
- *Fruit bony and nutlike; leaves not doubly toothed.* *Zelkova*

Leaves with at least 3 main veins arising at the base, hence not nearly parallel. Hackberry

ELM
(*Ulmus*)

Stately, imposing, and extremely decorative trees, among the finest for all landscape purposes, especially the American Elm which, unfortunately, is subject to the ravages of the Dutch elm disease. Leaves rather short-stalked, usually, and sometimes very strongly, oblique at the base, and inequilateral (see the pictures). Margins mostly doubly toothed, the main teeth themselves minutely toothed, the veins nearly parallel. Flowers in small clusters (not catkins), without petals, followed by a small, flat nutlet embedded in an almost papery wing.

It is impossible to suggest planting the American Elm, because most trees seem to be doomed by the spread of the Dutch elm disease. But some European elms and certain American ones do not appear to be so much infected by the disease, particularly the

Dutch Elm itself. Before planting, it is wise to secure current information from the state experiment station as to whether elm planting is safe in your vicinity. Most elms have relatively shallow roots, are gross feeders, but do well in many types of garden soils. Their culture is thus easy (except for the uncontrollable disease) and they make fairly rapid growth. At least 5 of the 18 known species are much planted and may be distinguished thus:

Flowers in bloom in the autumn. Chinese Elm
Flowers spring-blooming.
Flowers at the end of slender, drooping stalks. American Elm
Flowers in close, short-stalked clusters.
Leaves scarcely oblique, only once-toothed, not over 2 inches long. Siberian Elm
Leaves obviously oblique, always double-toothed, 2½–7 inches long.
Young twigs hairy; leaves rough above. English Elm
Young twigs not hairy; leaves usually smooth and shining above. Dutch Elm

CHINESE ELM (*Ulmus parvifolia*) p. 134

A small, quick-growing tree, inclined to fork, its foliage red or purple in the autumn in the North, but more or less persistent and green southward. **Size:** 30–50 feet and about a third as wide, its bark mottled, often exposing the lighter inner bark when mature. **Leaves:** More or less elliptic, 1–2½ inches long, shining green above, hairy beneath when young. **Flowers:** In small clusters, blooming in autumn, inconspicuous and greenish. **Fruit:** A small flat nutlet surrounded by a small papery wing. **Hardy:** From Zone 3 southward and apparently immune to the Dutch elm disease. **Varieties:** None for the East, but a nearly evergreen one in Calif. **Culture:** A quick-growing, valuable tree for places having a relatively poor soil; shallow-rooted and hence vulnerable to gales.

AMERICAN ELM (*Ulmus americana*) p. 134
The aristocrat of all elms, with its beautiful vase-shaped branching and dense shade, which make magnificent shaded aisles of many New England village streets. It is, however, not recommended for planting until the Dutch elm disease is conquered (see above). **Size:** 70–120 feet high and, in the crown, nearly half as wide, with light gray, deeply fissured bark. **Leaves:** Oval-oblong, 3–7 inches long, unequal at the base, smooth or roughish above, sometimes hairy beneath. **Flowers:** Small, greenish, 3 or 4 on a slender, drooping stalk, blooming before the leaves unfold. **Fruit:** A small nutlet, surrounded by a papery wing that is hairy on the margin. **Hardy:** Everywhere. **Varieties:** Several, of which *pendula,* the Weeping American Elm, is the most useful for its long, drooping branches. Others, often offered as immune to the disease, should be used with caution, for real immunity has yet to be demonstrated. **Culture:** See above.

Related plants (not illustrated):

Slippery Elm (*Ulmus rubra*). A close relative of the American Elm, 40–60 feet high, with leaves rough above, 5–9 inches long, its bruised twigs with a characteristic odor and flavor. Hardy from Zone 2 southward, and relatively immune to disease.

Rock Elm (*Ulmus thomasi*). Often called Cork Elm as its twigs are corky-winged; 60–90 feet high, and hardy from Zone 2 southward. Relatively immune to the disease.

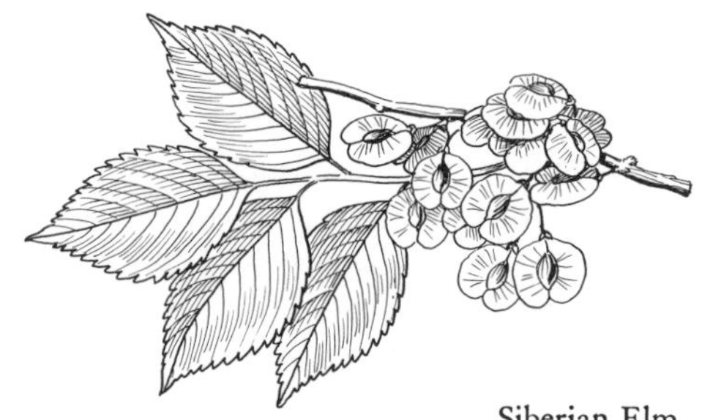

Siberian Elm
(*Ulmus pumila*), p. 119.

SIBERIAN ELM (*Ulmus pumila*) p. 118
A useful tree for the Middle West, where better elms do not always thrive. **Size:** 50–80 feet, about a third as wide. **Leaves:** 1½–3½ inches long, smooth both sides. **Flowers:** Small, greenish, inconspicuous, blooming as the leaves unfold. **Fruit:** A small nutlet, surrounded by a papery wing nearly 1 inch long. **Hardy:** From Zone 2 southward, and much used in the west for shelter-belt planting. **Varieties:** None. **Culture:** The most useful of all the elms because it will stand cold, bitter winds and poor soil.

ENGLISH ELM (*Ulmus procera*) p. 134
The European representative of our American Elm, but without its vaselike branching, and usually making a rounded, domelike canopy. Its bark is rough and fissured. **Size:** 100–150 feet high, its shade very dense, its twigs sometimes corky-winged. **Leaves:** Oval or elliptic, 2½–4 inches long, very oblique at the base, rough above, softly hairy beneath. **Flowers:** Small, greenish, blooming before the leaves unfold or with them. **Fruit:** Nearly round, about ¾ inch in diameter, not hairy on the margin. **Hardy:** From Zone 3 southward. **Varieties:** Several are listed in books, but most of them are doubtfully available here. **Culture:** See above.

Related plant (not illustrated):

Wych Elm (*Ulmus glabra*). A Eurasian tree, 80–120 feet high, with smooth bark, its twigs not corky-winged and leaves 3½–7 inches long. It has many varieties, among them the Camperdown Elm, which has drooping branches.

DUTCH ELM (*Ulmus hollandica*) p. 134
A beautiful hybrid elm much planted, especially in two of its varieties. Foliage very dense, shining green. **Size:** 60–100 feet, nearly half as wide. **Leaves:** Very oblique at the base, 3½–7½ inches long, shining green in the Huntingdon Elm, but somewhat rough above in the Belgian Elm. Mostly grown in these two **Varieties:** Huntingdon Elm (variety *vegeta*), a magnificent tree with a forking trunk; and the Belgian Elm (variety *belgica*), a tall tree with upright branches and a broad crown. Both these

trees, and the Dutch Elm itself, are relatively immune to the Dutch elm disease, and there are stately specimens in Md. and Va. **Hardy:** From Zone 3 southward. **Culture:** See above.

Related plant (not illustrated):

Smooth-leaved Elm (*Ulmus carpinifolia*). A Eurasian tree, rarely over 75 feet high, its glossy green leaves smooth both sides, 2–4 inches long. One of its varieties known as Buisman is reputedly immune to all elm diseases.

ZELKOVA SERRATA p. 134

An elm-like Japanese tree with rather small leaves and an open canopy, not very widely grown and the trunk forking. **Size:** 30–40 feet high, usually less as cultivated. **Leaves:** Only slightly unequal at the base, elliptic or oblongish, 1–2 inches long and with 8–14 pairs of straight, parallel veins, toothed but not doubly toothed. **Flowers:** Small, greenish, inconspicuous, blooming in early spring with the unfolding of the leaves. **Fruit:** Bony, hard, pea-size. **Hardy:** From Zone 3 southward. **Varieties:** None. **Culture:** Grown exactly as an elm (which see) and a possible substitute for it in regions where the Dutch elm disease is rampant.

HACKBERRY (*Celtis occidentalis*) p. 134

A large, native American elm-like tree, widely planted for its resistance to disease and unfavorable sites. **Size:** 80–100 feet, nearly half as wide in maturity, its trunk huge and with warty excrescences on many of the branches. **Leaves:** Ovalish or oblong, 3–5 inches long, somewhat inequilateral, with 3 main veins arising at the base, toothed on the margin, except at the base. **Flowers:** Small, greenish, inconspicuous, blooming in early spring. **Fruit:** Bony, hard, pea-size, only one to a short stalk, ultimately dark purple. **Hardy:** From Zone 2 southward. **Varieties:** None. **Culture:** Can be grown anywhere, in cities or towns, even with its roots in brackish water, and on wind-swept prairies. A valuable substitute for the elms in unfavorable places.

❦

Plants the Flowers of Which Have Obvious Petals

THE TREES and shrubs with obvious petals are rather dramatically separated from the coniferous evergreens and from the shade trees and shrubs, which have no petals. Besides having colored petals, these flowers rely on insects to carry pollen from flower to flower instead of being at the mercy of the wind as a pollen carrier, as in all the coniferous evergreens and most of the shade trees which immediately precede this section of the book.

Such a fundamental difference in plant structure needs a simple key for all the rest of the trees and shrubs in the book:

Plants with the petals separate, not united to form a cup-shaped, bell-shaped, or funnel-like corolla. See p. 123

Plants with a united, usually bell-shaped, cup-shaped, or funnel-like corolla. See p. 321

4. Trees and Shrubs with Separate Petals

THIS large group should be easy to identify, for anyone can see whether or not the flower has separate petals, as distinguished from one where the petals are all united. Identification would be easy, that is, if Nature had not endowed certain plants with separate petals that by their true relationship do not belong here at all. The resulting confusion would be complete unless one had been warned about these incongruous exceptions. They occur in some plants in the Heath Family (see p. 395), *Pterostyrax* (see p. 420) and the Ash (see p. 323).

In the trees and shrubs having separate petals in their flowers there is one group that differs from all the others in having flowers resembling the Sweet Pea. All belong to the Pea Family, and our first separation of plants with separate petals would hence be:

Flowers pea-like.	The Pea Family
Flowers never pea-like.	See p. 143

PEA FAMILY
(*Leguminosae*)

THE Pea Family is enormous, including nearly 14,000 species, many of them tropical trees, but in temperate regions including such garden favorites as the sweet pea, the common garden pea, lupines, the peanut, clover, alfalfa, and a large number of shrubs and trees, noted below. It also includes wisteria and the Kudzu-Vine (see p. 73 at VINES).

The leaves are always alternate and mostly compound, usually with many leaflets (but sometimes only 1 or 3). The leaflets are generally arranged feather-fashion, but in a few the leaflets are arranged finger-fashion. While most of the family have typically pea-like flowers (see the illustrations) one is an exception, as in so many things in nature. It does not have pea-like flowers, but has them arranged in a close ball-like cluster (see Silk Tree below). There are also a few of the Pea Family in which the flowers are not quite pea-like, and these include four small groups also noted below. In all of them the flower is followed by a splitting legume, familiar enough as the garden pea. Sometimes the pod is small and not easily recognized as a pea pod.

A simple scheme for the identification of the shrubs and trees of the Pea Family follows:

Flowers in a fuzzy, ball-like cluster. Silk Tree
Flowers pea-like or regular, if pea-like blooming before the leaves unfold.
 Leaflet 1, heart-shaped. Redbud
 Leaflets many.
  Leaves 20–36 inches long; flowers whitish. Kentucky Coffee-Tree
  Leaves smaller.
   Flowers yellow; an unarmed shrub. Bird-of-Paradise Bush
   Flowers greenish yellow; a thorny tree. Honey Locust
Flowers perfectly pea-like, usually blooming after the unfolding of the leaves.
 Leaflets 1 or 3, or leaves simple (rarely with 3 leaflets in Genista or Cytisus); or apparently leafless.
  Plants apparently leafless.
   Spiny shrub. Furze
   Spineless shrub. Spanish Broom
  Plants usually having leaves.
   Twigs round, green-striped. *Genista*
   Twigs obviously angled, not green-striped. Broom

Leaflets more than 1, often many more.
Leaflets 3.
Flowers yellow; mostly trees. — *Laburnum*
Flowers purple; low shrubs. — Bush Clover
Leaflets more than 3, always arranged feather-fashion.
Leaves always without an odd leaflet at the tip. — Pea-Tree
Leaves with an odd leaflet at the tip.
Flowers yellow. — Bladder-Senna
Flowers not yellow.
Twigs spiny. — Locust
Twigs not spiny.
Flowers white. — Yellow-Wood
Flowers not pure white.
Tree. — Pagoda Tree
Shrub. — Bastard Indigo

SILK TREE (*Albizzia Julibrissin*) p. 134
A beautiful flowering tree from Persia to central China, with a broad, spreading crown, lacy foliage and close, grayish bark. **Size:** 20–30 feet high, nearly as broad. **Leaves:** Twice-compound, 12–18 inches long, with 12–25 major divisions, each of which bears from 40–60 oblique leaflets that are scarcely ¼ inch long, the foliage graceful, drooping and feathery. **Flowers:** Light pink, very small, crowded in a dense, fuzzy, globe-shaped head, about 1–2 inches thick, usually in July. **Fruit:** A flat pod about 5 inches long and 1 inch wide. **Hardy:** From Zone 5 southward. **Varieties:** *rosea,* a form with deeper pink flowers and hardy up to Zone 4, thus the only form for gardens as far north as New York City and coastal Conn. **Culture:** Thrives in any ordinary garden soil, and is a quick-growing tree much planted from Md. southward and frequently, but incorrectly, called Mimosa (the true *Mimosa* is tropical American).

Redbud
(*Cercis*)

Medium-sized trees or shrubs attractive for their early bloom, which is conspicuous and profuse just before or with the unfolding of the leaves, which are not compound and generally heart-shaped, the main veins arising from the base of the blade. Flowers small, more or less pea-like, numerous, in small clusters, followed by a flat, narrow, slightly winged pod.

The redbud, especially the first species, is an attractive lawn plant, growing naturally under the canopy of the forest in the eastern U.S. and hence doing better in a half-shady place. It prefers sandy loams and should not be planted in heavy clay or in too moist a site. The two species, often called Judas-Trees, are:

Flowers less than ¾ inch long. American Redbud
Flowers more than ¾ inch long. Chinese Redbud

AMERICAN REDBUD (*Cercis canadensis*) p. 134
A medium-sized tree, often a shrub, much planted for its early bloom. **Size:** 15–30 feet, but often half this and shrubby, its crown spreading nearly half its height. **Leaves:** 3½–5 inches long, heart-shaped at the base, pointed at the tip. **Flowers:** Very numerous, crowded along the bare twigs, rosy pink and showy in Apr.–May. **Fruit:** Flat, narrowed, about 3 inches long. **Hardy:** From Zone 3 southward. **Varieties:** *alba* has white flowers; *plena* has double flowers but is not so desirable as the typical form. **Culture:** See above.

CHINESE REDBUD (*Cercis chinensis*) p. 130
The Asiatic representative of the American Redbud, differing only in having slightly larger flowers that are rosy purple. A fine shrub as grown here, but not safely hardy north of the southern part of Zone 4.

Related plant (not illustrated):

Judas-Tree (*Cercis Siliquastrum*). A small Eurasian tree or shrub, its leaves rounded or notched at the tip, with rose-purple flowers. Hardy from Zone 5 southward.

KENTUCKY COFFEE-TREE (*Gymnocladus dioica*) p. 135
An unusual tree of the eastern states, its leaves larger than almost any mentioned in this book, grown for its large flower cluster and persistent pods, the seeds of which were once used as a substitute for coffee by the pioneers. **Size:** 40–90 feet high, about a fourth as wide, and inclined to fork from near the base. **Leaves:** Very large, twice-compound, 20–36 inches long, split into 3–7 large divisions, each of which are composed of 6–14 leaflets that are more or less ovalish, 2–4 inches long and without marginal teeth. In the fall the yellow leaflets drop, and later the long leafstalk, the tree apparently thus shedding its twigs. **Flowers:** Small, nearly regular, white or greenish white, in a loose cluster, which, in the female flowers, may be 8–14 inches long. **Fruit:** A thick, flat, pulpy pod 8–12 inches long, its large black and flattened seeds embedded in the pulp, the brownish pod persisting until spring. **Hardy:** From Zone 3 southward. **Varieties:** None. **Culture:** Easy in most ordinary garden soils, particularly reasonably moist ones. It does not thrive in open windy places.

BIRD-OF-PARADISE BUSH (*Poinciana gillesi*) p. 135
A South American shrub or small tree, the branches straggling and sticky-hairy, grown in the South for its spectacular bloom. **Size:** 8–15 feet, and about a third as wide. **Leaves:** Alternate, twice-compound, the ultimate leaflets very numerous, and scarcely ¼ inch long, the foliage lacy, feathery, and very graceful. **Flowers:** Not pea-like, the showy yellow petals slightly unequal, the brilliant scarlet stamens protruding beyond the petals about 4–5 inches and very striking. The flowers are borne in an erect cluster 12–18 inches high; midsummer. **Fruit:** A pod, 3–4 inches long. **Hardy:** From Zone 6 southward, possibly in protected places near the southern edge of Zone 5. **Varieties:** None. **Culture:** Not difficult if young plants are set out while dormant, but difficult to move or plant when older; not particular as to soils.

HONEY LOCUST (*Gleditsia triacanthos*) p. 135
A tall tree of the eastern U.S., its trunk usually armed with many branched thorns which may be 3–5 inches long. **Size:** 70–130 feet

high, about a third as wide, its canopy open, hence casting only moderate shade. **Leaves:** Once- or twice-compound (often both on the same tree), the ultimate leaflets 20–30, and about 1¼ inches long, rather thick and leathery, all borne on thorny twigs. **Flowers:** Greenish yellow, not pea-like, borne in a small, narrow cluster, 2½–5 inches long; May. **Fruit:** A sickle-shaped and twisted pod, nearly 18 inches long, which persists for months, finally releasing its flattened seeds, which are extremely slow to germinate. **Hardy:** From Zone 3 southward. **Varieties:** *inermis,* a form with few, if any, thorns and offered under names like Moraine Locust and Imperial Locust. **Culture:** This does not compare with the true locust (*Robinia*), but is a useful tree in dry, windy places and is apparently indifferent as to soils.

FURZE (*Ulex europaeus*) p. 135

An extremely spiny, generally leafless Eurasian shrub, grown for its profuse bloom and often called Gorse and (in England) the Whin. **Size:** 2–3 feet high, nearly as wide, much-branched and twiggy. **Leaves:** Often wanting and replaced by a spiny leafstalk, or small and scalelike. **Flowers:** Pea-like, about ¾ inch long, fragrant, bright yellow, at the leaf joints or crowded at the tips of the twigs; Apr.–July. **Fruit:** A brown, hairy pod about ½ inch long. **Hardy:** From Zone 4 southward. **Varieties:** A double-flowered form (*plenus*) and an upright, columnar form (*strictus*) are offered, but the latter flowers sparingly. **Culture:** Prefers open, dry, sandy places, and in such sites it may spread so rampantly, especially in the South, as to become a spiny pest.

Related plant (not illustrated):

Ulex nanus. A dwarf European shrub, 12–20 inches high, its branches twiggy. Flowers golden yellow. Hardy from Zone 5 southward.

SPANISH BROOM (*Spartium junceum*) p. 135

Unfortunately this beautiful yellow-flowered shrub is not hardy over most of our range. Its stems are grooved and rushlike but not spiny as in the Furze. **Size:** 6–8 feet high, about half as wide. **Leaves:** Rarely seen, if produced simple, about ¾ inch long and bluish green. **Flowers:** Yellow, pea-like, fragrant, about 1 inch long, borne in a terminal, very showy cluster that may be 15 inches long; May–Sept. **Fruit:** A flattened pod, about 4 inches long and hairy. **Hardy:** From Zone 6 and possibly in protected places in Zone 5 southward. **Varieties:** None. **Culture:** Demands full sun and thrives on a variety of soils. Start with young plants as it is difficult to move when mature.

GENISTA
(*Genista*)

A confusing name to many as the "genista" of the florists does not belong here but to the closely related genus *Cytisus*. All the plants below are shrubs with round, green-striped twigs, rarely spiny, mostly with compound leaves (the leaflets only 1 occassionally) and yellow or white flowers in terminal clusters. Fruit a longish flattened pod.

The genistas require full sunshine, a light sandy soil and should be planted while young as mature bushes are moved with difficulty. Two species are prostrate vines and will be found at p. 49. Seven other species are in cultivation but only 3 can be included here, separated thus:

Leaves with 3 leaflets. — *Genista radiata*
Leaves simple, without leaflets.
    Shrub spiny. — Spanish Broom
    Shrub without spines. — Woadwaxen

GENISTA RADIATA p. 130

A European shrub, evergreen in the South but not northward, the twigs not spiny. **Size**: 1–3 feet high, nearly as wide. **Leaves:** Alternate, compound, silky, composed of 3 leaflets about ½ inch long. **Flowers:** Pea-like, yellow, crowded in dense clusters and showy; May–July. **Fruit:** A curved silky pod about ¼ inch long. **Hardy:** From Zone 5 southward. **Culture:** See above.

SPANISH BROOM (*Genista hispanica*) p. 130

A densely branched, spiny, low shrub, often leafless, but also bearing a single leaflet. For another plant also called Spanish Broom see p. 128. **Size:** 12–18 inches high, nearly as wide. **Leaves:** When present, reduced to a single ovalish leaflet about ½ inch long, usually soon falling. **Flowers:** Golden yellow, in a dense head of 2–10 flowers; May–June. **Fruit:** A hairy oblongish pod. **Hardy:** From Zone 4 southward. **Varieties:** *nana* is even smaller than the typical form. **Culture:** See above. It is one of the most showy of all the genistas.

Left: Spanish Broom (*Genista hispanica*), p. 129. Center: Chinese Redbud (*Cercis chinensis*), p. 126. Right: *Genista radiata,* p. 129.

WOADWAXEN (*Genista tinctoria*) p. 135

This Eurasian shrub is the most commonly cultivated of all the genistas for its profusion of flowers. **Size:** 2–3 feet high, nearly as wide, the twigs without any spines. **Leaves:** Reduced to a single oblongish leaflet ½–1½ inches long, quite smooth except on the hairy margin. **Flowers:** Pea-like, yellow, in profuse, sometimes branched clusters; June–Aug. **Fruit:** A small, nearly smooth, slightly arched narrow pod, about 1 inch long. **Hardy:** From Zone 3 southward. **Varieties:** *plena* has double flowers. **Culture:** See above. This is the most vigorous and best genista for the amateur's garden.

BROOM
(*Cytisus*)

A confusing group of shrubs, mostly from the Mediterranean region, often confused with the Spanish Broom, the Furze, and the genistas noted just above. Still more confusing is the fact that the "genista" of the florists is really *Cytisus canariensis,* a greenhouse plant excluded from this book.

The hardy brooms are shrubs the stems of which are angled and not green-striped. The leaves are compound and have 3 leaflets, except in the Warminster Broom. Flowers very profuse, showy, pea-like, rarely solitary and mostly in small clusters. Fruit a flat pod. One of the species is prostrate and will be found at p. 49 at VINES.

These shrubs demand full sunlight, are not particular as to soil, but do better on sandy ones. However, they transplant poorly when mature so that it is better to start with young plants. None of them will stand long-continued bitter winters. Of the 9 species in reasonably common cultivation, the following 4 are most likely to interest the amateur:

Flowers usually purple. — *Cytisus purpureus*
Flowers yellow or yellowish white.
- *Flowers bright yellow, solitary but profuse along the sides of the twigs.* — Scotch Broom
- *Flowers pale yellow or yellowish white.*
  - *Flowers of unpleasant odor.* — Warminster Broom
  - *Flowers essentially scentless.* — *Cytisus kewensis*

CYTISUS PURPUREUS p. 132

A showy, low shrub, often more or less sprawling and much cultivated as it has purple flowers, which is unusual in *Cytisus.* **Size:** 1–2 feet high, often as wide or more. **Leaves:** Leaflets 3, not over 1 inch long, stalked, dark green above. **Flowers:** Solitary or in pairs, profuse, pea-like, about ¾ inch long, purple; May–June. **Fruit:** A smooth pod, about 1¼ inches long. **Hardy:** From Zone 4 southward. **Varieties:** *albo-carneus* has pale pink flowers; *atropurpureus* has flowers still darker purple. **Culture:** See above.

SCOTCH BROOM (*Cytisus scoparius*) p. 135
The most widely cultivated of the group because of its profusion of rather large, pea-like yellow flowers and its ability to stand more cold than the other species of *Cytisus.* **Size:** 4–9 feet high, nearly three fourths as wide. **Leaves:** Leaflets 3 (rarely reduced to 1) ⅓–½ inch long, hairy when young, but becoming smooth later. **Flowers:** Solitary or in twos, mostly at the leaf joints, bright yellow and about 1 inch long, very profuse and showy; May–June. **Fruit:** A slender pod, nearly 2 inches long, usually hairy. **Hardy:** From Zone 3 southward. **Varieties:** Over 20, differing mostly in color or stature; of these *andreanus* has two crimson petals and is a very handsome plant; *sulphureus,* the Moonlight Broom, has sulphur-yellow flowers; *plenus* has double flowers, and there are many other named forms not always well defined. **Culture:** See above.

Left: *Cytisus purpureus,* p. 131. Center: *Cytisus kewensis,* p. 133. Right: Warminster Broom (*Cytisus praecox*), p. 133.

WARMINSTER BROOM (*Cytisus praecox*) p. 132
A bushy hybrid shrub, its branches at first erect, ultimately arching and even drooping at the tip. **Size:** 3–5 feet high. **Leaves:** Leaflet only 1, nearly linear, ½–¾ inch long, falling rather early. **Flowers:** Pea-like, of an unpleasant odor, pale yellow, borne in pairs or solitary at the leaf joints, very profuse and showy; May. **Fruit:** About 1 inch long, hairy. **Hardy:** From Zone 5 southward, possibly in protected parts of Zone 4. **Varieties:** None. **Culture:** See above.

CYTISUS KEWENSIS p. 132
A very desirable low hybrid shrub, suited mostly to the rock garden. **Size:** Hardly more than 12 inches high and often nearly prostrate. **Leaves:** The leaflets usually 3, very narrow, hairy. **Flowers:** Very numerous, solitary or in pairs at the leaf joints, pale yellow or yellowish white. **Fruit:** Unknown. **Hardy:** From Zone 4 southward. **Varieties:** None. **Culture:** See above. It is a fine plant for the rock garden or dry wall; and is best grown in a gritty or sandy soil.

LABURNUM
(*Laburnum*)

Very beautiful flowering trees, containing a poisonous juice (only if eaten), and widely planted for their profuse, drooping clusters of pea-like flowers. They are often called Golden Chain or Bean Trees. Leaves alternate, compound, with 3 nearly stalkless leaflets, the foliage thin and casting little shade. Flowers typically pea-like, borne in handsome, terminal and usually hanging clusters, followed by a flattened pod, its seeds also poisonous if eaten.

These most decorative trees flower in late May or early June and are outstanding lawn specimens. They thrive in a variety of garden soils, even on rocky and somewhat sandy sites. Much the best of the two below is the hybrid *Laburnum watereri*.

Flowers yellow, the hanging cluster 12–15 inches long. Golden Chain

Flowers golden yellow, the hanging cluster 15–18 inches long. *Laburnum watereri*

TREES AND SHRUBS, THE FLOWERS OF WHICH HAVE NO PETALS: 2–6, 8–9
PEA FAMILY (*Leguminosae*), 1, 7

1. **Silk Tree** (*Albizzia Julibrissin*) p. 125
   Universally, but incorrectly, called Mimosa in Washington and adjacent Md. A fine, showy tree.

2. **American Elm** (*Ulmus americana*) p. 118
   The finest of our native elms, apparently doomed over much of the East. Its vaselike outline unique.

3. **Paper Mulberry** (*Broussonetia papyrifera*) p. 115
   Not to be introduced into small gardens since it is very invasive. Useful in open, windy places with poor soil.

4. **Hackberry** (*Celtis occidentalis*) p. 120
   A native tree very tolerant of unfavorable sites, such as factory yards or edges of brackish marshes.

5. **English Elm** (*Ulmus procera*) p. 119
   A European counterpart of our American Elm, but without the vaselike outline.

6. ***Zelkova serrata*** p. 120
   A medium-sized, small-leaved, elm-like tree, not well known but useful where the American Elm is impossible.

7. **American Redbud** (*Cercis canadensis*) p. 126
   A shrublike small native tree, covered in early spring by showy flowers on the bare twigs.

8. **Chinese Elm** (*Ulmus parvifolia*) p. 117
   The hardiest, most drought-resistant of the elms, but the foliage not nearly so fine as in the others.

9. **Dutch Elm** (*Ulmus hollandica*) p. 119
   A magnificent lawn tree of hybrid origin. Foliage glistening green. So far immune to disease.

1
2
3
4
5
6
7
8
9

1
2
3
4
5
6
7
8

PEA FAMILY (*Leguminosae*), 1–9

1. **Bird-of-Paradise Bush** (*Poinciana gillesi*) p. 127
   Not for northern gardeners, this gorgeous South American shrub has yellow flowers with brilliant scarlet stamens.

2. ***Laburnum watereri*** p. 136
   A medium-sized hybrid tree with a profusion of hanging clusters of pea-like yellow flowers in spring.

3. **Kentucky Coffee-Tree** (*Gymnodadus dioica*) p. 126
   A native tree with enormous compound leaves, small greenish-white flowers, and a winter-persisting brown pod.

4. **Woadwaxen** (*Genista tinctoria*) p. 130
   A low, unarmed Eurasian shrub with a profusion of yellow, pea-like flowers, mostly in early summer.

5. **Black Locust** (*Robinia Pseudo-acacia*) p. 139
   The variety *descaisneana,* shown here, has pink instead of the usual white flowers of this sturdy, hard-wooded tree.

6. **Scotch Broom** (*Cytisus scoparius*) p. 132
   More hardy than most species of *Cytisus;* the profuse yellow bloom of this European shrub is most showy in June.

7. **Honey-Locust** (*Gleditsia triacanthos*) p. 127
   A spiny-trunked native tree with small, numerous leaflets, greenish-yellow flowers, and a twisted, persistent pod.

8. **Spanish Broom** (*Spartium junceum*) p. 128
   A usually leafless, green-twigged, unarmed shrub with yellow pea-like and fragrant flowers. Not hardy northward.

9. **Furze** (*Ulex europaeus*) p. 128
   A very spiny, showy shrub suited to dry, sandy places; long-flowering.

GOLDEN CHAIN (*Laburnum anagyroides*) p. 137
A small tree of European origin, with a stiff, upright habit, often branching close to the ground. **Size:** 20–30 feet high and about half as wide. **Leaves:** Compound, composed of 3 ovalish leaflets 1½–2½ inches long, nearly or quite smooth. **Flowers:** Yellow, about ¾ inch wide, in slender, hanging, very showy clusters that are 12–15 inches long; early June. **Fruit:** A thin, smooth pod with brown seeds which germinate easily in ordinary garden soil if planted in spring, about ½ inch deep. **Hardy:** From Zone 3 southward. **Varieties:** *pendulum* has even longer hanging flower clusters. **Culture:** See above.

Related plant (not illustrated):

Laburnum alpinum. A European tree, inferior to the one above or the one below, but one of the parents of *Laburnum watereri.* Its flower cluster is shorter and its stalks are silky-hairy. Often called Scotch Laburnum.

LABURNUM WATERERI p. 135
A very beautiful hybrid tree derived by crossing the Golden Chain with *Laburnum alpinum.* **Size:** 20–30 feet high, about half as wide. **Leaves:** Compound, composed of 3 elliptic or ovalish leaves 1½–3 inches long, grayish green and a little silky-hairy beneath when young. **Flowers:** Golden yellow, about 1¼ inch wide, the stunning, hanging cluster 15–18 inches long. **Fruit:** A small, sparingly hairy pod, with few or no seeds. **Hardy:** From Zone 5 southward, or in protected places in the southern part of Zone 4. **Varieties:** None. **Culture:** See above. Much the best of the laburnums for those within its climatic limits. Often listed as *Laburnum vossi.*

BUSH CLOVER (*Lespedeza bicolor*) p. 137
A small Asiatic shrub, valued chiefly for its pea-like late-flowering blossoms; flowers later than most shrubs. **Size:** 6–9 feet, half as wide. **Leaves:** Compound, composed of 3 ovalish leaflets ¾–1¼ inches long, grayish green beneath, the middle leaflet longer-stalked than the 2 side ones. **Flowers:** Pea-like, rosy purple, about ½ inch long, in small clusters 2–4 inches long, that are grouped in

a large, showy, branching cluster; Aug.–Sept. **Fruit:** A small 1-seeded pod. **Hardy:** From Zone 3 southward, but frequently dying to the ground northward and sending up new flowering shoots in the spring. **Varieties:** None. **Culture:** Easy in any light sandy soil in full sun, but it will also stand some shade.

Related plant (not illustrated):

Lespedeza thunbergi. An Asiatic relative of the above, its primary flower cluster longer and later-flowering (Sept.–Oct.). Hardy from Zone 4 (possibly Zone 3) southward.

Left: Golden Chain (*Laburnum anagyroides*), p. 136. Center: Pea-Tree (*Caragana arborescens*), below. Right: Bush Clover (*Lespedeza bicolor*), p. 136.

PEA-TREE (*Caragana arborescens*) above

An upright, Siberian, very hardy, sometimes spiny shrub or small tree with little to recommend it but its ability to thrive in regions of bitter winds and prolonged zero temperatures, hence much used in our northern prairie states and inhospitable sections of

Canada. **Size:** 10–20 feet, about a third as wide. **Leaves:** Compound, composed of 8–12 tiny leaflets about ⅓ inch long, without an odd one at the end. **Flowers:** Pea-like, yellow, about ¾ inch long, borne singly or in small clusters, rather profuse and hence showy; May. **Fruit:** A stalked pod, 1½–2 inches long. **Hardy:** From Zone 1 southward. **Varieties:** *lobergi,* with narrower leaflets and slightly larger flowers. **Culture:** Easy in any ordinary garden soil, in full sun. It is easily trimmed into a coarse but dense hedge and nothing exceeds it as a windbreak or snow trap in regions too severe for privet.

Related plant (not illustrated):

Caragana pygmaea. A low Asiatic shrub, 2–4 feet high, with yellow flowers and only 4 leaflets. Hardy from Zone 2 southward.

BLADDER-SENNA (*Colutea arborescens*) p. 140

A moderately desirable Eurasian shrub, chiefly valuable for its long-continuing bloom. **Size:** Usually 4 feet high, sometimes up to 10 feet, and half as wide. **Leaves:** Compound, composed of 9–13 leaflets, about 2 inches long, with an odd one at the tip, without marginal teeth. **Flowers:** Pea-like, yellow, about 1 inch long, in long-stalked clusters that arise at the leaf joints and blooming almost continuously from May to Sept. **Fruit:** Not a pea-like pod, but flat, papery, about 2 inches long and not splitting. **Hardy:** From Zone 4 southward. **Varieties:** None that are available. **Culture:** Easy in any ordinary garden soil, in full sun. It is disliked by some for its too vigorous growth and invasive habit, often crowding out better shrubs.

LOCUST
(*Robinia*)

Hard-wooded, thin-foliaged, spiny North American trees under which lawns will thrive, but in one species a weedy nuisance if allowed to spread. Leaves compound, the leaflets arranged feather-fashion, with an odd one at the tip. Flowers typically pea-like, borne in hanging, often showy clusters and much favored by bees. Fruit typically pea-like, but much larger, usually hanging, and with many seeds.

Except for the Black Locust, which is very widely planted, sometimes recklessly, the locusts are not well known to most gardeners in spite of their showy bloom. They thrive in a variety of soils, often in stony or sandy ones, and their culture is thus easy — too easy in the case of the invasive Black Locust, which has escaped from cultivation over much of the eastern states and in some places is a frankly weedy nuisance. Of the six species of *Robinia* that may be cultivated only 3 can be included here.

Flowers white; a tall tree. Black Locust
Flowers pink.
A shrub 3–4 feet high. Rose Acacia
A tree 30–40 feet high. Clammy Locust

BLACK LOCUST (*Robinia Pseudo-acacia*) p. 135
A vigorous tall tree, often harvested for its long-lasting and very hard wood, but also widely grown for its showy pea-like flowers. **Size:** 70–80 feet high, about a third as wide, its bark brownish and deeply furrowed. **Leaves:** Compound, composed of 7–19 ovalish leaflets that fold together at night in the so called "sleep movements" (cause unknown). **Flowers:** Pea-like, white, in many-flowered hanging clusters; June. **Fruit:** A smooth pod, 3–4 inches long. **Hardy:** From Zone 3 southward. **Varieties:** Many, of which the following are most likely to be useful: *bessoniana,* a form developing a rather ovoid outline; *descaisneana,* a variety with rose-colored flowers, differing from the Clammy Locust in not having red, sticky twigs; *umbraculifera,* a form with a dense, rounded outline; the Shipmast Locust (variety *rectissima*), an erect, columnar form chiefly valued for the lasting quality of its wood; a good street tree. **Culture:** See above; the only caution is not to plant it unless its subsequent spread can be controlled (by mowing in the early stages).

ROSE ACACIA (*Robinia hispida*) p. 140
A beautiful small shrub with brittle branches, its twigs, leaf-

stalks, and flowering stalks covered with red bristles. **Size:** 3–4 feet high, nearly as wide and often suckering freely. **Leaves:** Compound, composed of 7–13 oval leaflets that are about 1 inch wide. **Flowers:** Pea-like, pink, in few-flowered clusters; May–June. **Fruit:** A seldom-produced pod, 2–3 inches long, bristly. **Hardy:** From Zone 3 southward. **Varieties:** None. **Culture:** See above.

Left: Bladder-Senna (*Colutea arborescens*), p. 138.
Right: Rose Acacia (*Robinia hispida*), p. 139.

Related plant (not illustrated):

Robinia kelseyi. A shrub or small tree, 5–10 feet high, with 9–11 leaflets, pink flowers in bristly clusters, and a pod covered with purple hairs. Hardy from Zone 3 southward.

Clammy Locust
(*Robinia viscosa*), below.

CLAMMY LOCUST (*Robinia viscosa*) above
An extremely showy ornamental tree, its dark red twigs, leafstalks, and the stalk of the flower cluster sticky and hairy. **Size:** 30–40 feet high, about half as wide. **Leaves:** Compound, composed of 13–25 ovalish leaflets that are downy beneath when young. **Flowers:** Pea-like, pink, with a yellowish blotch, borne in many-flowered clusters that are 2–3 inches long, and not fragrant; May–June. **Fruit:** A somewhat sticky pod, 2–3 inches long. **Hardy:** From Zone 4 and protected sites in Zone 3 southward. **Varieties:** None. **Culture:** See above.

YELLOW-WOOD (*Cladrastis lutea*) p. 150
A graceful tree, rare in its natural habitat in the forests of southeastern U.S., but now widely planted for its wisteria-like clusters of flowers. To the pioneers its wood furnished an excellent, lasting yellow dye. **Size:** 30–50 feet high, about a third as wide, its smooth bark silvery gray. **Leaves:** Compound, composed of 7–9 ovalish leaflets, 3–4 inches long, without marginal teeth. **Flowers:** Pea-like, white, fragrant, about 1 inch long, borne in a showy, drooping cluster 10–20 inches long; June. **Fruit:** A pod 4–5

inches long, rather flat and with 3–6 seeds. **Hardy:** From Zone 3 southward. **Varieties:** None. **Culture:** Any ordinary garden soil suits it, and some trees have even thrived on city streets. An extremely handsome tree.

PAGODA TREE (*Sophora japonica*) p. 150
A Korean and Chinese tree, introduced into Japan and much planted there around Buddhist temples for its showy yellow flowers. Not as well known here as it should be. **Size:** 40–60 feet high, in maturity a round-headed tree nearly as wide. **Leaves:** Compound, composed of 7–17 ovalish, stalked leaflets, 1–2 inches long. **Flowers:** About ½ inch long, yellowish white, arranged in large, showy, terminal clusters 10–15 inches long; July–Sept. **Fruit:** A slender pod 2–3 inches long with 3–6 poisonous seeds (if eaten). **Hardy:** From Zone 3 southward. **Varieties:** *pendula* has drooping branches but is inclined not to flower. **Culture:** Needs full sun and some protection from wind, which seems more important than the soil type as it thrives on many different ones.

Related plant (not illustrated):

Sophora viciifolia. A Chinese shrub, 4–6 feet high, with 13–19 leaflets and bluish-violet flowers. Hardy from Zone 3 southward.

BASTARD INDIGO (*Amorpha fruticosa*) p. 150
A bushy shrub of the eastern U.S., with not particularly striking flowers and of minor use in the shrub border. **Size:** 8–18 feet, sometimes more, and about half as wide. **Leaves:** Compound, composed of 11–25 ovalish leaflets that are about 1½ inches long. **Flowers:** Pea-like, small, purplish or bluish, in a dense, terminal cluster about 6 inches long. **Fruit:** A curved pod 3½–4½ inches long. **Hardy:** From Zone 3 southward. **Varieties:** Several are noted in the literature, but they are of waning availability. **Culture:** Admirably suited to poor, dry, sandy or rocky soils. Often called False Indigo.

Related plant (not illustrated):

Lead Plant (*Amorpha canescens*). A very hardy relative

of the above, with hoary foliage, and more showy blue flowers. Hardy from Zone 2 southward.

Flowers Never Pea-like; Usually Perfectly Regular

This group of shrubs and trees which never have pea-like flowers, is so large that it is best divided upon the easily observed features of having opposite or alternate leaves, and whether the leaves are simple or compound. A brief key to the main groups of them is:

Leaves alternate. See p. 186

Leaves opposite (except in the upper leaves of crape myrtle, in two species of Rhamnus, for which see p. 220, and in Cornus alternifolia).

Leaves simple. See p. 167

Leaves compound.

Leaflets mostly 3. Bladder-Nut

Leaflets mostly 5–7, arranged finger-fashion. Horse-Chestnut

BLADDER-NUT (*Staphylea colchica*)
Family Staphyleaceae p. 150
A European shrub, rarely a small tree, suited to half-shady places, with not particularly showy flowers but rather striking, bladdery, fruit. **Size:** 8–12 feet high, about half as wide. **Leaves:** Opposite, compound, composed of 3 leaflets on flowering twigs, but often 5 on other twigs, the leaflets oblong or ovalish, 2–3 inches long, sharply toothed. **Flowers:** White or greenish white, of 5 separate petals, about ⅝ inch long, arranged in a terminal cluster. **Fruit:** A bladdery, rather leathery, inflated, more or less 3-sided pod, 1½–3½ inches long with bony seeds. **Hardy:** From Zone 3 southward. **Varieties:** None. **Culture:** Plant in a reasonably

moist, rich soil, in partial shade, and in a wind-free site. Not suited to rocky or sandy soils.

Related species (not illustrated):

Staphylea trifolia. A native American relative of the above with uniformly 3 leaflets and a small nodding flower cluster. Hardy from Zone 2 southward.

Horse-Chestnut
(*Aesculus*)
Family Hippocastanaceae

Showy shrubs, or more often tall trees, with opposite, compound, long-stalked leaves which in ours have 5–7 leaflets, all arranged finger-fashion. Besides the horse-chestnuts it includes the buckeyes. Flowers slightly irregular (*i.e.,* not symmetrical) in showy terminal clusters. Petals 4–5, slightly unequal, and with a longish shank. Fruit a large, ultimately splitting pod, sometimes smooth, but often with soft prickles on the outside, with usually one very large seed.

Of the 25 known species, only 4 can be included here. The tall trees cast such a dense shade that little or nothing will grow beneath them. They will thrive in a variety of garden soils but do not tolerate windy, dusty streets. The four species may be distinguished thus:

Flowers prevailingly white.
- *Tall tree.* — Horse-Chestnut
- *Shrub.* — *Aesculus parviflora*

Flowers not white.
- *Flowers pink or red.* — Red Buckeye
- *Flowers yellow.* — Sweet Buckeye

HORSE-CHESTNUT (*Aesculus Hippocastanum*) p. 145
A beautiful, tall, dome-shaped tree from the Balkans, with spectacular flowers and copious fruit, objected to by many because of the litter of its leaves, flowers, and fruit. **Size:** 70–100 feet,

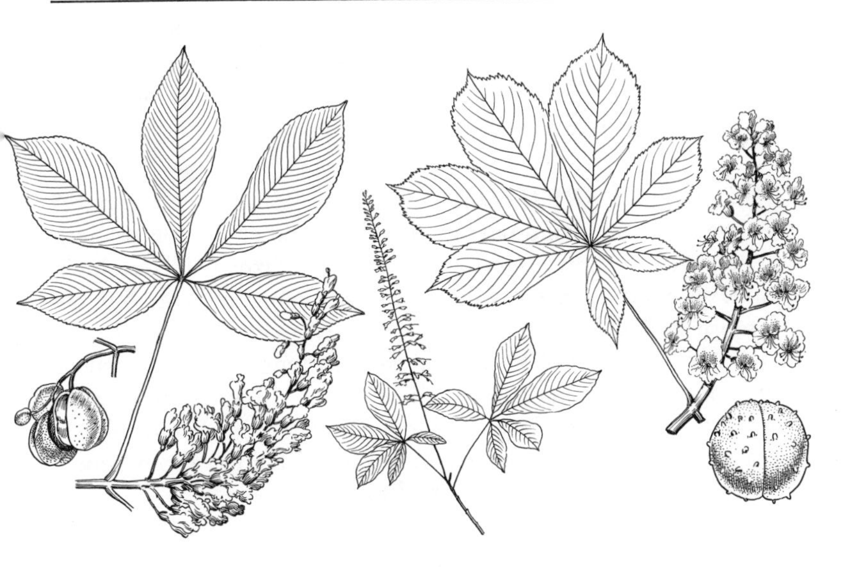

Left: Sweet Buckeye (*Aesculus octandra*), p. 146. Center: *Aesculus parviflora,* p. 146. Right: Horse-Chestnut (*Aesculus Hippocastanum*), p. 144.

more than half as wide, and casting a dense shade. **Leaves:** Opposite, compound (the winter buds very sticky), composed of 5–7 large leaflets arranged finger-fashion. Leaflets stalkless, more or less wedge-shaped, 5–9 inches long and with no autumnal color. **Flowers:** White, tinged with red, about ¾ inch wide, the very showy cluster 8–15 inches long; May–June. **Fruit:** A prickly bur, containing usually only one chestnut-like, large, shining, brown nut about 1½ inches wide. **Hardy:** From Zone 3 southward. **Varieties:** Many, the most important being *baumanni,* a double-flowered form that produces no fruit and is hence useful for those who object to the litter. It is often offered under the name *albo-plena.* **Culture:** See above. It dislikes the heat of pavements and should not be planted on city streets as the heat appears to encourage leaf scorch, to which the tree is subject. This is the "chestnut" of London and Paris.

Related species (not illustrated):

Red Horse-Chestnut (*Aesculus carnea*). A hybrid tree, very like the common Horse-Chestnut, but with red flowers and a somewhat greater tolerance of drought than the typical form.

AESCULUS PARVIFLORA p. 145
A late-flowering, handsome moundlike shrub of the southeastern states, with a showy, erect flower cluster. **Size:** 8–12 feet high, quite as wide or even wider. **Leaves:** Compound, the 5–7 oblongish leaflets 3½–8 inches long, a little broader toward the tip, stalkless. **Flowers:** White, about ½ inch long, with conspicuous and protruding pink stamens. The flower cluster is erect, cylindric, nearly a foot long and showy; Aug.–Sept. **Fruit:** Nearly egg-shaped, smooth, about 1¾ inches high. **Hardy:** From Zone 4 southward. **Varieties:** None. **Culture:** See above.

RED BUCKEYE (*Aesculus Pavia*) p. 150
A native shrub or small tree of the southeastern states, cultivated for its showy flowers. **Size:** As a shrub (its usual form) 5–15 feet high; as a tree (rare) 10–30 feet high. **Leaves:** Compound, its 5 short-stalked leaflets oblong, 3½–5 inches long, irregularly toothed on the margin. **Flowers:** Bright red, in a loose cluster 4–7 inches long; June. **Fruit:** Roundish or egg-shaped, smooth, 1¼–2 inches in diameter. **Hardy:** From Zone 4, southward. **Varieties:** None. **Culture:** See above.

SWEET BUCKEYE (*Aesculus octandra*) p. 145
A tall tree, valuable as the only yellow-flowered horse-chestnut and a native of the eastern states. **Size:** 40–60 feet high or even more, nearly half as wide. **Leaves:** Compound, the 5 leaflets elliptic, but broadest toward the tip, 4½–7 inches long. **Flowers:** Yellow, about 1¼ inches long, the open cluster 4½–7 inches long; May–June. **Fruit:** Nearly 2 inches in diameter, smooth, usually with 2 nuts. **Hardy:** From Zone 3 southward. **Varieties:** None. **Culture:** See above.

Related plant (not illustrated):

Ohio Buckeye (*Aesculus glabra*). A native tree, 15–30 feet high, with 5 leaflets, greenish-yellow flowers and a prickly fruit. Hardy from Zone 4 (possibly in sheltered parts of Zone 3) southward. The only horse-chestnut with effective fall foliage which is bright orange.

Trees and Shrubs Having Flowers with Separate Petals, Never Pea-like and Always with Opposite, Simple Leaves

AMONG this group are two exceptions to the rule of having simple leaves. They are the Box-Elder and *Acer griseum,* for which see pp. 169 and 170. All the rest can easily be divided upon the basis of whether the opposite, simple leaves have marginal teeth or lobes or have none:

Leaves with obvious marginal teeth or lobes that are often themselves toothed (NOTE: *The teeth are sometimes small and need close attention to be seen*). See p. 167
Leaves without marginal teeth, or, if any, they are remote and very small.
  Leaves not evergreen (except in 2 species of Hypericum). See p. 152
  Leaves evergreen. Box

BOX
(*Buxus*)
Family Buxaceae

This finest of all broad-leaved evergreens is so deservedly popular that people constantly attempt to grow it outside its normal hardiness range, usually with fatal results. The box in its innumerable horticultural forms may be anything from a dwarf shrub to a tree 20–25 feet high. Leaves evergreen, opposite, without marginal teeth, stalkless or with very short stalks. Flowers minute, greenish, without petals, usually passing unnoticed in Apr. or May, followed by very small, 3-horned pods, each with two shining seeds.

Box is so slow-growing and expensive that its cultivation should be confined to the region where it is climatically suited. It will not stand long, severe winters, nor will it stand intense, long-continued heat. Nowhere does it grow in such perfection as in tidewater Md., Del., and Va., where the air is warm but moist. North of Zone 4 it is extremely hazardous, and only the second species will thrive in the Deep South. In the prairie states its culture is next to impossible. Read carefully the hardiness notes below.

The plant is shallow-rooted, will grow in a variety of garden soils, and after planting should be watered frequently during the first year. It should never have its roots exposed to the air, and should not have them disturbed by surface cultivation or spading. It needs little, if any, fertilizer and if a mulch is applied (many think it is unnecessary) it should not be over 1½ inches deep, preferably of sawdust or peanut hulls. It roots readily from cuttings made early in Sept. and put outdoors in a mixture of sand and peat, out of the wind and in partial shade, if you are in Zones 5 or 6. Otherwise cuttings should be rooted in a cold frame.

The identification of the forms is puzzling, due to the many horticultural varieties:

Leaves more or less elliptic, broadest at or below the middle. Common Box

Leaves more or less round or oblong, broadest above the middle. *Buxus microphylla*

COMMON BOX (*Buxus sempervirens*) p. 150

The box of history and the one that enhances the splendor of many old estates in Md., Va., and Del. Quite unsuited to regions of bitter cold or stifling heat. **Size:** Impossible to specify as some varieties are reputedly permanent dwarfs, while others are treelike and up to 20–25 feet high. The usual form is 8–10 feet high and as wide, densely furnished with foliage from the ground up. **Leaves:** Opposite, evergreen, broadest at or below the middle, very short-stalked, ¾–1½ inches long, rounded or faintly notched at the tip, borne on 4-sided, slightly hairy twigs. The leaves, especially in a moist atmosphere, give off a slightly musky odor, pleasurable to some but definitely foxy to others. **Flowers and Fruit:** Negligible. **Hardy:** Definitely best suited to Zones 5 and 6, less so to Zone 4 where some winter protection is advisable, and hazardous or impossible from Zone 3 northward. **Varieties:** Nearly 30, many of them puzzling. Leaving out those with variegated or splotched leaves, the most important are:

angustifolia. A treelike form with narrower leaves than the type, the leaves being the largest of any of the varieties of the Common Box.

arborescens. The Tree Box, a treelike form with leaves like the typical box, but considerably more pointed. If sheared from youth and kept sheared twice a year it makes splendid, thick, bushy plants. Its twig growth may be 3–4 inches a year. This is the so-called American box.

handsworthi. A sturdy shrub with more or less upright branches, usually with darker green leaves notched at the tip.

myrtifolia. A small shrub, not over 4–5 feet high, its leaves scarcely ¾ inch long.

rotundifolia. A form with the leaves rounder than in the type, and the plant is a little more hardy than the common box.

suffruticosa. The common boxwood of gardens in Virginia, Maryland and Delaware. In spite of its Latin name (implying dwarfness), this plant is usually 8–10 feet high and as thick. It has faintly notched, rounded leaves and annual twig growth does not usually exceed an inch or so. It is the most important of the six varieties of *Buxus sempervirens*.

It is difficult or impossible to identify these varieties without mature specimens.

PEA FAMILY (*Leguminosae*), 1, 3, 9
2 and 4–8 in various families

1. **Bastard Indigo** (*Amorpha fruticosa*) p. 142
   A not particularly showy native shrub with compound leaves and purplish flowers in terminal clusters.

2. **Common Box** (*Buxus sempervirens*) p. 148
   Never grown for its inconspicuous flowers, this European evergreen shrub is the finest of all the varieties of box.

3. **Yellow-Wood** (*Cladastris lutea*) p. 141
   A medium-sized native tree with showy, hanging clusters of wisteria-like, white, and fragrant flowers.

4. **Gold-Flower** (*Hypericum moserianum*) p. 155
   A midsummer-blooming low hybrid shrub with simple leaves and yellow flowers nearly 2 inches wide.

5. ***Chimonanthus praecox*** p. 156
   Not for northern gardeners, this superlatively fragrant Chinese shrub blooms all winter on bare twigs.

6. **Bladder-Nut** (*Staphylea colchica*) p. 143
   European shrub with 3 leaflets, greenish-white flowers, and a bladdery, inflated, 3-sided pod.

7. **Red Buckeye** (*Aesculus Pavia*) p. 146
   A native shrub or small tree with 5 leaflets and showy clusters of red flowers in June.

8. **Rockrose** (*Cistus corbariensis*) p. 156
   A fine low hybrid shrub with half-evergreen foliage, with a white, yellow-centered flower in summer.

9. **Pagoda Tree** (*Sophora japonica*) p. 142
   An Asiatic midsummer-blooming, slow-growing tree with compound leaves and yellowish-white flowers.

1
2
3
5
4
6
7
8
9

1
2
3
4
5
6
7
8
9

MAPLE FAMILY (*Aceraceae*), 1, 6, 7
DOGWOOD FAMILY (*Cornaceae*), 8, 9
2–5 in various families

1. **Sycamore Maple** (*Acer Pseudo-platanus*) p. 171
   A large-leaved Eurasian tree with drooping clusters of yellowish-green flowers in May.

2. ***Philadelphus coronarius*** p. 158
   The commonest of the mock-oranges, usually, incorrectly, called Syringa; a disturbing fragrance.

3. **Crape Myrtle** (*Lagerstroemia indica*) p. 166
   Not for northern gardeners, this is the most gorgeous of summer-blooming shrubs.

4. **Strawberry-Shrub** (*Calycanthus floridus*) p. 165
   A highly aromatic native shrub, its dark purple flowers holding their fragrance even when faded.

5. **Pomegranate** (*Punica granatum*) p. 166
   Only a shrubby ornamental form of this fruit tree is much grown in the East. Not hardy northward.

6. **Norway Maple** (*Acer platanoides*) p. 171
   The most widely grown street tree in America, casting dense shade. Flowers early, yellow, and profuse.

7. **Japanese Maple** (*Acer palmatum*) p. 170
   A small tree grown for its extremely variable foliage, which may be dissected, and of various coloring.

8. **Flowering Dogwood** (*Cornus florida*) p. 162
   This pink-flowered variety is rare in nature but preferred by many to the common white sort.

9. **Tartarian Dogwood** (*Cornus alba*) p. 165
   Many grow this shrub not only for its white flowers but for its red twigs that make cheerful contrast against winter snow.

BUXUS MICROPHYLLA below

A Japanese shrub, two of its varieties of oustanding importance as they are hardier than the Common Box. **Size:** 3–6 feet high, nearly or quite as wide. **Leaves:** ⅓–1 inch long, broadest above the middle, its twigs angled and sometimes winged, perfectly smooth. **Flowers and Fruit:** Negligible. **Hardy:** Definitely more hardy than the Common Box, especially in the Korean Box, which is the hardiest of all the boxwoods, thriving as far north as Zone 4. **Varieties:** Korean Box (variety *koreana*), a short form rarely exceeding 3 feet, its leaves hairy on the midrib above; Japanese Box (variety *japonica*), a taller shrub, up to 6 feet high, and a little less hardy than the Korean Box. Both the varieties are a boon to New England gardeners.

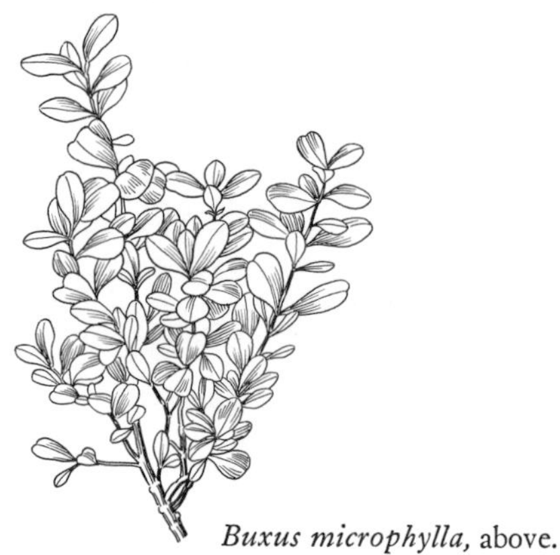

Buxus microphylla, above.

Leaves Not Evergreen

(*Except in 2 Species of Hypericum*)

Flowers white or yellowish.

Leaves with glandular dots, or aromatic when crushed, or both.

Leaves and twigs smooth. St. John's-wort

Leaves and twigs hoary. Rockrose

Leaves without glandular dots, not noticeably aromatic when crushed.

Petals 5 or more. *Chimonanthus*

Petals 4.

Flowers rather large, solitary, or in sparse clusters. Mock-Orange

Flowers small, crowded in a dense cluster. Dogwood and Cornel

Flowers neither white nor yellowish.

Flowers highly aromatic. Strawberry-Shrub

Flowers scentless or essentially so.

Flowers in a large terminal cluster. Crape Myrtle

Flowers solitary or in a sparse cluster. Pomegranate

St. John's-wort
(*Hypericum*)
Family Hypericaceae

Low shrubs with opposite, smooth leaves that are without marginal teeth and more or less spotted with usually yellowish glandular dots, often aromatic when crushed. Flowers prevailingly yellow, in small clusters or solitary. Stamens numerous. Fruit a dry capsule (in ours).

The St. John's-worts are of easy culture in most garden soils, but prefer a sandy loam. Their prevailingly yellow bloom, which comes in midsummer or later, and low stature make them useful garden plants. They are not much grown, however, and of the 18 species known to be in cultivation in the U.S., only the following can be included here. Several others, of which some are native, are found in the collections of fanciers. For the Rose-of-Sharon (*Hypericum calycinum*), which is an herb, see p. 131 of *The Guide to Garden Flowers* (Houghton Mifflin, 1958).

Leaves narrow, not ovalish; flowers 1 inch wide or less. *Hypericum kalmianum*

Leaves wider, ovalish; flowers considerably wider than 1 inch.

Shrub about 3 feet high; stems pinkish.

Hypericum patulum

Shrub about 12–15 inches high; stems reddish.

Gold-Flower

HYPERICUM KALMIANUM p. 155

A low evergreen or half-evergreen shrub, native in northeastern North America, grown for its late-blooming flowers. **Size:** 2–3 feet high, about half as wide, its twigs 4-angled. **Leaves:** Opposite, simple, without marginal teeth, narrowly oblong, 1½–2½ inches long. **Flowers:** Few, about 1 inch wide or less, yellow, and with 5 petals; Aug. **Fruit:** A dry capsule. **Hardy:** From Zone 3 southward. **Varieties:** None. **Culture:** See above.

Related species (not illustrated):

Hypericum frondosum. A native shrub, 2–3 feet high, with reddish, peeling bark and yellow flowers nearly 2 inches wide. Hardy from Zone 4 southward.

Bush Broom (*Hypericum prolificum*). An evergreen or half-evergreen native shrub, 4–5 feet high, its twigs 2-edged. Flowers about ¾ inch wide; July–Sept. Hardy from Zone 4 southward.

HYPERICUM PATULUM p. 155

An Asiatic evergreen or half-evergreen shrub, with large, nearly solitary flowers blooming in midsummer. **Size:** 2–3 feet high, about half as wide. **Leaves:** Opposite, simple, without marginal teeth, ovalish or oblong, 1½–2½ inches long. **Flowers:** Solitary or in sparse clusters, yellow, about 2 inches wide; July–Sept. **Fruit:** A dry capsule. **Hardy:** From Zone 5 southward. **Varieties:** *henryi,* a Chinese form more vigorous than the type, with larger flowers and hardy up to Zone 4, probably to Zone 3 in protected places. **Culture:** See above.

Related species (not illustrated):

Hypericum densiflorum. A native evergreen or half-evergreen shrub, 4–6 feet high, the branches 2-angled. Leaves

1–2 inches long. Flowers yellow, scarcely ½ inch wide, but in dense clusters. Hardy from Zone 4 southward.

Hypericum olympicum. A low Eurasian shrub, 8–12 inches high. Leaves narrow, 1–1½ inches long. Flowers nearly 2½ inches wide, showy, in terminal clusters. Hardy from Zone 6 southward.

Left: *Hypericum kalmianum,* p. 154.
Right: *Hypericum patulum,* p. 154.

GOLD-FLOWER (*Hypericum moserianum*) p. 150

A hybrid shrub and one of the best of the cultivated St. John's-worts. **Size:** 1–2 feet high, about half as wide. **Leaves:** Opposite, simple, without marginal teeth, ovalish, 1–2 inches long. **Flowers:** About 2½ inches wide, yellow, solitary or in a few-flowered cluster; midsummer. **Fruit:** A dry capsule. **Varieties:** *tricolor* has white-variegated leaves edged with red. **Culture:** See above.

Related plant (not illustrated):

Hypericum Coris. A European evergreen or half-evergreen shrub, 8–12 inches high, with very narrow leaves in clusters of 4–6, the leaves about 1 inch long. Flowers about ¾ inch wide, yellow. Hardy from Zone 5 southward.

ROCKROSE (*Cistus corbariensis*). Family Cistaceae p. 150
A beautiful little hybrid shrub grown for its summer-blooming flowers. **Size:** 18–30 inches high, quite bushy and nearly or quite as wide. **Leaves:** Opposite, simple, without marginal teeth but wavy-margined, evergreen or half evergreen, ovalish, 1–2 inches long, heart-shaped at the base, downy. **Flowers:** Solitary or few, white with a yellow center, about 1½ inches wide; midsummer. **Fruit:** A small, dry capsule. **Hardy:** From Zone 5 southward. **Varieties:** None. **Culture:** Not easy. They demand full sun, a sandy or gritty soil, preferably in a limestone region, and dislike winter slush.

Related plant (not illustrated):

Cistus ladaniferus. A European, sticky-stemmed shrub, 5–8 feet high, its lance-shaped leaves densely white-hairy beneath, 3–4 inches long. Flower solitary, white, but purple-blotched at the center. Hardy from Zone 6 southward.

CHIMONANTHUS PRAECOX.
Family Calycanthaceae p. 150
A Chinese shrub with very fragrant flowers that bloom before the leaves unfold; a favorite in the South. **Size:** 8–10 feet high, about half as wide, rather bushy in habit. **Leaves:** Lance-oval, long-pointed, somewhat rough, 4–6 inches long. **Flowers:** Very fragrant, solitary or in pairs, about 1 inch wide, the outer segments yellow, the inner striped purplish brown; blooming all winter in the South on the bare twigs. **Fruit:** Dry, egg-shaped, about 1½ inches long. **Hardy:** From Zone 5 southward. **Varieties:** *grandiflora* has somewhat larger but less fragrant flowers. **Culture:** Can be grown in any ordinary garden soil, but in the northern part of its range it is well to grow it in a place sheltered from north winds.

MOCK-ORANGE
(*Philadelphus*)
Family Saxifragaceae

Extremely fragrant shrubs (with a sinister odor to some), commonly called syringa, but *Syringa* is the correct Latin name of the lilac. Some of the species have peeling bark. Leaves opposite, mostly without marginal teeth, or with a few very small and remote teeth. Flowers solitary or in clusters, white, the petals 4. Fruit a small capsule. The mock-oranges are sometimes mistaken for *Deutzia,* but that group has hollow twigs and 5-petaled flowers, while the twigs of *Philadelphus* have a white pith, and its flowers have 4 petals.

These shrubs are so popular that of the 40 known species 15 were included in the fourth edition of *Taylor's Encyclopedia of Gardening.* Of these only 5 of the most popular can be included here and a simple key to them follows. They are of the easiest culture as they tolerate all ordinary garden soils. Because of the many hybrid forms, the identification of the species is rather puzzling.

Flowers in obvious clusters, often numerous.
Flowers always double. — *Philadelphus virginalis*
Flowers single.
Leaves with tufts of hair at the vein joints on under side. — *Philadelphus coronarius*
Leaves without such tufts of hair. — *Philadelphus lemoinei*
Flowers solitary or few, rarely in large clusters.
Flowers 3–9 in each cluster. — *Philadelphus cymosus*
Flowers 1–3 in each cluster. — *Philadelphus inodorus*

PHILADELPHUS VIRGINALIS p. 158

A hybrid shrub, usually with brown, peeling bark, or with grayish bark only slightly peeling. **Size:** 4–6 feet high, about one half as wide. **Leaves:** Opposite, simple, ovalish, 2½–3 inches long, slightly toothed on the margin, soft-hairy beneath. **Flowers:**

Double, white, about ½ inch wide, in clusters of 3–7; June. **Fruit:** A small dry capsule. **Hardy:** From Zone 3 southward. **Varieties:** Many, of which the following are the most grown. Argentine, with very double flowers; Glacier, with very peeling bark and semidouble flowers in clusters of 5–7; Virginal, with peeling bark and semidouble flowers in clusters of 5–7. These and several others are difficult to identify. **Culture:** See above.

Left: *Philadelphus virginalis,* p. 157.
Right: *Philadelphus lemoinei,* p. 159.

PHILADELPHUS CORONARIUS p. 151

The most widely cultivated of the mock-oranges, very fragrant, but John Gerard (in 1597) wrote that he found the odor "too sweet, troubling and molesting the head," and many agree with him today. Bees will not touch it. **Size:** 6–10 feet high, half as wide, its bark peeling. **Leaves:** Opposite, simple, ovalish or oblong, pointed, 1½–4 inches long, with a few remote marginal teeth or with none, and with tufts of hair at the veins beneath. **Flowers:** Single, white, about 1½ inches wide, in clusters of 5–7; June. **Fruit:** A small dry capsule. **Hardy:** From Zone 2 southward. **Varieties:** 8 or 10, of which the following are most likely to interest the grower; *dianthiflorus* and *flore-pleno* have double flowers; Golden Mock-orange (variety *aureus*) has bright yellow leaves in youth; *variegatus,* with white-bordered leaves; *pumilus,* a dwarf form with smaller leaves.

Related plant (not illustrated):

Philadelphus nivalis. A hybrid shrub, its white flowers 1–1½ inches wide, in clusters of 5–9, the leaves with marginal teeth, slightly soft-hairy beneath. Hardy from Zone 2 southward.

PHILADELPHUS LEMOINEI p. 158

A very popular hybrid shrub, perhaps the most fragrant of all the mock-oranges. **Size:** 4–6 feet high, about three fourths as wide, its brown bark peeling. **Leaves:** Opposite, simple, ovalish or narrower, pointed, 1½–2½ inches long, toothed on the margin, smooth above, stiff-hairy beneath, noticeably smaller and with fewer teeth on flowering twigs. **Flowers:** Showy, about 1½ inches wide, in clusters of 3–7. **Fruit:** A small dry capsule. **Hardy:** From Zone 2 southward. **Varieties:** At least 17, most of them rather ill-defined. Two of the best are Boule d'argent and Avalanche. **Culture:** See above.

Related plants (not illustrated):

Philadelphus pekinensis. A Chinese and Korean shrub, 4–6 feet high, its leaves with purple stalks. Flowers cream-white, in clusters of 5–9. Hardy from Zone 3 southward.

Philadelphus microphyllus. A Rocky Mountain shrub, 3–4 feet high, with usually solitary flowers having a delightful pineapple fragrance. Hardy from Zone 4 southward.

PHILADELPHUS CYMOSUS p. 160

A hybrid mock-orange with flowers larger than most of the others, and very fragrant. **Size:** 4–6 feet high, about half as wide, its bark peeling. **Leaves:** Opposite, simple, ovalish, 2½–3 inches long, remotely toothed with minute teeth, usually somewhat hairy beneath. **Flowers:** White, single or double, about 2½ inches wide, in clusters of 3–9 and hence showy; June. **Hardy:** From Zone 3 southward. **Varieties:** Bannier, has semidouble flowers; Norma, with double or single flowers. **Culture:** See above.

PHILADELPHUS INODORUS below

The only mock-orange here included without any fragrance and a favorite with those who dislike the usual odor; a native of the southeastern states. **Size:** 6–8 feet high, about half as wide. **Leaves:** Ovalish or broader, smooth and shiny, 3–5 inches long, occasionally toothed on the margin. **Flowers:** Nearly cuplike, but of 4 separate petals, usually solitary, but sometimes in clusters of 3, nearly 2 inches wide. **Fruit:** A small dry capsule. **Hardy:** From Zone 4 southward. **Varieties:** None, but see the next item. **Culture:** See above.

Related plant (not illustrated):

Philadelphus grandiflorus. Perhaps only a variety of *Philadelphus inodorus,* and differing chiefly in having larger flowers and leaves. Hardy from Zone 3 southward, where it is often known as "English dogwood," although it is neither English nor a dogwood.

Left: *Philadelphus cymosus,* p. 159.
Right: *Philadelphus inodorus,* above.

Dogwood and Cornel
(*Cornus*)
Family Cornaceae

Trees and shrubs of quite diverse habit and aspect, having opposite, simple leaves without marginal teeth and often brightly colored bark, most decorative in winter. Flowers of three kinds: (1) small flowers in profuse, usually flat-topped clusters (mostly shrubs); (2) small flowers (greenish and inconspicuous) having below them large, conspicuous bracts, often wrongly called petals (Flowering Dogwood and related trees); and (3) yellow flowers blooming before the leaves expand (Cornelian Cherry). Petals always 4 (5 in shrubby viburnums which are sometimes confused with *Cornus*), followed by a fleshy, berrylike fruit with a single stone. To the confusion of nearly everyone the Blue Dogwood, instead of having opposite leaves, has alternate ones.

The culture of the shrubby species below is simple in any ordinary garden soil, as they transplant easily. This is not true of the dogwood and its tree relatives which must be moved with care, always with a large ball of earth tightly done up in bagging. Water them freely the first year of planting, and if possible choose a sheltered, wind-free site. At least six species are in common cultivation. For the Bunchberry (*Cornus canadensis*), a practically herbaceous relative of the Flowering Dogwood, see *The Guide to Garden Flowers,* p. 125.

Trees.
  Flowers yellow, appearing weeks before the leaves unfold. Cornelian Cherry
  Flowers small, greenish, having below them 4–6 conspicuous bracts.
    Bracts usually notched at the tip. Flowering Dogwood
    Bracts not notched at the tip. *Cornus Kousa*
Shrubs.
  Leaves alternate. Blue Dogwood
  Leaves opposite.

Twigs not bright red or pink. Gray Dogwood
Twigs bright red or pink. Tartarian Dogwood

CORNELIAN CHERRY (*Cornus Mas*) p. 163
A small Eurasian tree, sometimes shrubby, much grown for its early-blooming flowers which cover the twigs weeks before the leaves expand. **Size:** 15–25 feet high, about half as wide, its bark close, scaly and dark brown. **Leaves:** Simple, opposite, without marginal teeth, 3–4 inches long. **Flowers:** Minute, yellow, in small, headlike clusters, so profuse that they nearly cover the bare twigs; Mar.–Apr. (earlier in the South). **Fruit:** Berrylike, scarlet, edible but acid, ripening in Aug. **Hardy:** From Zone 3 southward. **Varieties:** None. **Culture:** See above; it stands city conditions better than most trees.

FLOWERING DOGWOOD (*Cornus florida*) p. 151
A stunning native tree, always naturally an inhabitant of the undercanopy of the forest, and hence doing better in the garden where there is at least partial shade and freedom from wind. **Size:** 20–30 feet high, a little more than half as wide. **Leaves:** Opposite, simple, without marginal teeth, ovalish, 3–5 inches long, abruptly pointed at the tip, beautifully scarlet in the fall and so profuse that the tree casts a dense shade. **Flowers:** Minute, greenish, and inconspicuous, set in the midst of 4 large, showy, white, notched, petal-like bracts (often mistaken for petals); May. **Fruit:** Berrylike, scarlet in Oct., but not persistent. **Hardy:** Not quite hardy north of Zone 4, and most profuse south of this. **Varieties:** *rubra,* a similar tree with red or pink bracts, which has been in cultivation since 1731, generally under the name of Red Dogwood; there are also a weeping and a "double-flowered" form (actually with 6–8 bracts). **Culture:** See above; it is not easy to transplant.

CORNUS KOUSA p. 163
An Asiatic representative of our native Flowering Dogwood and very similar. Its chief difference is that its white bracts are not notched, and its pinkish fruits are in a beadlike cluster; it is also not quite so tall. Its chief virtue is that it flowers nearly a month

later than the native flowering dogwood, and its variety *chinensis* has longer and more showy bracts. **Hardy:** From Zone 4 southward. **Culture:** See above.

Related plant (not illustrated):

Cornus nuttalli. A smaller tree than the native Flowering Dogwood, and the Pacific Coast representative of it, with 6 petal-like, white or pinkish bracts. Doubtfully hardy above Zone 6.

Left: Cornelian Cherry (*Cornus Mas*), p. 162. Right: *Cornus Kousa,* p. 162.

BLUE DOGWOOD (*Cornus alternifolia*) p. 164

A shrubby exception to the rest of the dogwoods in having alternate leaves; native in eastern North America, its twigs greenish. **Size:** 8–15 feet high, and about half as wide, usually a shrub (rarely a small tree). **Leaves:** Simple, alternate, ovalish or elliptic, without marginal teeth, 2–6 inches long. **Flowers:** Very small, white, crowded in a loose, moundlike cluster about 2½ inches wide; May–June. **Fruit:** Berrylike, blue, with a bloom, not persistent. **Hardy:** Everywhere. **Varieties:** *argentea* has white-variegated leaves, and is often offered under the name *variegata.* **Culture:** See above.

GRAY DOGWOOD (*Cornus racemosa*) below

A native shrub often offered under the incorrect name of *Cornus paniculata,* grown for its white flowers and fruit, its twigs gray. **Size:** 6–10 feet high, about half as wide. **Leaves:** Simple, opposite, without marginal teeth, elliptic or narrowly oval, 2–4 inches long, tapering at the tip, but wedge-shaped at the base. **Flowers:** Very small, white, in a moundlike, branched cluster that is about 2 inches wide; June–July. **Fruit:** Berrylike, white, on red stalks, pea-size, not persistent. **Hardy:** From Zone 3 southward. **Varieties:** None. **Culture:** See above.

Left: Gray Dogwood (*Cornus racemosa*), above.
Right: Blue Dogwood (*Cornus alternifolia*), p. 163.

TARTARIAN DOGWOOD (*Cornus alba*) p. 151

The finest of all the shrubby dogwoods for winter effects as its bright red twigs and branches make dramatic contrasts against the snow. It is a medium-sized Asiatic shrub, perhaps the commonest *Cornus* in cultivation. **Size:** 6–10 feet high, inclined to spread and hence often as wide. **Leaves:** Simple, opposite, without marginal teeth, ovalish, 3–5 inches long, bluish green beneath, red in the fall. **Flowers:** Very small, white, in numerous clusters that are about 2 inches wide and showy; May–June. **Fruit:** Berrylike, bluish white, not persistent. **Hardy:** Everywhere. **Varieties:** Siberian Dogwood (variety *siberica*) has coral-red twigs; there are several other varieties with variegated or otherwise blotched leaves. **Culture:** See above.

Related plants, all with red or pink twigs (not illustrated):

Red Osier (*Cornus stolonifera*). A native shrub, 4–6 feet high, spreading widely, with white flowers and bluish-white fruit. Hardy everywhere.

Silky Cornel (*Cornus Amomum*). A native shrub, 6–8 feet high, its leaves silky beneath. Flowers white. Fruit blue. Hardy everywhere.

Cornus baileyi. A native shrub, 6–9 feet high, closely related to the Red Osier, with white fruit. Hardy everywhere and used in seaside gardens.

Red Dogwood (*Cornus sanguinea*). A Eurasian dogwood, 6–10 feet high, closely related to the Tartarian Dogwood, but with black fruit. Hardy everywhere.

STRAWBERRY-SHRUB (*Calycanthus floridus*)

Family Calycanthaceae p. 151

A very aromatic native shrub, so spicy in odor that it is also called Carolina Allspice; densely hairy throughout. **Size:** 4–6 feet high, about half as wide. **Leaves:** Opposite, simple, without marginal teeth, oval or elliptic, 3–5 inches long, pale on the under side. **Flowers:** Solitary, dark purplish brown, very fragrant, about 2 inches wide; June–July. **Fruit:** Urn-shaped, capsule-like, about 2½ inches long. **Hardy:** From Zone 4 southward. **Varieties:** None. **Culture:** Preferably planted in a rich, reasonably moist soil, but with no standing water at its roots. It will, however, do reasonably well in any ordinary garden soil.

Related plant (not illustrated):

Calycanthus fertilis. A native shrub, perfectly smooth in all its parts, also aromatic, but not so much so as the Strawberry-Shrub. Hardy from Zone 4 southward.

CRAPE MYRTLE (*Lagerstroemia indica*).
Family Lythraceae p. 151

Without much doubt the most magnificent midsummer flowering shrub in cultivation, but unhappily not for northern gardeners. It is a native of China and very widely grown throughout the South. **Size:** 6–25 feet high, about a third as wide, its stems gray and smooth, the bark shredding in midsummer. In the northern part of its range it is often cut down to the ground, as it will bloom from the current year's shoots. Such annual shoots are likely to be winter-killed in the North. **Leaves:** Simple, opposite (some upper ones may be alternate), without marginal teeth, elliptic to oblong, 1–2 inches long, nearly stalkless and turning rusty red in the fall. **Flowers:** Pink, the 6 petals frilled or crinkled, and with a long shank. They are borne in a dense cluster 4–9 inches long, lilac-like but much more floriferous, so profuse as to nearly obscure the foliage; July 15 to Sept. 1, according to variety. **Hardy:** Safely from Zone 5 southward, but often grown slightly beyond this and usually winter-killed there. **Varieties:** *alba* has white flowers; *purpurea* has purple flowers; *rubra* has red, almost scarlet flowers. **Culture:** A little difficult to get started and always late in leafing out in the spring. It will thrive, when once established, in almost any good garden soil, preferably a reasonably rich one.

POMEGRANATE (*Punica Granatum*).
Family Punicaceae p. 151

A cultivated fruit in southern Europe and Asia, but as here considered, a showy ornamental shrub or small tree, usually a little spiny. **Size:** 10–20 feet high, about half as wide. **Leaves:** Opposite, simple, without marginal teeth, oblongish, short-stalked, 1½–

3 inches long. **Flowers:** See below at Varieties. **Fruit:** Practically never maturing within the area of this book. **Hardy:** From Zone 5 southward. **Varieties:** Quite a number as an ornamental. The flowers, borne in all of them in small clusters at the leaf joints, are comprised of 5–7 separate, wrinkled red petals about 1½ inches wide. The most showy of the varieties have double scarlet flowers which never produce fruit, blooming mostly in June. **Culture:** Any good garden soil will do. By careful winter pruning the plant can be made into an attractive hedge.

Trees or Shrubs with Opposite, Simple (2 Maples Are Exceptions) Leaves, the Margins Always Toothed or Lobed, Sometimes Both. Flowers with Separate Petals, Never Pea-like

HERE come a group quite unrelated to each other, but readily separated thus:

Mostly tall trees, or tree-like shrubs:
- *Fruit a winged "key" (see illustrations).* — Maple
- *Fruit a many-seeded pod.* — Katsura Tree

Shrubs of medium size.
- *Flowers small, inconspicuous, prevailingly greenish, yellowish, or purplish.*
  - *Twigs round.* — Japanese Laurel
  - *Twigs usually 4-angled.* — Spindle-Tree
- *Flowers large and showy, prevailingly white.*
  - *Twigs hollow.* — *Deutzia*
  - *Twigs not hollow.*
    - *Flowers in large clusters.* — *Hydrangea*
    - *Flowers solitary.* — Jetbead

MAPLE

(*Acer*)

Family Aceraceae

Tall trees, rarely shrubs, the leaves all opposite, the veins arranged finger-fashion, and simple except in the Box-Elder where the leaves are compound. Most of the maples have lobed or toothed leaves, or both. Flowers small, each inconspicuous, some of them without petals, but en masse quite showy, especially in those that bloom before the leaves unfold. Fruit 2-winged, keylike.

All those below are relatively tall trees or treelike shrubs except the Japanese Maple, which as usually grown is a small tree. Two native shrubs, the Striped Maple (*Acer pensylvanicum*) and the Mountain Maple (*Acer spicatum*) are woodland plants of the North, not much cultivated, and hence excluded here. The trees are very easily grown in most garden soils, and the Box-Elder is particularly suited to windy prairie states where better species will not thrive. Of the 20 species of maple in cultivation only the 7 below can be included here:

Leaves compound. Box-Elder
Leaves simple.
  Flower clusters blooming with or after the leaves unfold.
    Small-sized trees or treelike shrubs; flowers greenish, purple, or yellowish white.
      June-flowering; the flowers purple; a tree. Japanese Maple
      May-flowering; treelike shrubs. *Acer Ginnala*
    Tall trees; flowers greenish yellow.
      Sap of leafstalk milky. Norway Maple
      Sap of leafstalk not milky.
        Lobes of the leaf coarsely toothed; flower cluster 3–6 inches long. Sycamore Maple

Lobes of the leaf with few or no teeth; flower cluster not usually more than 2 inches long. Sugar Maple

Flower cluster on naked twigs, blooming weeks before the leaves unfold. Red Maple

Left: *Acer Ginnala,* p. 170.
Right: Box-Elder (*Acer Negundo*), below.

BOX-ELDER (*Acer Negundo*) above

A relatively worthless native tree in the East, where it seeds so freely it can become a nuisance, but extremely useful in the West where it will stand drought and wind that make better maples impossible. **Size:** 40–70 feet high, half as wide, not casting a dense shade, and a rapid grower. **Leaves:** Opposite, compound, comprising 3–5 bright-green leaflets 2–4 inches long, coarsely toothed or the terminal one 3-lobed. **Flowers:** Yellowish green, blooming before the leaves unfold, some in flat-topped clusters, others in hanging clusters; Mar.–Apr. **Fruit:** A winged "key," the wings usually incurved. **Hardy:** From Zone 1 southward, and it will stand wind, smoke, dust, and drought. **Varieties:** Many, of which *variegatum* has white-variegated foliage. The others are of indifferent value. **Culture:** See above.

Related plant (not illustrated):

Paperback Maple (*Acer griseum*). A Chinese tree with paper-thin, peeling gray bark, compound leaves with 3 leaflets and short-stalked hairy clusters of small flowers. Foliage orange-red in autumn. Hardy from Zone 5 southward.

JAPANESE MAPLE (*Acer palmatum*) p. 151

A slow-growing Japanese and Korean tree valued chiefly for its thin, colorful, and decorative foliage. **Size:** 10–25 feet high, nearly three fourths as wide, but often shrubby and lower. **Leaves:** Opposite, simple, 5–9-lobed and variously colored (see below), pointed, smooth, and often cut nearly to the base, the lobes toothed; the green varieties turning bright red in autumn. **Flowers:** Small, purple, in flat-topped clusters; June. **Fruit:** Keylike, the wings incurved. **Hardy:** From Zone 3 southward. **Varieties:** Over 40, and the grower is urged to visit a well-labeled arboretum or nursery before making a choice. The varietal names, some of them Japanese, are in much confusion, but the following appear to be reasonably reliable: *dissectum* has deeply divided, narrow, green leaf lobes; *atropurpureum* has deep red foliage and only 7 lobes to the leaf; *ornatum* has deeply dissected leaves, red at first, ultimately bronzy; *sanguineum* has red leaves a little smaller than the typical Japanese Maple; *aureum* has golden-yellow leaves. **Culture:** See above.

Related plant (not illustrated):

Acer japonicum. A Japanese shrub or small tree, 10–30 feet high, its leaves 7–11-lobed, 3–5½ inches long. Flowers purple, in flat-topped clusters; May. Hardy from Zone 3 southward.

ACER GINNALA p. 169

A medium-sized tree or tall shrub from eastern Asia, its fall foliage scarlet, as are the fruits. **Size:** 10–20 feet high, about half as wide. **Leaves:** Opposite, simple, 3-lobed, 1½–4 inches long, the central lobe larger than the side ones and toothed. **Flowers:** Small, yellowish white, in dense clusters, fragrant; May. **Fruit:**

Keylike, red, and maturing while the leaves are still green. **Hardy:** From Zone 3 southward. **Varieties:** None. **Culture:** See above.

NORWAY MAPLE (*Acer platanoides*) p. 151

Perhaps the most widely planted lawn and street tree in America, splendidly dome-shaped and casting so dense a shade that nothing will grow under this outstanding European tree. **Size:** 60–100 feet high, its canopy nearly as wide in maturity. **Leaves:** Opposite, simple, long-stalked, the leafstalks with a milky sap (not always easy to see), 5-lobed, 4–7 inches across, the lobes with marginal teeth; foliage bright yellow in autumn. **Flowers:** Small, yellow, or greenish yellow, in erect, many-flowered, flat-topped clusters; Apr.–May, just before or with the unfolding of the leaves. **Fruit:** Keylike, drooping, its wings spreading almost horizontally. **Hardy:** Everywhere. **Varieties:** Schwedler's Maple (variety *schwedleri*) has leaves red in the spring, green by midsummer; Crimson King is a form wherein the red leaves reputedly stay so all summer; *columnare* is a narrow-headed, more erect form useful for street planting; *globosum* is an even more round-headed, but shorter variety. There are also variegated, golden-leaved, and dissected-leaved varieties. **Culture:** See above; in spite of its perhaps excessive popularity it is still the best maple for lawn and street planting (if the streets are wide).

Related plant (not illustrated):

Hedge Maple (*Acer campestre*). A Eurasian tree or large shrub, often trimmed as a hedge in Europe, 10–30 feet high, its leaves 3-lobed. Flowers greenish, in May. Hardy from Zone 2 southward.

SYCAMORE MAPLE (*Acer Pseudo-platanus*) p. 151

The largest-leaved maple in common cultivation, its leafstalk also very long. It is a Eurasian tree grown for centuries for its handsome foliage and drooping clusters of flowers. **Size:** 60–100 feet high, about half as wide. **Leaves:** Opposite, simple, coarse, 5-lobed, 4–8 inches wide, green above, paler beneath, the leafstalk 4–8 inches long, the lobes toothed; with no fall color.

Flowers: Yellowish green, small, but in relatively showy, drooping clusters that are 4–5 inches long; May and later. **Fruit:** Keylike, its wings spreading at an acute or right angle. **Hardy:** From Zone 2 southward. **Varieties:** Several, but ill-defined, comprising forms with purplish foliage and one (variety *erectum*) with narrower, more erect habit. **Culture:** See above. It is one of the best of the maples for seaside gardens.

Left: Sugar Maple (*Acer saccharum*), below.
Right: Red Maple (*Acer rubrum*), p. 173.

SUGAR MAPLE (*Acer saccharum*) above

A native tree with the finest autumnal color of all the maples and perhaps of any cultivated tree, extensively grown also for its yield of maple syrup. **Size:** 75–120 feet high, nearly three fourths as wide in maturity, its bark furrowed and shaggy in age. **Leaves:** 3–5-lobed, resembling the Norway Maple, but without the milky juice in the leafstalks, 4–5½ inches wide, the lobes scarcely or not at all toothed, turning first yellow then through orange to scarlet in the fall. **Flowers:** Greenish yellow, appearing just before or with the unfolding of the leaves; Apr. **Fruit:** Keylike, its wings widely divergent. **Hardy:** From Zone 2 southward. **Varieties:**

monumentale is a more upright, columnar form. **Culture:** See above. It is deservedly one of the finest street trees, especially in the north.

RED MAPLE (*Acer rubrum*) p. 172

In late Feb., early Mar., and on into Apr., the bare twigs of this native tree are so covered with myriads of tiny red flowers that it is well named Red Maple. **Size:** 75–120 feet high, about three fourths as wide. **Leaves:** Simple, opposite, 3–5-lobed, 2–4 inches long, shiny above but paler and with a bloom beneath. **Flowers:** Small, red, very profuse, blooming weeks before the leaves unfold; Feb.–Apr. **Fruit:** Keylike, red, often ripe and falling off before the leaves expand. **Hardy:** From Zone 2 southward. **Varieties:** *columnare* has a more upright habit. **Culture:** See above; it grows naturally in moist or wet places and will do best in similar sites although it thrives in most ordinary garden soils.

Related plant (not illustrated):

Silver Maple (*Acer saccharinum*). A native maple with a deservedly bad reputation for shedding twigs and branches in only half a gale. It is related to the Red Maple but has sharper lobes to the leaf and deeper indentations between the lobes. Hardy from Zone 2 southward.

KATSURA TREE (*Cercidiphyllum japonicum*)

Family Trochodendraceae p. 182

An Asiatic foliage tree, its flowers negligible, but useful where a wide-spreading low tree is needed. **Size:** 60–100 feet high (in the wild) never attaining this stature as cultivated; usually with several trunks and wide-spreading. **Leaves:** Opposite or nearly so, mostly borne on short spurs, nearly round, heart-shaped at the base, 2–4½ inches long, shallowly and bluntly toothed, purplish when unfolding, yellow or scarlet in the fall. **Flowers:** Negligible. **Fruit:** A splitting pod with many seeds, persistent through most of the winter. **Hardy:** From Zone 4 southward. **Varieties:** *sinense* is inclined to have only one trunk. **Culture:** It will thrive in any ordinary garden soil, and hence is easy to grow.

JAPANESE LAUREL (*Aucuba japonica*)

Family Cornaceae p. 182

An Asiatic evergreen shrub with striking foliage and hardy only in the southern part of our range. **Size:** 4–15 feet high, about a third as wide. **Leaves:** Opposite, simple, broadly lance-shaped or ovalish, evergreen, 4–8 inches long, rather distantly toothed. **Flowers:** Male and female on separate plants, small, green, red, or purplish. **Fruit:** A usually scarlet berry (rarely produced in cultivated plants and never without both sexes in proximity). **Hardy:** From Zone 5 southward. **Varieties:** Gold-dust tree (variety *variegata*), the most usual form in gardens, has the leaves plentifully spotted with golden dots and is a handsome foliage plant. **Culture:** Needs a rich soil, freedom from wind, plenty of moisture, and a half-shady place. Frequently grown northward in the cool greenhouse.

SPINDLE-TREE
(*Euonymus*)
Family Celastraceae

Valuable shrubs or small trees, some of them evergreen or nearly so, with usually 4-angled twigs; grown mostly for their fine foliage and showy fruits, as the flowers are mostly small and inconspicuous. Leaves opposite in those below, persistent and nearly evergreen in some but losing the leaves in winter in others, the margins with sharp or rounded teeth. Flowers small, usually greenish or yellowish green, followed by a generally lobed pod, which as it splits discloses 1–2 seeds covered with a red or orange fleshy coat, hence very showy. Of over 100 species about 14 are in reasonably common cultivation, and of these the following 5 are most likely to interest the amateur. One of the species is a handsome, lustrous, evergreen vine and for this see p. 60 at VINES.

The cultivation of *Euonymus* is not at all difficult, but some of them are not hardy everywhere, as noted at the species. The evergreen or half-evergreen sorts should be moved only with a ball of earth (or in pots), as they do not thrive if the roots are exposed to the sun and wind. The 5 may be distinguished thus:

Leaves evergreen or nearly so.
Leaves thin, nearly evergreen; flowers in August. *Euonymus kiautschovicus*
Leaves evergreen and thick; flowers in June. *Euonymus japonicus*
Leaves not evergreen or hardly so.
Twigs usually corky-winged. Winged Spindle-Tree
Twigs not corky-winged.
Leaves 3–5½ inches long. *Euonymus yedoensis*
Leaves 1½–3 inches long. *Euonymus europaeus*

EUONYMUS KIAUTSCHOVICUS p. 176

A handsome Chinese evergreen or half-evergreen shrub, grown more for its fine foliage than for anything else. **Size:** 4–8 feet high, the lower branches sometimes prostrate and rooting. **Leaves:** Opposite, simple, oblongish, 4–6 inches long, bluntly fine-toothed. **Flowers:** Negligible, but as they are plentiful they give "the bush a filmy appearance that is very pleasing"; Aug. **Fruit:** Nearly round, pinkish, the interior orange red, about ⅝ inch wide, often failing to develop, especially in the northern part of our range. **Hardy:** From Zone 5 southward, and in Zone 4 with protection. **Varieties:** None. **Culture:** See above.

EUONYMUS JAPONICUS p. 176

A very popular Japanese evergreen shrub grown mostly for its fine, dense foliage. **Size:** 8–15 feet high, about a third as wide, as its slightly 4-angled branches are stiffly erect. **Leaves:** Simple, opposite, evergreen, narrowly elliptic or broadest toward the tip, 1–3 inches long and bluntly toothed. **Flowers:** Greenish orange, inconspicuous; June. **Fruit:** Nearly round, pinkish, the interior orange, the fruit often failing to develop. **Hardy:** From Zone 5 southward, and with protection, in the southern part of Zone 4. **Varieties:** Over 20, mostly variations in color of foliage, such as yellow-leaved, silvery-leaved, blotched, etc. One important variety

Left: *Euonymus japonicus,* below.
Right: *Euonymus kiautschovicus,* p. 176.

is *microphyllus,* which has smaller leaves and dwarf stature, making it an acceptable substitute for the edging box where that plant is not hardy. This small-leaved form of *Euonymus japonicus* will stand subzero temperatures. **Culture:** See above.

WINGED SPINDLE-TREE (*Euonymus alatus*) p. 182
An Asiatic shrub, its foliage not evergreen, its twigs prominently corky-winged. **Size:** 6–9 feet high, about half as wide. **Leaves:** Opposite, simple, elliptic to ovalish, 1½–3 inches long, turning scarlet in autumn. **Flowers:** Negligible. **Fruit:** A collection of usually 4 pods, for which the shrub is largely grown, as they are quite showy, disclosing the bright scarlet interior as they split; persistent into Nov. **Hardy:** From Zone 3 southward. **Varieties:** *compactus,* a valuable dwarf form, densely branched, easily clipped into a hedge and furnishing brilliant scarlet autumn color. It needs clipping only once a year, whenever convenient. **Culture:** See above.

Related plants (not illustrated):

Burning Bush (*Euonymus americanus*). A sparingly

branched native shrub, 5–8 feet high, with wavy-toothed, oblongish leaves 1½–4 inches long. Fruit pinkish and warty. Hardy from Zone 4 southward.

Euonymus nanus. A low Eurasian shrub, 2–3 feet high, its narrow leaves with slightly rolled margins. Fruit pinkish, 4-lobed. Hardy from Zone 2 southward.

EUONYMUS YEDOENSIS below

An Asiatic shrub with the largest leaves of any of the *Euonymus* here considered, not evergreen, but with a dense habit and fine foliage. **Size:** 4–10 feet high, about half as wide. **Leaves:** Opposite, simple, oblongish, 3–5½ inches long, bluntly fine-toothed and turning brilliant scarlet in the fall. **Flowers:** Yellowish and inconspicuous. **Fruit:** A 4-lobed pinkish pod, the interior (after splitting) bright orange. **Hardy:** From Zone 2 southward. **Varieties:** None. **Culture:** See above.

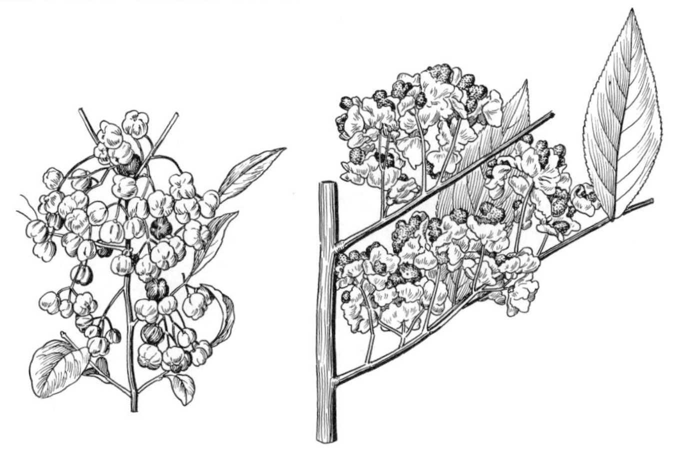

Left: *Euonymus europaeus,* below. Right: *Euonymus yedoensis,* above.

EUONYMUS EUROPAEUS above

A Eurasian shrub or small tree, often called Spindle-Tree or Prick-Timber, grown mostly for its colorful fall fruits. **Size:** 9–20 feet high, about half as wide. **Leaves:** Opposite, simple, oblong, 1½–3 inches long, wavy-toothed, wedge-shaped at the base, not evergreen, but reddish in the fall. **Flowers:** Negligible. **Fruit:** 4-lobed, about ¾ inch wide, red or pink, when splitting disclosing the bright orange interior; persistent into Nov. **Hardy:** From Zone 3 southward. **Varieties:** *aldenhamensis* has bright pink, long-stalked fruit. There are also several not readily available varieties. **Culture:** See above.

DEUTZIA
(*Deutzia*)
Family Saxifragaceae

These, among the most widely cultivated of all flowering shrubs, do not have a valid common name. All of the 50 species are shrubs, and of these nearly half are cultivated in the U.S., but only 4 can be included here. Bark usually shreddy, the twigs hollow. Leaves opposite, the margins always toothed. Flowers prevailingly white, but pinkish on the outside in some forms, the 5 separate petals often overlapping and the flower appearing (falsely) as though bell-shaped. Fruit a dry pod, with many small seeds. There are a great number of horticultural forms, some of them double-flowered.

The ease of cultivation of *Deutzia* is notorious, for they root readily and luxuriate in any ordinary garden soil, always in full sun. They are among the cheapest and best of all ornamental shrubs, although their fruits and fall color are of little value.

Low shrubs, 18–30 inches high.
- *Flowers white.* — *Deutzia gracilis*
- *Flowers pinkish.* — *Deutzia rosea*

Taller shrubs, 4–8 feet high.
- *Leaves green and smooth on upper side.* — *Deutzia lemoinei*
- *Leaves hairy on both sides.* — *Deutzia scabra*

DEUTZIA GRACILIS p. 182

A low, graceful, very old garden favorite, its profusion of flowers astonishing in such a small shrub; a native of Japan. **Size:** 18–30 inches, nearly as wide. **Leaves:** Simple, opposite, oblong or narrower, 1½–3 inches long. **Flowers:** White, very numerous, about ¾ inch wide, the clusters dense; May–June. **Fruit:** A small dry capsule. **Hardy:** From Zone 3 southward. **Varieties:** None, except a pink-flowered form properly assigned to the next species. **Culture:** See above; so easy that this and *Deutzia scabra* are among the most widely grown of all shrubs.

DEUTZIA ROSEA

A hybrid shrub, considered by many as scarcely distinct from *Deutzia gracilis,* of the same stature, hardiness, and popularity, its only significant difference being its pink flowers.

Deutzia lemoinei, below.

DEUTZIA LEMOINEI above

A hybrid shrub, originating in France and, with the next, a widely grown plant much taller than *Deutzia gracilis.* **Size:** 5–7 feet high, about as wide. **Leaves:** Opposite, simple, elliptic or narrower, 3–4 inches long, sharply toothed, green and smooth on the upper side. **Flowers:** White, about ¾ inch wide, very numerous in flattish or pyramidal clusters 2–4 inches wide; June. **Fruit:** A small dry capsule. **Hardy:** From Zone 3 southward. **Varieties:**

compacta, a low more dense form; Boule de Neige, with larger flowers in denser clusters. **Culture:** See above.

Related plant (not illustrated):

Deutzia kalmiaeflora. A hybrid shrub, 4–6 feet high, with white flowers, flushed carmine on the outside, in rather sparse clusters. Hardy from Zone 4 southward.

DEUTZIA SCABRA p. 182

A very widely cultivated tall Asiatic deutzia, especially in its variety Pride of Rochester (see below). **Size:** 5–8 feet high, nearly as wide, its shreddy bark reddish brown. **Leaves:** Opposite, simple, ovalish, 2–3 inches long, hairy both sides. **Flowers:** White or pinkish outside, nearly 1¼ inches wide, in showy, spirelike clusters 3–5 inches long; June–July. **Hardy:** From Zone 4 southward. **Varieties:** At least 15, of which the best is Pride of Rochester, a superb shrub with double, pinkish flowers; the rest are mostly variegated-leaved forms or dwarf plants. **Culture:** See above. It is the best deutzia for city conditions.

Related plant (not illustrated):

Deutzia magnifica. A hybrid shrub, not very different from *Deutzia scabra,* but its flower cluster not so spirelike, shorter and broader; June. Hardy from Zone 4 southward.

Hydrangea

(*Hydrangea*)

Family Saxifragaceae

Very popular medium-sized shrubs, the bark usually shreddy and the twigs not hollow but with a whitish pith. Leaves opposite, always toothed on the margin (lobed in one species) and with an obvious leafstalk. Flowers small, prevailingly white (blue or pink in *Hydrangea macrophylla*), in large, showy clusters, the marginal flowers in some clusters larger than the central ones and sterile. Fruit a dry pod with many small seeds. Over 30 species are known, but the 4 below are of chief garden interest. Another species, the climbing *Hydrangea petiolaris,* will be found at p. 61 at VINES.

The cultivated hydrangeas fall into two distinct categories, the obviously hardy shrubs such as *Hydrangea paniculata grandiflora,* and the more tender *Hydrangea macrophylla,* commonly called Hortensia, which needs special conditions noted at that species. The others are almost weedy in their ease of growth, in sun or shade and in any kind of soil.

Leaves lobed. *Hydrangea quercifolia*
Leaves not lobed, merely toothed.
Flowers usually blue or pink, rarely white. Hortensia
Flowers always white.
Flower cluster flat. Wild Hydrangea
Flower cluster not flat. Common Hydrangea

Left: *Hydrangea quercifolia,* p. 184.
Right: Wild Hydrangea (*Hydrangea arborescens*), p. 184.

SAXIFRAGE FAMILY (*Saxifragaceae*), 5, 7–9
1–4 and 6 in various families

1. **Winged Spindle-Tree** (*Euonymus alatus*) p. 176
   A corky-winged Asiatic shrub, its foliage scarlet in autumn. Flowers negligible. Fruit scarlet, persistent.

2. **Sweet Gum** (*Liquidambar Styraciflua*) p. 187
   A native tree with gorgeous autumnal color, negligible flowers, and a ball-like, rather prickly brownish fruit.

3. **Jetbead** (*Rhodotypos scandens*) p. 185
   An Asiatic shrub grown as much for its jet-black shiny fruit as for the white, waxy flower that blooms in May.

4. **Katsura Tree** (*Cercidiphyllum japonicum*) p. 173
   A little-known, wide-spreading Asiatic tree with negligible flowers but yellow or scarlet foliage in the fall.

5. **Hortensia** (*Hydrangea macrophylla*) p. 184
   The finest of the garden hydrangeas with globelike flower clusters that may be white, blue, or pink.

6. **Gold-Dust Tree** (*Aucuba japonica variegata*) p. 174
   A yellow-speckled variety of the green Japanese laurel, grown as a fine foliage plant. Not hardy northward.

7. ***Deutzia gracilis*** p. 178
   The finest of the low shrubby deutzias, with profuse bloom of white flowers in May or early June.

8. ***Deutzia scabra*** p. 180
   An Asiatic deutzia, 5–8 feet high and one of the finest of the taller deutzias. Flowers white, in spirelike clusters.

9. ***Hydrangea paniculata grandiflora*** p. 185
   The hydrangea of the uninitiated and anathema to some, but its profuse bloom and ease of culture make it ever popular.

1
2
3
4
5
6
7
8
9

1
2
3
4
5
6
7
8
9

ROSE FAMILY (*Rosaceae*), 3, 6, 8
LINDEN FAMILY (*Tiliaceae*), 7, 9
1, 2, 4, 5 in various families

1. **Tulip-Tree** (*Liriodendron tulipifera*) p. 189
   Not for small gardens, this magnificent native tree may be 150 feet high and nearly as wide.

2. **Sassafras** (*Sassafras albidum*) p. 189
   A medium-sized tree with very variable foliage, aromatic bark, and yellow flowers followed by black fruit.

3. **Double-Flowered Peach** (*Amygdalus persica duplex*) p. 204
   Stunning, waxy, pink or red double flowers cover the twigs of this relative of the edible peach in early spring.

4. **Stewartia** (*Stewartia Pseudo-Camellia*) p. 193
   A handsome shrubby Japanese tree, with red bark and white, waxy flowers in summer. Not hardy northward.

5. **London Plane** (*Platanus acerifolia*) p. 188
   This hybrid tree is perhaps the most widely planted — and certainly the best — city street tree in the world.

6. ***Malus sargenti*** p. 196
   A low, moundlike Japanese shrub, covered in May with a profusion of small white flowers.. Fruit red.

7. **Small-Leaved European Linden** (*Tilia cordata*) p. 192
   Perhaps the most satisfactory linden for fairly large gardens. In June its fragrant flowers are much loved by bees.

8. **Showy Crabapple** (*Malus floribunda*) p. 200
   A smallish, round-headed tree, easily grown anywhere, and covered in early May with a profusion of white flowers that become pinkish.

9. **Silver Linden** (*Tilia tomentosa*) p. 191
   Smaller than No. 7, and suited to more restricted gardens. Leaves silver beneath and handsome.

HYDRANGEA QUERCIFOLIA p. 181
A distinctive native shrub in the southeastern states, its leaves lobed like those of a white oak, and grown for its showy flower clusters. **Size:** 4–6 feet high, about as wide, the twigs reddish and hairy. **Leaves:** Opposite, simple, 3–7-lobed, the lobes toothed, white-hairy beneath, 4–7 inches long. **Flowers:** A pyramidal cluster 4–10 inches high, composed of many sterile white flowers that are 2 inches wide, all pure white at first, fading to pinkish. **Fruit:** A small dry capsule. **Hardy:** From Zone 5 southward and (less surely) in protected sites in Zone 4. **Varieties:** None. **Culture:** See above.

HORTENSIA (*Hydrangea macrophylla*) p. 182
A Japanese shrub, widely forced by florists for winter bloom, but perfectly hardy outdoors as outlined below. **Size:** 6–12 feet in the wild, but as usually grown outdoors (and often winter-killed) not over 3–4 feet, which is the usual length of the growth of a single season, after winter-killing to the ground. **Leaves:** Opposite, simple, broadly oval or nearly round, 3–6 inches long, coarsely toothed. **Flowers:** Normally pink or blue (rarely white), all sterile, in the usual globe-shaped clusters that may be 4–6 inches in diameter; only the marginal flowers sterile in the much more rare form with a flat-topped cluster. **Fruit:** Usually lacking. **Hardy:** From Zone 4 southward, and regularly killed to the ground in winter in Zone 4, often so in Zone 5, but south of this the stems persisting over the winter, and sometimes northward in sheltered sites. **Varieties:** Several, mostly color variations. Pink-flowered forms may sometimes be changed to blue by putting bits of iron or alum in the soil. **Culture:** See above.

WILD HYDRANGEA (*Hydrangea arborescens*) p. 181
A poor relation of the next species and a native shrub in the woods from N.Y. to Iowa and southward. **Size:** 3–5 feet, and nearly as wide, as the bush is open and more or less straggling. **Leaves:** Simple, opposite, toothed, ovalish, more or less rounded or heart-shaped at the base, 3–6 inches long. **Flowers:** White, small, in a flat-topped or slightly roundish cluster, 2½–5 inches

wide, only some of the marginal ones sterile. **Fruit:** Negligible. **Hardy:** From Zone 3 southward. **Varieties:** Hills-of-Snow (the variety *grandiflora*) is the only form worth growing and is deservedly popular as it blooms in June–July, and all the flowers in the ball-like cluster are sterile. **Culture:** See above.

COMMON HYDRANGEA (*Hydrangea paniculata grandiflora*) p. 182

So common is this ubiquitous shrub that it is anathema to many and the trade terms for it are simply P.G. or peegee (both attempts to shorten *paniculata grandiflora*). This Asiatic shrub, in spite of its perhaps too profuse use, is the best of all the hydrangeas for the amateur, for it is practically foolproof, will grow in any kind of soil and never fails to produce its large trusses of flowers. **Size:** 8–25 feet high, usually shrubby but sometimes tree-like. **Leaves:** Elliptic or ovalish, rounded or wedge-shaped at the base, 3–5 inches long and toothed on the margin. **Flowers:** White, fading to pink or even brownish, in a very large, close, pyramidal or oblongish cluster 8–12 inches long, usually long-persistent, all or nearly all the flowers sterile; Aug.–Sept. **Fruit:** Usually lacking. **Hardy:** From Zone 2 southward. **Varieties:** The Common Hydrangea is itself a variety, as the parent *Hydrangea paniculata* is seldom grown. For those who want it there is a tree form. **Culture:** See above. It will stand city conditions and grow well in exposed, wind-swept seaside gardens.

JETBEAD (*Rhodotypos scandens*). Family Rosaceae p. 182

An attractive shrub from eastern Asia, grown for its handsome flowers and shining coal-black fruit. **Size:** 4–6 feet high, nearly as wide. **Leaves:** Opposite, simple, doubly toothed, oblongish, 3–4 inches long. **Flowers:** Solitary, about 2 inches wide, white, the 4 rounded petals suggesting a single rose; May. **Fruit:** Berrylike, comprising a collection of usually 4, jet black and shining drupes, each about ⅜ inch thick. **Hardy:** From Zone 3 southward. **Varieties:** None. **Culture:** Easy in any ordinary garden soil, preferably in full sun, although it will stand some shade.

Leaves Alternate, Flowers with Separate Petals, Not Pea-like

From pp. 143 to 185 all the shrubs and trees have their leaves oppositely arranged (except the Blue Dogwood, which is with the other species of *Cornus* at p. 161).

The diversified and quite unrelated groups below have only this one distinguishing feature — all of them have alternate leaves, whether simple or compound, and some have marginal teeth or lobes and others have none.

Hence, to separate such a large group it is necessary to look for the following features:

Leaves compound. p. 290

Leaves simple.

Leaves without marginal teeth or lobes. See p. 257

Leaves with marginal teeth or lobes or both. The groups immediately below

Shrubs and Trees with Alternate Simple Leaves with Marginal Teeth or Lobes or Both

Medium-sized shrubs (rarely tall and treelike in some hollies). See p. 215

Trees (usually tall) or treelike shrubs. (The differences between a small tree and a large shrub are sometimes elusive, and in cases of doubt it is well to look at the key below and at the illustrations.)

Plants unarmed or with a few weak spines or prickles.

Leaves conspicuously lobed (except in some unlobed leaves in the sassafras).

Fruit a round, prickly ball. Sweet Gum
Fruit not round or, if round, not prickly.
Bark in large plates, usually peeling. Plane
Bark not peeling, close.
Tip of leaf not notched. Sassafras
Tip of leaf with a conspicuous notch.
Tulip-Tree
Leaves not conspicuously lobed, but with marginal, sometimes small or blunt teeth. See p. 190
Plants with often branched, formidable thorns.
Hawthorn; see p. 208

SWEET GUM (*Liquidambar Styraciflua*)
Family Hamamelidaceae p. 182
A tall native tree, perhaps the most gorgeously colored of any cultivated tree in the fall. **Size:** 70–120 feet high, at least half as wide or even more in vigorous specimens, its twigs and young branches corky-winged. **Leaves:** Alternate, simple, 3–7-lobed, the leaves starlike and somewhat resembling a maple, 5–7 inches long, the lobes finely toothed and pointed. The leaves turn a brilliant scarlet in autumn. **Flowers:** Negligible. **Fruit:** A globe-shaped collection of shining brown capsules, each tipped with a prickle, the whole ball hence prickly and about 1½ inches in diameter. **Hardy:** From Zone 3 southward. **Varieties:** None. **Culture:** An inhabitant of cool, moist woods, it prefers such sites, but will grow in ordinary garden soil in full sun if carefully transplanted with a ball of earth tightly wrapped in burlap. Not easy to get established, but growing satisfactorily when thoroughly recovered from the move (in about a year). Water it if drought intervenes.

Plane
(*Platanus*)
Family Platanaceae

The plane trees, variously called Sycamore, Buttonwood or Buttonball Tree, contain the much-planted London Plane, as well as our native Sycamore. In all species the bark peels off in large plates, disclosing the much lighter inner bark. Leaves alternate, long-stalked, deeply lobed, the lobes and veins arranged finger-fashion. Flowers small and inconspicuous, followed in the female flowers by a ball-like mass of small nutlets which are not spiny, the fruiting balls solitary, or in pairs or even more than two.

The London Plane is *the* street tree in the U.S. as well as in London, Paris, Vienna, and over most of the north temperate zone. It is so because it will stand more dust, wind, smoke, and hot pavements than any other tree. One of its parents is our native Sycamore, which is more suited to cool, moist sites in the country than to city or suburban conditions. It can become our largest deciduous tree. The two species should be planted only in the spring, and while much alike can be distinguished thus:

Fruiting balls solitary. Sycamore
Fruiting balls 2 or sometimes more. London Plane

SYCAMORE (*Platanus occidentalis*) below
The largest native tree that drops its leaves in the fall, cultivated specimens being known with trunk diameters of 8–10 feet and a wild one at Worthington, Ind., having a diameter of over 14 feet; often called Buttonwood. **Size:** 70–130 feet high, nearly or quite as wide at maturity if grown in the open. **Leaves:** Alternate, simple, usually 3-lobed (rarely 5-lobed), coarsely toothed, 5–9 inches wide, long-stalked, lobed and veined finger-fashion, and with no autumnal color. **Flowers:** Negligible. **Fruit:** A ball-like cluster, usually solitary (rarely 2), about 1½ inches in diameter,

Sycamore (*Platanus occidentalis*), above.

not prickly, its stalk about 2 inches long, persistent until early winter. **Hardy:** From Zone 3 southward. **Varieties:** None. **Culture:** See above. A fine tree for the country, but not suited to city streets, or any street if it is narrow.

LONDON PLANE (*Platanus acerifolia*) p. 183
A hybrid tree, and perhaps the best of all street trees, originating from a cross of our Sycamore and *Platanus orientalis* before 1800. Its chief difference from the Sycamore is that it is not so wide-spreading and its fruiting balls are in groups of 2–5 and persist a little longer in winter. Superficially the trees are much alike. Widely planted throughout the U.S. and Europe. **Hardy:** From Zone 3 southward. **Varieties:** Several, but scarcely available and having only minor differences from the typical London Plane. **Culture:** See above. The tree can be clipped to form a tree hedge, as was done in the pleached allée in the Schönbrunn Palace in Vienna, but it takes a long time and annual pruning of unwanted branches from the beginning.

SASSAFRASS (*Sassafras albidum*). Family Lauraceae p. 183
A pleasantly aromatic native tree (only if the foliage is crushed), of little garden value except for its brilliant scarlet autumnal foliage. **Size:** 20–60 feet high, rarely more in the wild, about half as wide. **Leaves:** Alternate, simple, 3–5 inches long and very variable. Some are unlobed, others lobed on one side, still others have irregular lobes on both sides of the blade, the tip rounded, the lobes without teeth. **Flowers:** Small, yellow, delightfully fragrant, blooming before the leaves unfold; Apr.–May. **Fruit:** Berrylike, bluish black, with a bloom, the stalks red. **Hardy:** From Zone 3 southward. **Varieties:** None. **Culture:** Not easy. While it grows in a variety of rocky or sandy soils it should be moved only with a ball of earth tightly wrapped in burlap. Once established it does well.

TULIP-TREE (*Liriodendron Tulipifera*)
Family Magnoliaceae p. 183
No small garden has room for this magnificent forest tree which is a native of eastern North America, and, in the forest, may have no branches nearer than 40–60 feet from the ground. In the open

it may branch somewhat lower. **Size:** 80–150 feet high, nearly as wide in maturity, its bark close, dark brown. **Leaves:** Alternate, simple, saddle-shaped, 3–5½ inches long, lobed, the broad tip deeply notched, turning bright yellow in the fall. At each leaf, as it opens, are 2 conspicuous stipules which are long-persistent. **Flowers:** Solitary, resembling a tulip, about 2½ inches wide, the 6 petals greenish, but with an orange band at the base; June. **Fruit:** A conelike mass of long-persistent fruits. **Hardy:** From Zone 3 southward. **Varieties:** *fastigiatum,* a form with upright branches forming a narrower head than the typical form. **Culture:** Not easy. Only spring planting is safe, preferably a young specimen with the roots in ball and burlap, as they will not stand exposure to sun and wind. Choose a deep rich, cool, moist site and water if there is a drought the first year. Once established they grow rather quickly.

Leaves Not Conspicuously Lobed, but with Marginal, Sometimes Blunt or Small Teeth

Flowers not borne on a winged bract. p. 193
Flowers borne on a winged bract. Linden

Linden
(*Tilia*)
Family Tiliaceae

Beautiful tall lawn trees, casting a dense shade, commonly called linden here, but in Europe called lime trees. The alternate leaves are more or less heart-shaped at the base, and one side of the blade is a little longer than the other; always toothed and with little or no fall color. Flowers extremely fragrant, beloved by bees, rather small, yellowish white, borne in small clusters the stalk of which arises from the midde of a long, leaflike bract that is not toothed. Fruit roundish, nutlike.

The lindens are among our most popular shade trees, both for lawn specimens and for broad streets (not in crowded cities). While relatively indifferent as to soil, they appear to thrive best in cool moist sites and, for such valuable trees are fairly quick-growing. Over 30 species are known, but of these the following 4 are most likely to be grown.

Leaves white-hairy beneath. Silver Linden
Leaves not white-hairy beneath.
  Leaves hairy beneath and sometimes above. Large-leaved Linden
  Leaves smooth beneath, except for tufts of hair at the vein joints.
    Leaves green both sides. European Linden
    Leaves pale beneath. Small-leaved European Linden

SILVER LINDEN (*Tilia tomentosa*) p. 183
A Eurasian tree, useful in the lawn, but it casts so dense a shade that most plants will not thrive under it. **Size:** 50–90 feet high, of pyramidal outline and hence three fourths as wide. **Leaves:** Alternate, simple, 2–5 inches long, roundish, heart-shaped at the base or not noticeably so, sharply toothed (very rarely lobed), the under surface silvery and downy. **Flowers:** Whitish, in clusters of 3–10; July. **Fruit:** Hard, pea-size. **Hardy:** From Zone 3 southward. **Varieties:** None. **Culture:** See above.

Related tree (not illustrated):

Tilia petiolaris. Closely related to the Silver Linden, but having drooping branches. Hardy from Zone 4 southward.

LARGE-LEAVED LINDEN (*Tilia platyphyllos*) p. 192
A European tree, much planted there and here for its handsome foliage which is dense enough to cast a heavy shade. **Size:** 70–120 feet high, its lower branches often sweeping the ground if the tree is in the open, hence needing much space. **Leaves:** Alternate, simple, roundish-oval, heart-shaped at the base, 2½–5 inches long, hairy on the veins and midrib beneath and sometimes above. **Flowers:** Whitish, small, in hanging 3-flowered

clusters; June. **Fruit:** Hard, pea-size. **Hardy:** From Zone 3 southward. **Varieties:** *rubra* has red twigs; *fastigiata* is a form with erect branches and hence a narrower canopy. **Culture:** See above. A favorite street tree.

Related tree (not illustrated):

American Linden (*Tilia americana*). Often called Basswood, this native tree is inferior for lawn or street planting to the large-leaved lindens, from which it differs in having the under side of the leaves mostly without hairs. It is similarly large-leaved.

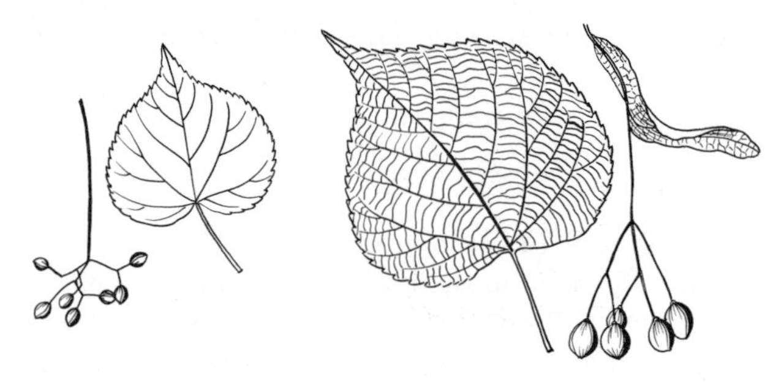

Left: European Linden (*Tilia europaea*), below.
Right: Large-leaved Linden (*Tilia platyphyllos*), p. 191.

EUROPEAN LINDEN (*Tilia europaea*) above

A hybrid tree resembling both the Large-leaved Linden and the Small-leaved European Linden, which were its parents, and differing from the one above in having smaller leaves, and from the Small-leaved European Linden in having the leaves green both sides. **Hardy:** From Zone 3 southward, but not commonly cultivated in this country. **Varieties:** None. **Culture:** See above.

SMALL-LEAVED EUROPEAN LINDEN
(*Tilia cordata*) p. 183

A splendid lawn tree, its lower branches sweeping the ground if grown in the open, and one of the finest of shade trees, its outline broadly oblong and casting a dense shade. **Size:** 90–100 feet high, at least three fourths as wide. **Leaves:** Alternate, simple, roundish, 1½–3 inches long, heart-shaped at the base, dark green above, paler or even whitish beneath, often with tufts of brownish hairs at the vein joints beneath. **Flowers:** Yellowish, 5–7 in each cluster, very fragrant; June. **Fruit:** Hard, pea-size. **Hardy:** From Zone 3, and possibly Zone 2, southward. **Varieties:** None. **Culture:** See above. This is one of the best of all lindens if there is space for it.

Related tree (not illustrated):

Tilia euchlora. A hybrid tree, not over 50 feet high and hence useful in smaller places. Flowers in June, fragrant. Hardy from Zone 3 southward.

Flowers Not Borne on a Winged Bract as They Are in the Lindens

Flowers relatively inconspicuous or insignificant. See p. 213
(NOTE: *The dove tree has showy bracts often mistaken for the flower.*)

Flowers conspicuous, usually very showy.

Flowers silky on the outside. Stewartia

Flowers not silky on the outside. Rose Family

STEWARTIA (*Stewartia Pseudo-camellia*).

Family Theaceae p. 183

A very beautiful Japanese tree or treelike shrub with waxy white summer-blooming flowers. **Size:** 10–25 feet high as usually cultivated, but twice this height in the wild, its handsome red bark peeling off in large flakes. **Leaves:** Alternate, simple, more or less elliptic, 2–4 inches long, pointed, rather thick, turning crimson in autumn, somewhat remotely blunt-toothed. **Flowers:** Solitary, cup-shaped, but of 5 petals, white, mostly at the leaf joints, 2–3 inches wide, silky on the outside; July–Aug. **Fruit:** A 5-angled woody capsule, about $\frac{5}{8}$ inch long. **Hardy:** From Zone 5 southward. **Varieties:** None. **Culture:** Plant in partial shade, out of the wind, in a deep, rich loam, preferably somewhat acid. Keep a mulch of peanut hulls or pine needles over the soil.

Related plants (not illustrated):

Mountain Camellia (*Stewartia ovata*). A native shrub, 10–15 feet high. Flowers nearly 3 inches wide, white, but the stamens orange. Its variety *grandiflora* has still larger flowers. Hardy from Zone 3 southward.

Stewartia koreana. A Korean tree 20–30 feet high, the white flowers nearly 4 inches wide. Hardy from Zone 5 southward, and perhaps not distinct from *Stewartia Pseudo-camellia.*

ROSE FAMILY

(*Rosaceae*)

Only three groups of the important trees of the Rose Family are considered here, the cherries and plums, the flowering crabs or crabapple, and the hawthorns, which bear formidable thorns. For other plants in this family and a general account of the Rose Family, which contains many fine garden shrubs and small trees see pp. 242 to 256.

The three groups below are all trees or shrubs, the latter often treelike in maturity. Among them are the incomparable Japanese flowering cherries and the equally fine flowering crabapples. If commercial fruit trees were in this book, here would come the peach, nectarine, plum, pear, apple, loquat, and quince, but only the 3 ornamentals are here included. (For culture of the fruit trees see *Fruit in the Garden,* by Norman Taylor. D. Van Nostrand Co., Princeton, N.J., 1954.)

The plants below all have alternate, simple leaves, always with marginal teeth. Flowers typically with 4–5 petals, but in some horticultural, double-flowered forms there may be many petals. Fruit inedible in most of those below, apple-like in the ornamental crabapples, like a miniature plum in some, and wanting in nearly all the double-flowered horticultural varieties. The 3 genera are:

Plants without formidable thorns.

Petals roundish; anthers often reddish; fruit like a miniature apple, several-seeded, or wanting in some horticultural forms. Crabapple

Petals narrower; anthers usually yellowish; fruit like

a small cherry or plum, 1-seeded or wanting in some horticultural hybrids. Plum and Cherry

Plants with formidable, often branched thorns. Hawthorn

CRABAPPLE
(*Malus*)

There are about 25 species of these fine flowering trees (rarely shrubs), but many more hybrids and horticultural forms. The difficulty of identification is thus considerable. While they all have a superficial resemblance to the common apple tree, the distinctions between the different sorts is rather elusive. All except one are medium-sized trees with alternate, simple, always toothed leaves. The crushed foliage is not unpleasantly scented as in so many of the cherries, due to the fact that *Malus* contains no prussic acid. Flowers always in small clusters, usually very numerous. Petals nearly round in outline. Stamens many, their anthers often reddish. Fruit (often withered or dried) like a miniature apple, several-seeded, and in some, still more apple-like, fleshy, edible, and very tart.

The ornamental crabapples are among our finest spring-blooming trees. So popular are they that there are many hybrids, prevailingly white, pink, or even reddish. While they are easily grown in a variety of soils, they thrive best in a soil that is slightly alkaline. Otherwise their culture is simple. Of the 25 species known to be in cultivation only the following can be admitted here and they are frankly only a meager representation of the genus. There are many more horticultural forms and for their identity the reader should consult *Ornamental Crabapples,* by A. den Boer, published by the American Association of Nurserymen, Washington, D.C. (1959). Still better, if one is to select plantings from trees in bloom (the only safe way, for there are over 130 named horticultural varieties) is to visit a well-labeled arboretum in the spring. Many trade names are apt to be unreliable.

Shrub, not over 6 feet high. *Malus sargenti*

Trees, in maturity 15–40 feet high.

Young twigs and young leaves purple. *Malus purpurea*
Young twigs and leaves not purple.
  Twigs always smooth. Siberian Crabapple
  Twigs hairy in youth or permanently so.
    Twigs more or less densely hairy, especially when young. American Crabapple
    Twigs somewhat hairy only when young.
      Flower buds pink or red, the flowers fading to lighter pink or white. Showy Crabapple
      Flowers always pink. *Malus theifera*

MALUS SARGENTI p. 183

A low, bushy, much-branched Japanese shrub, moundlike in outline and very floriferous. **Size:** 3–5 feet high, as wide or sometimes twice as wide. **Leaves:** Alternate, simple, ovalish, or elliptic, 2–3 inches long, short-pointed, turning orange and yellow in autumn. **Flowers:** Pure white, about 1 inch wide, in clusters of 5–6, the petals overlapping; May. **Fruit:** Nearly round, ½ inch or less in diameter, dark red and with a slight bloom. **Hardy:** From Zone 3 southward. **Varieties:** *rosea* has pink flowers. **Culture:** See above.

Related plant (not illustrated):

Malus toringoides. A Chinese shrub or small tree, 10–25 feet high, its yellow or reddish fruit nearly ¾ inch in diameter and persistent into Nov. Hardy from Zone 3 southward.

MALUS PURPUREA p. 197

A hybrid crabapple mostly grown in the variety *aldenhamensis,* and a medium-sized tree. **Size:** 15–25 feet, about three quarters as wide, its young twigs decidedly purple. **Leaves:** Alternate, simple, ovalish, 1½–3 inches long, the young growth purple, fading to green, without autumnal color. **Flowers:** Wine red, the petals oblong; May. **Fruit:** About 1 inch in diameter, purplish red, persistent throughout Oct. **Hardy:** From Zone 4 southward. **Varieties:** *aldenhamensis,* the most widely popular, is a small

Left: *Malus purpurea,* p. 196.
Right: American Crabapple (*Malus coronaria*), p. 200.

tree, the veins of the leaf purple, and with partially double wine-red flowers. **Culture:** See above.

SIBERIAN CRABAPPLE (*Malus baccata*) p. 201
A round-headed Asiatic tree, with many smooth, slender twigs, and extremely floriferous. **Size:** 20–40 feet high, nearly as wide. **Leaves:** Alternate, simple, thin, smooth, ovalish, 2½–4 inches long, pointed at the tip. **Flowers:** White, about 1½ inches wide, very profuse and fragrant; early May. **Fruit:** Yellow or red, about ¾ inch in diameter, persistent until late Oct. **Hardy:** From Zone 2 southward. **Varieties:** Several, two of the best being *gracilis,* with slightly larger flowers and the tips of the branches drooping; and *mandshurica,* which has still larger flowers that bloom in late April and is hence the first of the crabapples to flower. **Culture:** See above.

Related plant (not illustrated):

Toringo Crabapple (*Malus sieboldi arborescens*). A Japanese tree, 20–30 feet high, with coarsely toothed or lobed leaves, the flower buds pink, fading to white; late May. Fruit ½ inch in diameter, yellowish red. Hardy from Zone 3 southward.

ROSE FAMILY (*Rosaceae*), 1–5, and 7–9
DOGWOOD FAMILY (*Cornaceae*), 6

1. **Japanese Flowering Cherry** (*Prunus serrulata* or *P. yedoensis*) p. 207
   A confused group of hybrid trees with pink or white, often double flowers before or with the expansion of the leaves.

2. **Flowering Almond** (*Prunus triloba*) p. 205
   Sometimes mistaken for No. 9, but taller and its leaf teeth also coarsely toothed. Flowers pinkish.

3. **Rosebud Cherry** (*Prunus subhirtella*) p. 207
   A striking medium-sized Japanese tree, its profusion of light pink flowers blooming before the leaves unfold or just after.

4. **Cockspur Thorn** (*Crataegus crus-galli*) p. 210
   A shrub or small tree, very thorny, with white apple-like flowers and bright red winter-persisting fruit.

5. **Cherry Plum** (*Prunus cerasifera pissardi*) p. 204
   A form of the cherry plum with showy leaves that are purple at first, and white flowers. A medium-sized tree.

6. **Dove Tree** (*Davidia involucrata*) p. 214
   A Chinese relative of the dogwood, but with showy white bracts of unequal length. Flowers without petals.

7. **English Hawthorn** (*Crataegus Oxyacantha*) p. 212
   A medium-sized, very thorny tree, its profusion of flowers white, pink, or even red. Called "May" in England.

8. **Washington Thorn** (*Crataegus Phaenopyrum*) p. 212
   A round-headed native tree with slender thorns and orange or scarlet fall foliage. Fruit scarlet, persistent.

9. **Flowering Almond** (*Prunus glandulosa*) p. 206
   Related to No. 2, but a lower shrub and the marginal leaf teeth are not themselves toothed.

1
2
3
4
5
6
7
8
9

1
2
3
4
5
6
7
8
9

HOLLY FAMILY (*Aquifoliaceae*), 2, 3, 5, 9
WITCH-HAZEL FAMILY (*Hamamelidaceae*), 4, 6, 8
1, 7 in other families

1. ***Idesia polycarpa*** p. 214
   A little-known Asiatic tree, with small greenish-yellow flowers in a showy, hanging cluster.

2. **Japanese Holly** (*Ilex crenata*) p. 218
   A completely unarmed evergreen shrub, with box-like leaves. Useful in regions unfit for box.

3. ***Ilex cornuta*** p. 220
   The finest evergreen holly to come from China and faster-growing than the English Holly. A superb shrub.

4. ***Parrotia persica*** p. 214
   A small Persian tree grown mostly for its mottled, showy bark and scarlet, orange, and yellow fall foliage.

5. **American Holly** (*Ilex opaca*) p. 219
   Our native relative of the English Holly and not so fine, but much more hardy and easier to grow.

6. **Witch-Alder** (*Fothergilla monticolor*) p. 225
   A striking shrub, without petals, but its profusion of white stamens makes the flowers very showy in May.

7. **Buckthorn** (*Rhamnus cathartica*) p. 221
   Not of much garden interest since its flowers and fruit are inconspicuous, but it is found in many old gardens.

8. ***Hamamelis vernalis*** p. 223
   A winter-blooming native witch-hazel, its reddish or yellow flowers blooming from Jan. to Mar.

9. **English Holly** (*Ilex Aquifolium*) p. 219
   The holly of history, difficult to grow in the East and found in many horticultural forms.

AMERICAN CRABAPPLE (*Malus coronaria*) p. 197
A native tree found wild from Ontario to Ala. and west to Mo. **Size:** 20–30 feet high, its branches twiggy and stiffish. **Leaves:** Alternate, simple, ultimately smooth, thin, ovalish, 2–3 inches long, sharp-toothed and sometimes slightly lobed. **Flowers:** About 1 inch wide or a little less, pink at first, changing to white; late May. **Fruit:** About 1 inch in diameter, green and of little decorative value. **Hardy:** From Zone 3 southward. **Varieties:** *charlottae* has semidouble white or pinkish flowers. **Culture:** See above.

Related plant (not illustrated):

Prairie Crabapple (*Malus ioensis*). Chiefly noted for its variety known as Bechtel's Crab, which is a tree 20–30 feet high with a mass of rose-tinted or white, always double flowers. Hardy from Zone 3 southward.

SHOWY CRABAPPLE (*Malus floribunda*) p. 183
The most widely planted of all crabapples, thought to be of hybrid origin, but possibly Chinese. It is sometimes bushy but most often a tree. **Size:** 15–25 feet high, forming a round-headed, wide-spreading tree needing considerable space. **Leaves:** Alternate, simple, thin, oval, 2–3 inches long, toothed, very rarely lobed. **Flowers:** Buds pink or rose red, fading to lighter pink or ultimately white, very profuse, and single, about 1½ inches wide. **Fruit:** About ⅓ inch in diameter, red, persistent until late Oct. **Hardy:** From Zone 3 southward. **Varieties:** Innumerable and much confused; see above. **Culture:** See above.

Related plants (not illustrated):

Malus spectabilis. A Chinese tree, 15–25 feet high, its flowers deep rose red in bud, fading to blush, 1½–2 inches wide. Fruit nearly 1 inch in diameter yellowish and bitter. Hardy from Zone 3 southward.

Malus Zumi. Similar to the Toringo Crabapple (*Malus sieboldi arborescens*), but with red fruit. Hardy from Zone 3 southward.

Carmine Crabapple (*Malus atrosanguinea*). A hybrid

bushy shrub, resembling the Showy Crabapple, but its flowers not fading to white. Hardy from Zone 4 southward.

MALUS THEIFERA

below

An extremely handsome Asiatic tree, often offered as *Malus hupehensis,* widely popular for its profusion of white or pinkish, fragrant flowers. **Size:** 10–20 feet high, with stiff, spreading branches, hence nearly as wide. **Leaves:** Alternate, simple, 2–4 inches long, with sharp marginal teeth, pointed at the tip. **Flowers:** 3–7 in each cluster, fragrant, pink or pinkish; early May. **Fruit:** About ⅓ inches in diameter, greenish yellow, with a red cheek. **Hardy:** From Zone 3 southward. **Varieties:** None. **Culture:** See above.

Related plant (not illustrated):

Kaido Crabapple (*Malus micromalus*). An upright tree, 15–25 feet high, with nearly oblong leaves 2–4 inches

Left: Siberian Crabapple (*Malus baccata*), p. 197.
Right: *Malus theifera,* above.

long, deep pink flowers about 1½ inches wide and nearly red fruit that is hollowed at the base. Hardy from Zone 4 southward.

PLUM AND CHERRY
(*Prunus*)

The genus *Prunus,* comprising about 200 species, is of the utmost importance to all gardeners, for besides the edible cherries and plums, which cannot be treated here, it contains the Japanese flowering cherries and many more superb flowering shrubs and trees. For the edible cherries and plums see *Fruits in the Garden,* by Norman Taylor (D. van Nostrand Co., Princeton, N.J. 1954). Nearly all those below have inedible fruit or none at all in the double-flowered forms.

All have alternate, simple leaves, the margins toothed. In several species the leafstalk has one or two obvious glands on the stalk, and similar glands are found on the leaf teeth of some species. The crushed foliage of many species has an unpleasant odor, due to the prussic acid in the sap. Flowers with usually 5 more or less narrow petals, but in the double-flowered horticultural varieties the number of petals is much more than 5. Fruit (if present) fleshy, with a single stone, which is nearly round in the cherries, but flattened in the plums. In many of the most ornamental species there is no fruit or it is an apparently withered replica of a cherry or plum, and inedible.

The trees and shrubs of this group are not very long-lived, and 20–25 years is about the limit of their growth. Some of the Asiatic species like the Rosebud Cherry bloom just before the leaves expand, while others are apt to bloom just with or slightly after the unfolding of the leaves.

The number of hybrid horticultural forms is so great that specific identity of each is quite complicated. In the key below the characters are based on those found in the *species* of *Prunus,* leaving to the individual species the notes on the horticultural varieties derived from them. All of them thrive in any ordinary garden soil, preferably in a reasonably wind-free site.

There is a single very handsome small tree, the Double-flowered Peach (included in *Prunus* by some), with superb waxy red, white, or pink flowers. It is the first species in the list, but is excluded from the key as it belongs to the genus *Amygdalus* (the peach and almond). The 8 species in the key comprise less than a quarter of those in reasonably common cultivation. They were selected as typical of those most likely to interest the amateur, and for the 28 other species the seeker should consult more technical works such as Alfred Rehder's *Manual of Cultivated Trees and Shrubs* (Macmillan, New York, 1940).

Flowers in long, drooping, fingerlike clusters. — *Bird-Cherry*
Flowers not so arranged.
  Fruit fleshy, plum-like. — *Cherry Plum*
  Fruit often wanting, if present cherry-like.
    Shrubs, rarely over 6–10 feet high.
      Leaves doubly toothed, i.e., *the teeth themselves toothed.* — Flowering Almond
      Leaves singly toothed.
        Leaf teeth sharp; twigs densely hairy. — Nanking Cherry
        Leaf teeth incurved, blunt; twigs smooth or slightly hairy. — Flowering Almond
    Trees 20–40 feet high at maturity.
      Flowers blooming before the leaves unfold. — Rosebud Cherry
      Flowers blooming with the unfolding of the leaves.
        Leaves hairy on the veins beneath. — Flowering Cherry
        Leaves not hairy on the veins beneath, or only slightly so. — Japanese Flowering Cherry

NOTE: *It is obvious that the identification of the 8 species above cannot be even reasonably certain, without leaves, flowers, and fruit, which of course do not come at the same time. For complete identification it is necessary to make notes on the flowers and wait for the fruit, if any, to mature. Also, to still further complicate the problem, there are several wild cherries in our range, and many cultivated fruit cherries have been spread by birds.*

DOUBLE-FLOWERED PEACH
(*Amygdalus Persica duplex*) p. 183
A very beautiful spring-flowering tree, resembling the peach but with worthless fruit and a profusion of double showy flowers. It was derived from the peach, which is a native of China. **Size:** 10–20 feet high, about as wide. **Leaves:** Alternate, simple, peach-like, narrowly oblongish or narrower, 4–7 inches long, finely toothed, the leafstalks with obvious glands. **Flowers:** Solitary, double, 1½–2½ inches wide, pink, white, or maroon (according to variety), very showy, and in such profusion as to nearly cover the bare twigs, just before or with the unfolding of the leaves; Apr. **Fruit:** Peach-like, but worthless and often not produced. **Hardy:** From Zone 4 southward. **Varieties:** The Double-flowered Peach is itself a variety of the edible peach, and comes in mostly unnamed color forms, notably pink, white, and an especially deep maroon form that is extremely handsome. **Culture:** See above.

BIRD CHERRY (*Prunus Padus*) p. 206
A Eurasian tree of no particular decorative value, but the only one of the cherries here considered that has a fingerlike, hanging cluster of flowers. **Size:** 20–40 feet high, about half as wide. **Leaves:** Alternate, simple, elliptic or oblong, 3–5½ inches long, sharply toothed, grayish beneath. **Flowers:** Fragrant, small, white, in finger-shaped, hanging clusters that are 5–7 inches long. **Fruit:** Cherry-like, black, about ½ inch in diameter, relatively worthless. **Varieties:** Several, but none of the easily available ones any better than the typical form. **Culture:** See above.

CHERRY PLUM (*Prunus cerasifera*) p. 198
An Asiatic tree, closely related to the edible plum, often a little spiny, and frequently called the Myrobalan Plum. **Size:** 15–25 feet high, rather slender and hence about a third as wide. **Leaves:**

Alternate, simple, thin, bluntly oval, finely toothed, 1½–2 inches long. **Flowers:** White, solitary or in small clusters, about ¾ inch wide, very profuse, blooming as the leaves unfold; Apr.–May. **Fruit:** A small plum, red or yellow, juicy, sweet, about 1 inch in diameter. **Hardy:** From Zone 3 southward. **Varieties:** *divaricata* has smaller but still more profuse flowers; *pissardi* has purple leaves (in youth) and larger, pink flowers. It is one of the best of the early-flowering ornamental plums. There is also a form with drooping branches. **Culture:** See above.

Related plants (not illustrated):

Beach Plum (*Prunus maritima*). A native maritime shrub, 4–6 feet high, with white flowers and delicious black or purple bloomy fruit. Suited only to dune conditions or sandy soil in Zones 3 and 4.

Sloe (*Prunus spinosa*). A somewhat spiny Eurasian shrub or small tree, 8–12 feet high, its wood the blackthorn of the Irish, with white flowers and red tart fruit, used to flavor sloe gin. Hardy from Zone 3 southward.

FLOWERING ALMOND (*Prunus triloba*) p. 198

A shrub or small tree from China and one of the finest of the genus *Prunus* where small specimens are needed. **Size:** 6–10 feet high, about three fourths as wide. **Leaves:** Alternate, simple, broadly oval, sometimes 3-lobed but usually coarsely double-toothed, 1½–2½ inches long, sometimes a little hairy beneath. **Flowers:** Extremely profuse, pinkish, about 1½ inches wide, solitary or in clusters of 2, essentially stalkless; Apr. **Fruit:** Hairy, red, about ½ inch thick, often wanting. **Hardy:** From Zone 3 southward. **Varieties:** The most commonly cultivated is the variety *flore-pleno,* a double-flowered pink form of great beauty. **Culture:** See above.

NANKING CHERRY (*Prunus tomentosa*) p. 206

An Asiatic shrub, cultivated chiefly for ornament although it bears edible fruit. **Size:** 5–10 feet high, nearly as wide, the twigs densely hairy. **Leaves:** Alternate, simple, numerous, rather crowded, more or less elliptic, 2½–3½ inches long, unequally but

Left: Nanking Cherry (*Prunus tomentosa*), p. 205.
Right: Bird Cherry (*Prunus Padus*), p. 204.

singly toothed. **Flowers:** 1 or 2 in a cluster, about 1 inch wide, white or pinkish, blooming before the leaves unfold; Apr. **Fruit:** Nearly round, red, about ¾ inch in diameter and edible. **Hardy:** From Zone 1 southward. **Varieties:** None. **Culture:** See above.

FLOWERING ALMOND (*Prunus glandulosa*) p. 198
A low Asiatic shrub, one of the most widely cultivated of any species of *Prunus* because of the profusion of its early bloom. **Size:** 3–5 feet high, nearly as wide. **Leaves:** Alternate, simple, ovalish oblong, 1½–4 inches long, its blunt marginal teeth incurved. **Flowers:** Extremely profuse, borne singly or in twos, completely covering the bare twigs, about ¾ inch wide, white or pinkish; Apr. **Fruit:** Red, about ½ inch thick. **Hardy:** From Zone 3 southward. **Varieties:** *rosea* has pink flowers; *sinensis* has double pink flowers; *albo-plena* has double white flowers. **Culture:** See above.

Related plants (not illustrated):

Sand Cherry (*Prunus besseyi*). A low native shrub, 2–4 feet high, with white flowers and black fruit. It is extremely hardy and often planted in the prairie states.

Prunus cistena. A hybrid shrub with reddish leafstalks and leaves, single white flowers and blackish-purple fruit (a cherry). Hardy from Zone 1 southward.

Prunus japonica. Perhaps scarcely different from *Prunus glandulosa,* but it has more finely toothed leaves and even more profuse pink or white flowers. Hardy from Zone 3 southward.

ROSEBUD CHERRY (*Prunus subhirtella*) p. 198

One of the most showy of all Japanese trees, especially in a weeping variety with hanging branches. **Size:** 20–30 feet high, about half as wide, rarely shrubby. **Leaves:** Alternate, simple, ovalish, 1½–3 inches long, often doubly toothed, hairy on the veins beneath. **Flowers:** Almost incrediby numerous, light pink, in clusters of 2–5, about 1 inch wide, the petals notched. They bloom in mid-Apr., before the leaves unfold or just after. **Fruit:** About ⅓ inch in diameter, black. **Hardy:** From Zone 3 southward. **Varieties:** *pendula,* one of the finest of weeping trees, with long, hanging branches. It comes in both single or double-flowered forms; *autumnalis,* which produces its semidouble flowers both in the spring and fall. **Culture:** See above.

FLOWERING CHERRY (*Prunus yedoensis*) p. 198

This and the next species are both commonly called Japanese flowering cherries and differ from each other only in that this one has the leaves hairy on the veins beneath, while the next one has few or no hairs. It is of chief interest as the parent of many hybrids, which because they mostly belong to the next species are treated there. Hardy from Zone 3 southward.

JAPANESE FLOWERING CHERRY
(*Prunus serrulata*) p. 198

An enormously popular ornamental tree originally from Japan, China, and Korea, most widely planted in some of its innumerable horticultural forms, nearly all produced by the Japanese and making that country spectacularly beautiful in Apr. **Size:** 20–30 feet high, about half as wide, rarely smaller and shrubby. **Leaves:**

Alternate, simple, ovalish or narrower, long-pointed, 2½–5½ inches long, toothed or doubly toothed, the teeth short-bristly. **Flowers:** Typically white and single (but see below), opening just with or slightly after the unfolding of the leaves; Apr.–May. **Fruit:** Black (often wanting). **Hardy:** From Zone 3 southward. **Varieties:** Over a hundred, mostly produced in Japan, but some here and in Europe. The confusion in their names is boundless — some Japanese, British, Dutch, American, etc. The only safe way to select one of them is to visit a well-labeled arboretum or nursery in the spring and pick out the variety of your choice. The best collection of these is that at the Tidal Basin in Washington, D.C., presented by the City of Tokyo in 1912. It includes pink, white, and double-flowered forms and is usually at its best in mid-Apr. **Culture:** See above.

Related plant (not illustrated):

Prunus sargenti. A handsome Japanese tree, 50–60 feet high, differing from the Japanese Flowering Cherry in having the leaf teeth without bristles. Flowers rose-pink, about 1½ inches wide, very numerous. Fruit purplish black, about ½ inch in diameter. Hardy from Zone 4 southward.

Plants with Often Branched, Formidable Thorns

HAWTHORN
(*Crataegus*)

A huge group of thorny shrubs or trees, difficult and often unnecessary to identify, but containing handsome garden subjects suited to a variety of sites. Of 900 rather dubious species perhaps 7 are commonly cultivated and all of these, except the English Hawthorn and the hybrids, are natives of eastern North America.

The always toothed or lobed simple leaves are alternate, never evergreen, and often with gorgeous autumnal coloring. Flowers prevailingly white (pink or red in some horticultural varieties), usually in small but very profuse clusters. Fruit often brightly colored, resembling a miniature apple, but with 1–5 bony seeds. Of easy culture in any garden soil but growing better in limestone areas than elsewhere. Called, also, Thorn and Haw.

The identification of the few cultivated species here treated is not easy (there is no way of identifying immature specimens), but they may be divided thus:

Shrubs or small trees (not usually over 10 feet high). *Crataegus intricata*
Trees or tall shrubs 15–35 feet high.
  Veins of the leaf ending only at the tips of the marginal teeth or lobes of the leaf.
    Leaves wedge-shaped at the base, singly toothed. Cockspur Thorn
    Leaves broader at the base, doubly toothed or even lobed.
      Fruit dull red or orange; thorns often branched *Crataegus punctata*
      Fruit crimson. *Crataegus arnoldiana*
      Fruit ashy red. *Crataegus nitida*
  Veins of the leaf ending at the tips of the marginal teeth or lobes, and at the notches between them.
    Thorns stout, not over 1 inch long. English Hawthorn
    Thorns slender, 2–3 inches long. Washington Thorn

For accurate identification it is essential to have leaves, thorns, flowers and fruit.

CRATAEGUS INTRICATA p. 210

An irregularly branched shrub with slender curved thorns, native in eastern U.S. **Size:** 4–9 feet high and as wide. **Leaves:** Thin, wedge-shaped, 1–3½ inches long, rather sparse. **Flowers:** Profuse, but in clusters of only 3–7, white; May. **Fruit:** Dull reddish brown, football-shaped or rounder; Oct., not persistent. **Hardy:** From Zone 4 southward. **Varieties:** None. **Culture:** See above.

Related plant (not illustrated):

Crataegus prunifolia, 10–15 feet high, with rounded clusters of white flowers in May.

Left: *Crataegus intricata,* p. 209.
Right: *Crataegus punctata,* p. 210.

COCKSPUR THORN (*Crataegus crus-galli*) p. 198

A large shrub or small tree, with many short, slender thorns, usually not over 2 inches long; native in eastern North America and one of the most popular of American thorns because of its profuse flowers, fine autumnal color, and persistent fruit. **Size:** 15–30 feet high, about 8 feet wide. **Leaves:** Wedge-shaped, without marginal teeth at the base, but shallowly toothed toward the tip, 1–3½ inches long, bright green, but orange and scarlet in the fall. **Flowers:** White, in clusters, 2–3 inches wide, in May–June, profuse. **Fruit:** Nearly round, about ½ inch wide, bright red; Oct. and winter-persistent. **Hardy:** From Zone 3 southward. **Varieties:** None. **Culture:** See above.

Related plant (not illustrated):

Crataegus lavallei. A hybrid thorn resembling the Cockspur Thorn but with longer thorns, and orange-red, winter-persisting fruit.

CRATAEGUS PUNCTATA p. 210

A tree with spreading branches, rather open foliage and short, stout, sometimes branched thorns, but often lacking all thorns; native of eastern North America. **Size:** 20–30 feet high, half as wide. **Leaves:** Lobed or coarsely double-toothed, 2–3½ inches long, the veins impressed. **Flowers:** Showy, white, in profuse close clusters that are 2–3 inches wide; May. **Fruit:** Dull red or orange, dotted, nearly round, about ¾ inch wide; Oct., not persistent. **Hardy:** From Zone 3 southward. **Varieties:** Some with yellow, cherry-red, or yellow-streaked fruits are even finer than the typical form. **Culture:** See above.

Related plant (not illustrated):

Red Haw (*Crataegus mollis*). A tree with short, stout thorns, or sometimes thornless, the branches wide-spreading and forming a round-topped tree with profuse white flowers in May, and crimson fruit.

CRATAEGUS ARNOLDIANA below

A handsome lawn tree with zigzag branches and plentiful thorns 2–3 inches long; native in northeastern U.S. **Size:** 20–30

Left: *Crataegus arnoldiana,* above.
Right: *Crataegus nitida,* p. 212.

feet high, about 10–15 feet wide. **Leaves:** Broadly wedge-shaped, 2½–4 inches long, the stalks softly hairy, but not white-hairy. **Flowers:** Numerous, showy, in lax clusters, white; their stalks hairy; May. **Fruit:** Nearly round, about ½ inch thick, bright crimson; Aug.–Sept., not persistent. **Hardy:** From Zone 3 southward. **Varieties:** None. **Culture:** See above.

CRATAEGUS NITIDA p. 211
A round-headed tree with rather sparse foliage, often unarmed or with a few slender thorns 1–1½ inches long; native to the central U.S. **Size:** 20–35 feet high, about 10–15 feet wide. **Leaves:** Coarsely double-toothed and with a few shallow, sharp-pointed lobes, oblongish, 2–4 inches long. **Flowers:** Numerous, showy, in close clusters, white, in May. **Fruit:** Ashy red, nearly round, about ½ inch wide; Oct., and winter-persistent. **Hardy:** From Zone 3 southward. **Varieties:** None. **Culture:** See above. Its orange-scarlet autumnal foliage and winter-persisting fruit makes this a valuable garden tree.

ENGLISH HAWTHORN (*Crataegus Oxyacantha*) p. 198
Commonly called May in England and a showy, round-headed tree with many short, stiff thorns and with some twigs ending in a sharp spine; native of Europe and northern Africa. **Size:** 20–25 feet high, 10–15 feet wide. **Leaves:** Deeply 3–5-lobed, but sparingly double-toothed, generally triangular or ovalish, about 2 inches long. **Flowers:** Typically white, in May (but see below), in profuse clusters of 6–12, very showy. **Fruit:** Red, nearly round, about ½ inch thick; Sept.–Oct., not persisting. **Hardy:** From Zone 3 southward. **Varieties:** Many, some with pale pink, others with scarlet flowers, some double-flowered, others with yellow fruit, and one with weeping habit and rose-colored flowers. Another tree also known as English Hawthorn is *Crataegus monogyna,* which is less thorny, has smaller, less deeply lobed leaves, and is a smaller tree. It is little known in this country, while *Crataegus Oxyacantha* is everywhere.

WASHINGTON THORN (*Crataegus Phaenopyrum*) p. 198
A round-headed tree, perfectly smooth in all its parts, and with

many slender thorns 2-3 inches long; native to the eastern U.S. and a desirable lawn specimen. **Size:** 20–30 feet high, 12–15 feet wide. **Leaves:** More or less triangular in outline, shallowly but sharply lobed and double-toothed, 1½–3 inches long, turning orange and scarlet in autumn. **Flowers:** White, in May, in profuse but close clusters. **Fruit:** Scarlet, nearly round, about ½ inch thick, shining; Oct., the numerous clusters persistent for part of the winter. **Hardy:** From Zone 3 southward. **Varieties:** None. **Culture:** See above. A deservedly popular garden tree.

There are over 100 species of *Crataegus* in eastern North America, any of which may be dug from the wild. For their enumeration and description the seeker should see the treatment of this difficult genus by Ernest J. Palmer in the second volume of the new Britton and Brown *Illustrated Flora* by H. A. Gleason, published by the New York Botanical Garden in 1952.

Flowers Relatively Inconspicuous or Insignificant

(NOTE: The Dove Tree has showy bracts often mistaken for flowers.)

THE three groups below are apt to be confusing for, as with so many of nature's "rules," they are exceptions to the general rule of plants that have obvious separate petals, for none of them has any petals at all. The true flowers are hence inconspicuous or insignificant, but in the Dove Tree, just beneath the flower cluster are two unequal, showy white bracts. The distinctions are:

Flower with 2 unequal, showy white bracts beneath it. Dove Tree
Flowers without 2 showy bracts.
  Leaves 6–12 inches long; fruit fleshy, orange-red. *Idesia polycarpa*
  Leaves 2–4 inches long; fruit a dry pod. *Parrotia persica*

DOVE TREE (*Davidia involucrata*).
Family Cornaceae p. 198
A Chinese curiosity and a relative of the dogwood, but having no petals and yet a quite striking tree when in bloom. Called also Ghost Tree in England. **Size:** 30–50 feet high, about half as wide, branching like a linden. **Leaves:** Alternate, simple, broadly oval, 4–6 inches long, short-stalked, toothed, and silky-hairy beneath. **Flowers:** Small, inconspicuous, without petals, yellowish from the profusion of stamens that are gathered in a ball-like cluster about 1 inch in diameter. Just beneath are two creamy-white bracts, the lower one about 7 inches long and hanging, the other shorter and not hanging, making the tree very showy; May. **Fruit:** Fleshy, pear-shaped, about 1½ inches long, green and with a bloom. **Hardy:** From Zone 5 southward, and in protected places in Zone 4, but it may not flower there. **Varieties:** *vilmoriniana* is quite similar, but has the leaves smooth and blue green beneath. **Culture:** Put in a wind-free, sheltered place, in cool, moist soil, preferably in partial shade. If there is summer drought it may fail to bloom for years, but trees along the coast appear to bloom more or less regularly.

IDESIA POLYCARPA. Family Flacourtiaceae p. 199
Not a very well-known Asiatic tree, but its autumn fruit is so handsome that it ought to be more widely planted than it is. **Size:** 25–40 feet high, about half as wide, its bark close and grayish white. **Leaves:** Alternate, simple, long-stalked, ovalish, 5–9 inches long, the margins wavy-toothed. **Flowers:** Small, greenish-yellow, grouped in a hanging cluster 4–9 inches long; May–June. **Fruit:** A fleshy berry about ½ inch in diameter, orange red or brownish, grouped in a hanging, decidedly showy cluster, persisting into Nov. **Hardy:** From Zone 5 southward, probably from Zone 4 if in a protected site. **Varieties:** None. **Culture:** Plant in a sandy, well-drained loam, in full sun, otherwise it appears to have few cultural preferences.

PARROTIA PERSICA. Family Hamamelidaceae p. 199
A small Persian tree, often branching from near the base, grown not for its insignificant flowers but for its showy bark and

splendid fall foliage. **Size:** 8–20 feet high, nearly as wide from its low branching, its bark mottled gray and flaking off. **Leaves:** Alternate, simple, ovalish or oblong, 2–4 inches long, coarsely toothed toward the tip, hairy on both sides, turning scarlet, orange, or yellow in the fall and quite persistent. **Flowers:** Negligible, in Mar.–Apr., on bare twigs. **Fruit:** A dry 2-beaked capsule about ½ inch high. **Hardy:** From Zone 3 southward. **Varieties:** None. **Culture:** In any ordinary garden soil, in full sun. A showy autumnal feature because of its brilliant foliage.

Shrubs of Moderate Size, if Tall, Rarely Tree-like (Except in Some Hollies)

The distinction between tree-like shrubs and shrublike trees is never easy, especially in young or undeveloped trees. For that reason it may, in some cases, be necessary to consult not only the key below but also the one on p. 186 which deals with tall trees, some of which, in youth, might be mistaken for some of the plants below.

THE shrubs below are mostly of moderate size, but in a few cases they are above medium height, as in the buckthorns and witch-hazels, but not treelike, *i.e.*, with a single trunk, except in some hollies. Leaves alternate, simple, always toothed or lobed. (Exceptions are 2 buckthorns which have opposite leaves, and 2 hollies that have no marginal teeth or lobes.)

All those below are quite unrelated botanically, and are best separated thus:

Leaves very large and deeply lobed. See p. 256

Leaves of moderate size, merely toothed on the margin, or if lobed, shallowly so.

Flowers obviously showy. See p. 222

Flowers greenish or greenish yellow, small and inconspicuous.

Leaves prevailingly evergreen, often spiny-margined. Holly

Leaves neither evergreen nor spiny-margined. Buckthorn

HOLLY
(*Ilex*)
Family Aquifoliaceae

Prevailingly evergreen and very handsome shrubs (rarely tree-like in the English and American Hollies), with alternate, often spiny-margined leaves. Flowers small, greenish yellow, usually in small clusters or solitary. Fruit berrylike, often brightly colored. Male and female flowers on separate plants so that to ensure berries both male and female plants must be in close proximity. Some nurserymen have female plants upon which male shoots have been grafted. This is impossible to tell from foliage characters, so the buyer must depend upon the reliability of the nurseryman. A few hollies apparently bear male and female flowers on the same tree, but the sexes are always separate.

As most of the more desirable evergreen species will not luxuriate in places with extreme heat, drying winds, and deficient rainfall, hollies do not thrive in the prairie states, and some of them are impossible in the North (see hardiness notes below). All the evergreen sorts should be moved only with a tightly wrapped ball of earth, as the roots must not be exposed to the sun and wind. In planting it is better to knock off most of their leaves and prune severely. They grow well in a variety of soils, always slowly, and in the American Holly appear to do best in a somewhat acid soil.

Over 300 species are known, of which about 20 are in cultivavation in the U.S., but there are scores of horticultural varieties. Here only 6 species can be included, all evergreen except the Black Alder.

Leaves not evergreen, never spiny-margined. Black Alder
Leaves evergreen, often spiny-margined.
Leaves never spiny-margined.
Low shrub, not usually over 6 feet high. Inkberry
Taller shrubs, 8–20 feet high. Japanese Holly
Leaves always more or less spiny-margined.
Leaves distinctly paler beneath than above. American Holly
Leaves nearly uniformly green and shining both sides.
Leaves ovalish, not rectangular, with several spiny teeth. English Holly
Leaves more or less rectangular, with 3 strong spines toward the tip. *Ilex cornuta*

BLACK ALDER (*Ilex verticillata*) p. 218
This native, never evergreen shrub is useful in the shrubbery for its profusion of bright scarlet, winter-persistent fruit. **Size:** 6–8 feet high, nearly as wide. **Leaves:** Alternate, simple, not evergreen, ovalish or narrower, wedge-shaped at the base, 1½–2½ inches long, very finely toothed. **Flowers:** Insignificant. **Fruit:** Berrylike, pea-size, essentially stalkless, very profuse, scarlet, persistent for most of the winter. **Hardy:** From Zone 3 southward. **Varieties:** None. **Culture:** See above. In the wild the plant grows mostly in swamps or moist woods and in the garden should have a moist site.

INKBERRY (*Ilex glabra*) p. 218
A native bog shrub, evergreen in the South, but only half evergreen northward, and turning a rusty green in the fall. **Size:** Usually 3–4 feet high, but wild plants as much as 6 feet, the foliage sparse. **Leaves:** Alternate, simple, oblongish, but broadest toward the tip, wedge-shaped at the base, 1–2½ inches long.

Left: Black Alder (*Ilex verticillata*), p. 217.
Right: Inkberry (*Ilex glabra*), p. 217.

Flowers: Negligible but much liked by bees. **Fruit:** Stalked, pea-size, black. **Hardy:** From Zone 3 southward. **Varieties:** None. **Culture:** Although a bog shrub, it will grow well in a reasonably sandy loam, in full sun, preferably with a mulch of leaves.

JAPANESE HOLLY (*Ilex crenata*) p. 199

An immensely valuable Japanese evergreen shrub with box-like habit and foliage, and much faster-growing than the box. **Size:** From 2–3 up to 15–20 feet high, according to the variety, the foliage not quite so dense as that of the box, but an acceptable substitute for it in regions where the box will not thrive. **Leaves:** Alternate (opposite in the box), simple, oblongish, but broadest toward the tip, wedge-shaped at the base, 1–2 inches long, very finely toothed. **Flowers:** Insignificant. **Fruit:** Pea-sized, stalked, solitary, black. **Hardy:** From Zone 4 southward, and in protected places in the southern part of Zone 3. **Varieties:** Many, of which the most important are *convexa,* which has the leaves convex above and concave beneath; *microphylla,* which has smaller leaves than the typical form and is a little hardier; *latifolia,* a form with elliptic leaves and probably the most common variety in cultivation. Nurserymen offer at least 30 other "varieties" under many rather dubious names, but the three above are the most likely to interest the grower. **Culture:** See above. It is one of the best of all spineless hollies.

AMERICAN HOLLY (*Ilex opaca*) p. 199
A valuable shrub or even tree of the eastern U.S., most often shrubby, but on the author's farm near Princess Anne, Md., there is a tree 50 feet high, with a trunk diameter of 18 inches. **Size:** 8–30 feet, usually shrubby but often treelike and very slow growing. **Leaves:** Alternate, simple, evergreen, elliptic, 1¾–3 inches long, dull green above, distinctly paler beneath, the marginal teeth remote and spiny. **Flowers:** Negligible. **Fruit:** Pea-sized, usually solitary, red, persistent beyond New Year's. **Hardy:** From Zone 4 southward. **Varieties:** Over 70, mostly with slight differences in size of the leaves or fruit. There is no safe way to select them without visiting a well-labeled nursery or arboretum. The largest collection in the country is at the U.S. National Arboretum at Washington, D.C. **Culture:** Grows best in a sandy loam, somewhat acid, in sun or partial shade, and a mulch of pine needles or cottonseed meal is valuable.

ENGLISH HOLLY (*Ilex Aquifolium*) p. 199
How this evergreen acquired the name English Holly is something of a mystery as it is native in many parts of Eurasia and northern Africa. **Size:** 3–4 feet as a shrub (according to variety) but in maturity usually a tree 30–50 feet high. **Leaves:** Simple, alternate, short-stalked, ovalish or oblong, the margin wavy and with several large, triangular spiny teeth, shining green above and beneath, 1½–2½ inches long. **Flowers:** Negligible. **Fruit:** Pea-size, bright red, usually in clusters, persistent through Jan. or beyond. **Hardy:** Precariously hardy in Zone 4, and hardy from Zone 5 southward, but it does not like hot, dry summers. **Varieties:** Legion (at least 100 or more) and the only safe way to select them is to visit a well-labeled nursery or arboretum. The best

collection in the country is at the U.S. National Arboretum, Washington, D.C. **Culture:** See above.

Related plant (not illustrated):

Ilex altaclarensis. Resembling the English Holly, but the leaves are less wavy-margined. Hardy from Zone 4 southward.

ILEX CORNUTA p. 199

Many enthusiasts rate this the finest holly in cultivation, superior to the English Holly in foliage and in its larger, more showy fruit. It is a Chinese evergreen relative of the English Holly. **Size:** A shrub 7–10 feet high, as wide or wider, the foliage dense. **Leaves:** More or less rectangular, 2–3 inches long, dark green, with 1–2 large spines on each side and one at the tip, the lateral spines recurved. **Flowers:** Negligible. **Fruit:** Oblong, ⅓–½ inch long in stalked clusters of 5–8, persistent and very showy. **Hardy:** From Zone 4 southward. **Varieties:** *burfordi,* similar in size, but with smaller, less rectangular leaves that have only a single terminal spine (occasionally none). **Culture:** See above.

Related plant (not illustrated):

Ilex perneyi. Closely related to *Ilex cornuta,* but taller and with nearly stalkless paired fruit. Hardy from Zone 4 southward.

BUCKTHORN

(*Rhamnus*)

Family Rhamnaceae

Rather uninteresting shrubs (rarely small trees) of little decorative value, but found in many old gardens. Leaves simple, toothed, but never spiny-margined or evergreen, alternate in most species, but opposite in the Buckthorn. Flowers small, greenish, inconspicuous, followed by a small, fleshy fruit. They are of easy culture anywhere. Only two are likely to be grown, one of them having leaves without marginal teeth.

Leaves relatively long-stalked.	Buckthorn
Leaves almost stalkless.	Alder Buckthorn

BUCKTHORN (*Rhamnus cathartica*) p. 199
A sometimes thorny Eurasian shrub, of little decorative value but useful as a thick screen, and found in many old gardens. **Size:** 10–25 feet high, nearly as wide. **Leaves:** Simple, opposite, ovalish, 2–3 inches long, the margins with rounded teeth. **Flowers:** Negligible. **Fruit:** About ½ inch in diameter, black. **Hardy:** Everywhere. **Varieties:** None. **Culture:** See above. It has escaped from gardens in the eastern U.S. and is often found wild in thickets.

Related plant (not illustrated):

Rhamnus davurica. An Asiatic relative of the Buckthorn, with larger, narrower leaves. Hardy everywhere.

Alder Buckthorn (*Rhamnus Frangula*), below.

ALDER BUCKTHORN (*Rhamnus Frangula*) above
The only buckthorn with any autumn coloration, its foliage turning bright yellow in the fall. It is a Eurasian shrub or small tree of little garden interest otherwise. **Size:** 10–18 feet high, about a third as wide. **Leaves:** Simple, alternate, ovalish, or broadest toward the tip, 1½–2½ inches long, without marginal teeth. **Flowers:** Negligible. **Fruit:** Red at first, ultimately black-

ish. **Hardy:** From Zone 2 southward. **Varieties:** None. **Culture:** See above.

Related plant (not illustrated):

Rhamnus Alaternus. A European evergreen shrub 10–15 feet high, its leaves remotely or not at all toothed, with small black fruit. Hardy from Zone 6 southward.

Flowers Obviously Showy, Not Greenish and Inconspicuous

Petals more than 4. See p. 225
Petals 4 (or none in the Witch-Alder). Witch-Hazel Family

WITCH-HAZEL FAMILY
(*Hamamelidaceae*)

Shrubs (in ours), often blooming in fall, winter, or very early spring, usually while the plants are leafless or just before leaf-fall. Leaves alternate, simple, often hairy. Flowers with 4 showy petals in the witch-hazel, but without petals in the Witch-Alder, which is showy because of its protruding white stamens. Fruit a dry capsule, splitting explosively in the witch-hazel and shooting the two black seeds for a considerable distance. The 2 genera below are easily distinguished:

Flowers yellow, petals 4. Witch-Hazel
Flowers white, petals none. Witch-Alder

Witch-Hazel
(*Hamamelis*)

Most of the cultivated witch-hazels are tallish shrubs, very rarely small trees, of easy culture in any garden soil, but generally found wild in cool, damp woods. Leaves usually turning a soft yellow in the autumn. The four yellow, narrow petals are showy, the flowers blooming in late fall, in winter, or very early spring. The explosive discharge of the two black seeds from the capsule is unique among shrubs and trees. At least 4 species are in pretty common cultivation, and may be distinguished thus:

Flowering in the fall. — *Hamamelis virginiana*
Flowering in the winter or very early in the spring.
- *Leaves hairy both sides.* — *Hamamelis mollis*
- *Leaves mostly hairless, except for occasional hairs on the veins below.*
  - *Leaves with mostly 5 pairs of veins.* — *Hamamelis vernalis*
  - *Leaves with mostly 7 pairs of veins.* — *Hamamelis japonica*

HAMAMELIS VIRGINIANA p. 224

A rather coarse native shrub of no value until its fall flowers begin to open late in Oct. or Nov., about the time its leaves are falling. **Size:** 8–15 feet high, nearly as wide. **Leaves:** Alternate, simple, coarse in texture, elliptic or broadest toward the tip, 4–6 inches long, coarsely toothed. **Flowers:** Showy, bright yellow, about ¾ inch wide, the 4 petals strap-shaped, the flowers essentially stalkless on the bare twigs. **Fruit:** An explosively splitting capsule. **Hardy:** From Zone 2 southward. **Varieties:** None. **Culture:** See above.

HAMAMELIS MOLLIS p. 224

A Chinese relative of the next species, and the only witch-hazel here treated which has fragrant flowers. **Size:** 10–25 feet high, often treelike in the larger size, about half as wide. **Leaves:** Simple, alternate, roundish or broadest toward the tip, 3½–7 inches long, finely toothed and hairy both sides. **Flowers:** Golden yellow but reddish at the base, about ¾ inch long, essentially stalkless, delightfully and aromatically fragrant; Feb.–Mar. **Fruit:** An explosively splitting pod. **Hardy:** From Zone 3 southward. **Varieties:** None. **Culture:** See above.

HAMAMELIS VERNALIS p. 199

A many-stemmed shrub native in the central U.S., its winter-blooming flowers unique in our native flora. **Size:** 4–6 feet high, about half as wide. **Leaves:** Alternate, simple, oblongish or broadest toward the tip, 3–5 inches long, coarsely toothed above the middle and with mostly 5 pairs of veins. **Flowers:** About ½ inch long, the 4 strap-shaped petals dark yellow or reddish toward the base, the flower essentially stalkless; Jan.–Mar. **Fruit:** An explosively splitting capsule. **Hardy:** From Zone 4 southward. **Varieties:** None. **Culture:** See above.

HAMAMELIS JAPONICA below

A Japanese shrub or small tree valued for its midwinter bloom. **Size:** 10–25 feet high, with spreading branches and hence nearly as wide. **Leaves:** Alternate, simple, roundish or broadly oval, 3–4 inches long, with mostly 7 pairs of veins. **Flowers:** Essentially stalkless, bright yellow, about ¾ inch long; Jan.–Mar. **Fruit:** An explosively splitting capsule. **Varieties:** *arborea* is generally more treelike. **Culture:** See above.

Left: *Hamamelis japonica,* above. Upper Center: *Hamamelis virginiana,* p. 223. Right: *Hamamelis mollis,* p. 223.

WITCH-ALDER (*Fothergilla monticola*) p. 199
For a shrub with no petals the Witch-Alder is a conspicuous addition to the shrubbery border because of its masses of white stamens. **Size:** 4–6 feet high, nearly as wide. **Leaves:** Alternate, simple, broadly oval, 3–4 inches long, hairy on the veins beneath, turning yellow to scarlet in the autumn. **Flowers:** In terminal, spirelike clusters 2–3 inches long, each flower without petals, but its many, white, protruding stamens conspicuous; May. **Fruit:** A capsule with a single brown seed. **Hardy:** From Zone 4 southward. **Varieties:** None. **Culture:** Best planted in a low, moist, half-shady place in the garden with as little wind as possible.

Related plants (not illustrated):

Fothergilla gardeni. Not over 3 feet high, and blooming long before the leaves unfold. Hardy from Zone 4 southward.

Fothergilla major. Usually 7–10 feet high, its white flowers blooming with the unfolding of the leaves. Hardy from Zone 4 southward.

Petals More Than 4

Petals prevailingly 5 (except in some double-flowered horticultural forms). See p. 229
Petals 6. Barberry

Barberry
(*Berberis*)
Family Berberidaceae

For low or medium-sized shrubs, few plants excel the barberries, most of which are evergreen, and all of them spiny. The wood and inner bark are apt to be yellow. Leaves simple, alternate, usually toothed, some of the leaves replaced by 3-pronged spines. Flowers prevailingly yellow, borne in small clusters. Petals 6. Fruit a black, blue, or red berry, often showy and winter-persistent.

The barberries are of extremely easy culture in any ordinary garden soil, but some of the evergreen species are not hardy everywhere so that hardiness notes below should be studied before planting. The cultivation of some barberries, especially the common native one (*Berberis canadensis*) and its European relative (*Berberis vulgaris*), is forbidden in cereal states because they serve as an alternate host for a serious rust on some cereals. In any case these 2 are of slight garden interest. All of the species below are immune to the rust and may be planted anywhere. Of over 175 species, at least 55 are known to be in cultivation here, but of these only the following can be included:

Leaves evergreen.
  Leaves without marginal teeth. — Magellan Barberry
  Leaves with marginal teeth.
    Twigs more or less warty. — *Berberis verruculosa*
    Twigs not warty.
      Twigs a little angled. — Wintergreen Barberry
      Twigs decidedly grooved. — *Berberis mentorensis*
Leaves not evergreen. — Japanese Barberry

Magellan Barberry
(*Berberis buxifolia*), p. 227.

MAGELLAN BARBERRY (*Berberis buxifolia*) p. 226

A spiny evergreen Chilean shrub, much grown for its attractive foliage and purplish fruits but precariously hardy in the East. **Size:** 4–7 feet high, about half as wide. **Leaves:** Alternate, simple, wedge-shaped at the base, ½–1¼ inches long, without marginal teeth but prickle-tipped. **Flowers:** Only 1 or 2 in a cluster, orange yellow; June. **Fruit:** Nearly round, pea-size, dark purple, not winter-persistent. **Hardy:** Precariously so in Zones 5 or 4, but it does not thrive in hot, dry places. **Varieties:** *nana* is a compact shrub, scarcely 2 feet high, grown for its foliage as it rarely flowers or fruits. **Culture:** See above. The plant is best suited to the Pacific Coast, especially from northern Calif. to Wash. and Ore.

Related plant (not illustrated):

Berberis stenophylla. A hybrid shrub 6–8 feet high with evergreen, hair-fringed, but not spiny-margined leaves, and golden-yellow flowers in clusters of 2–6. Hardy from Zone 5 southward, but not thriving in the East.

BERBERIS VERRUCULOSA p. 230

This and its relatives, some of which are hybrids, comprise some of the best of the evergreen barberries. It is a native of China and spiny. **Size:** 2–3 feet high, moundlike, compact, and hence nearly as wide, the twigs warty. **Leaves:** Alternate, simple, ovalish or elliptic, ¾–1½ inches long, glossy green above, whitish beneath, the margins spiny-toothed. **Flowers:** Nearly ¾ inch wide, golden yellow; May–June. **Fruit:** Pea-size, bluish black, with a bloom, not winter-persistent. **Hardy:** From Zone 4 southward. **Varieties:** None. **Culture:** See above.

Related plants (not illustrated):

Berberis chenaulti. A hybrid shrub, more arching in habit and with larger flowers than *Berberis verruculosa,* blooming in June. Hardy from Zone 4 southward.

Berberis darwini. An evergreen Chilean shrub, 5–8 feet high, with clusters of golden-yellow flowers and dark purple fruit. Precariously hardy in Zone 6 in the East and best suited to the Pacific Coast.

Berberis candidula. A low Chinese evergreen shrub,

scarcely 2 feet high, its leaves with rolled margins that nearly conceal the few spiny teeth. Flowers yellow; June. Fruit pea-size, blue, with a bloom. Hardy from Zone 5 southward.

WINTERGREEN BARBERRY (*Berberis julianae*) p. 230

Perhaps the most popular of all the Chinese evergreen barberries for its ease of growth and hardiness. **Size:** 4–6 feet high, nearly as wide, its twigs a little angled. **Leaves:** Alternate, simple, narrowly elliptic, 1½–3 inches long, the margins toothed and spiny. **Flowers:** Bright yellow, in close clusters, each flower about ⅓ inch in diameter; May. **Fruit:** Pea-size, bluish black, with a bloom, not winter-persistent. **Hardy:** From Zone 4 southward. **Varieties:** One listed by several nurserymen as variety *nana* is reputedly lower, but the variety is unknown to science. **Culture:** See above.

BERBERIS MENTORENSIS p. 230

A hybrid evergreen barberry, much resembling the Wintergreen Barberry, which is one of its parents and from which it differs chiefly in having definitely grooved twigs and the ability to stand drought. Hardy from Zone 4 southward, and forming a dense spiny hedge.

JAPANESE BARBERRY (*Berberis thunbergi*) p. 230

Of all the barberries this one that drops its leaves in winter is by far the commonest — so common that it is anathema to some. Its autumn foliage and bright fruit make it the most showy in late fall and early winter. **Size:** 4–6 feet high, nearly as wide, the twigs spiny. **Leaves:** Alternate, simple, not evergreen, without marginal teeth, usually broadest toward the tip, ½–1½ inches long, turning a brilliant scarlet in the fall. **Flowers:** Yellow, but red-tinged on the outside, about ⅓ inch wide, in clusters of 2–5; May. **Fruit:** Bright red, football-shaped, about ½ inch long, persisting through the winter, and so profuse as to make the plant a cheerful sight against the snow. **Hardy:** From Zone 3 southward. **Varieties:** Many, mostly with different-colored foliage or habit. The most important are the Box Barberry (variety *minor*), which is a permanent dwarf and good for edging or low hedges; *atropurpurea,* with reddish foliage; *columnaris,* an erect form most easily made into hedges; *erecta,* a form with rigidly upright branches, the plant thus narrower than the type. **Culture:** See above. It is the easiest of all the barberries to grow.

Related plants (not illustrated):

Berberis koreana. A Korean shrub, not evergreen, with red twigs. It grows up to 8 feet high, has thin leaves 1½–3 inches long, turning red in the fall. Flowers yellow, in dense, nearly stalkless clusters. Fruit bright red, persisting until spring. Hardy from Zone 4 southward.

Berberis aggregata. A densely branched Chinese shrub, not evergreen, 4–7 feet high, its spines 3-pronged. Leaves oblongish, about 1 inch long. Flowers yellow, in stalkless clusters. Fruit egg-shaped, nearly ½ inch long, red, but with a bloom, persistent for part of the winter. Hardy from Zone 4 southward.

Petals Prevailingly 5 (Except in Some Double-flowered Horticultural Forms)

Flowers never yellow or greenish yellow. See p. 235

Flowers yellow or greenish yellow (red in one species of Ribes).

Flowers appearing in early spring, on leafless twigs. *Corylopsis*

Flowers appearing on leafy branches, after or with the unfolding of the leaves.

Fruit a berry, flowers less than ½ inch wide. Flowering Currant

Fruit dry; flowers (often double) at least 1 or 1½ inches wide. *Kerria*

BARBERRY FAMILY (*Berberidaceae*), 1, 2, 9
TEA FAMILY (*Theaceae*), 6, 8
SAXIFRAGE FAMILY (*Saxifragaceae*), 3, 4
5, 7 in other families

1. **Wintergreen Barberry** (*Berberis julianae*) p. 228
   A splendid evergreen Chinese shrub, rather spiny, with yellow flowers in spring. *B. mentorensis* is a related hybrid.

2. **Japanese Barberry** (*Berberis thunbergi*) p. 228
   The brilliant autumn color of its foliage and its winter-persistent scarlet fruit makes this the most used of all barberries.

3. **Flowering Currant** (*Ribes sanguineum*) p. 233
   A stout Pacific Coast shrub, with hanging clusters of showy red flowers. Not hardy northward.

4. **Golden Currant** (*Ribes aureum*) p. 234
   An easily grown shrub from the western United States, with yellow fragrant flowers.

5. **"Japanese Rose"** (*Kerria japonica pleniflora*) p. 235
   A Chinese shrub, usually grown in its double-flowered form, the flowers ball-like and nearly 2 inches wide.

6. ***Camellia Sasanqua*** p. 237
   An evergreen Asiatic shrub, in many fine varieties and much easier to grow than the common camellia.

7. ***Corylopsis spicata*** p. 232
   A Japanese relative of the witch-hazels, its yellow flowers blooming on the bare twigs in Feb. or Mar.

8. ***Camellia japonica*** p. 236
   The camellia to most gardeners, now found in scores of varieties, mostly winter-blooming. Not hardy northward.

9. ***Berberis verruculosa*** p.227
   A low, moundlike evergreen Chinese shrub, somewhat spiny, the golden-yellow flowers blooming in the spring.

1
2
3
4
5
6
7
8
9

1
2
3
4
5
6
7
8
9

ROSE FAMILY (*Rosaceae*), 1, 2, 5–8
3, 4, 9 in other families

1. **Bridal Wreath** (*Spiraea prunifolia plena*) p. 245
   One of the most widely planted of all spireas, this Asiatic shrub has a tremendous profusion of doubled flowers in early spring.

2. ***Spiraea Bumalda*** p.251
   The variety Anthony Waterer is the best of the summer-blooming spireas and bears profuse red clusters in July.

3. **Franklinia** (*Gordonia altamaha*) p. 238
   A remarkable native shrub or small tree, its white or pinkish waxy, fragrant flowers blooming in early fall. Difficult to grow.

4. **Blue-Blossom** (*Ceanothus thyrsiflorus*) p. 241
   A nearly evergreen Pacific Coast shrub, often sprawling, with showy, midsummer blue flowers. Not hardy northward.

5. **Cream Bush** (*Holodiscus discolor*) p. 244
   A spirea-like shrub from the western United States, its branching cluster of small whitish flowers blooming in summer.

6. ***Spiraea vanhouttei*** p. 248
   Early-blooming and often called Bridal Wreath (see No. 1), this medium-sized shrub has only single flowers.

7. ***Stephanandra incisa*** p. 243
   Not showy except in the fall, when its finely dissected leaves are reddish. Flowers white or greenish white in June.

8. **Ninebark** (*Physocarpus opulifolius*) p. 243
   A native spirea-like shrub with white flowers and bark that peels off in long ribbons. Not very showy.

9. **Rose-of-Sharon** (*Hibiscus syriacus*) p. 239
   A very popular Chinese shrub now found in many color forms that are more attractive than the typical species.

CORYLOPSIS
(*Corylopsis*)
Family Hamamelidaceae

Tall shrubs, relatives of the witch-hazel, but with 5 petals instead of 4. Like the witch-hazel they bloom only on the bare twigs and thrive best in cool moist sites. The flowers appear in very early spring before the leaves unfold. Their culture is relatively easy. Only 2 species are here considered:

Young twigs hairy. *Corylopsis spicata*
Young twigs not hairy or only faintly so. *Corylopsis pauciflora*

CORYLOPSIS SPICATA p. 230
A rather coarse Japanese shrub of value only because of its very early bloom. **Size:** 4–6 feet high, nearly as wide, the young twigs hairy. **Leaves:** Alternate, simple, toothed, ovalish, but obliquely heart-shaped at the base, 4–6 inches long, of no autumnal color. **Flowers:** Yellow, in nearly stalkless clusters, blooming in Feb.–Mar. on bare twigs. The flowers are about ¾ inch long. Petals 5. **Fruit:** A 2-beaked pod, with 2 black seeds. **Hardy:** From Zone 4 southward. **Varieties:** None. **Culture:** See above.

CORYLOPSIS PAUCIFLORA below
A smaller Japanese version of the above, but its young twigs are hairless or nearly so. **Size:** 3–5 feet, nearly as wide. **Leaves:** Rather coarse, toothed, ovalish, but obliquely heart-shaped at the base, 2–3 inches long. **Flowers:** Yellow, in nearly stalkless clusters,

Corylopsis pauciflora, above.

blooming on the bare twigs in Feb.–Mar. Petals 5. **Fruit:** A 2-beaked pod, with 2 black seeds. **Hardy:** From Zone 4 southward. **Varieties:** None. **Culture:** See above.

Flowering Currant
(*Ribes*)
Family Saxifragaceae

A group of over 150 species of low shrubs of the cooler parts of the world, of which over 50 are in cultivation in the U.S. Besides the ornamentals the genus is important to the fruit grower, for it contains the gooseberry and the red and black currants (which are excluded from this book). All the ornamental species, usually called Flowering Currant, are low shrubs with alternate, stalked, simple leaves which are usually somewhat lobed. Flowers in small clusters, the small or even minute petals 5. Fruit a juicy, usually tart berry. Some plants of this group are alternate hosts of the white pine blister rust. If you are near any white pines it is necessary to consult with your county agent, as the cultivation of some species of *Ribes* is unlawful in many states (nearly all on the Atlantic Coast and many in the Midwest).

Many of the flowering currants are of little garden merit, but the 3 here included are well worth growing and of extremely easy culture in any garden soil. They are:

Flowers red. Flowering Currant
Flowers yellow or greenish yellow.
  Flower cluster erect. Mountain Currant
  Flower cluster drooping. Golden Currant

FLOWERING CURRANT (*Ribes sanguineum*) p. 230
A handsome red-flowered shrub from the Pacific Northwest much cultivated for its showy bloom. **Size:** 8–10 feet high, nearly as wide, the twigs not spiny. **Leaves:** Simple, alternate, roundish or kidney-shaped, 2–5 inches long, 3–5-lobed, the lobes irregularly toothed, white-felty beneath. **Flowers:** Red, the sticky, many-flowered, drooping clusters very showy; May. **Fruit:** A bluish-black berry, with a bloom. **Hardy:** From the southern part of

Zone 5 southward. **Varieties:** Several, of which the best are: *albescens,* with white flowers; *splendens,* with darker red flowers; *atrorubens,* with smaller, but many, dark red flowers; there are also double-flowered forms. **Culture:** See above.

Related plant (not illustrated):

Ribes gordonianum. A hybrid shrub, 4–6 feet high, with a many-flowered cluster of sterile yellow blooms that are tinged red outside, and no fruit. Hardy from Zone 4 southward.

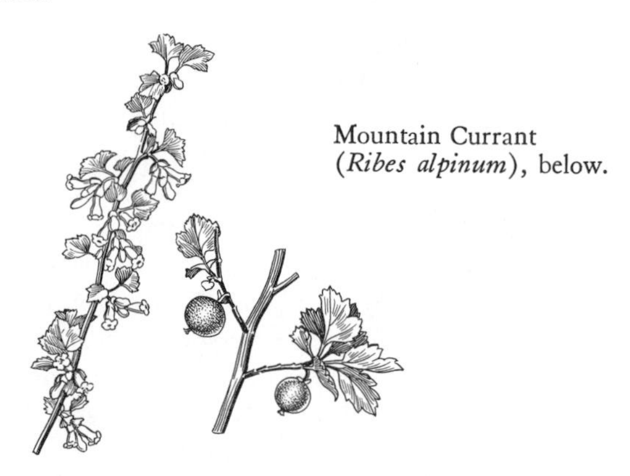

Mountain Currant (*Ribes alpinum*), below.

MOUNTAIN CURRANT (*Ribes alpinum*) above

A European ornamental shrub with upright clusters of flowers, the male and female on different plants. **Size:** 5–8 feet high, nearly as wide, of compact growth and making a good hedge plant. **Leaves:** Alternate, simple, roundish, 1½–2½ inches long, 3-lobed (sometimes 5-lobed), the lobes toothed. **Flowers:** Greenish yellow, the clusters erect; May. **Fruit:** A scarlet berry (on female plants only). **Hardy:** From Zone 5 southward. **Varieties:** *aureum,* a dwarf form with yellowish leaves. **Culture:** See above.

GOLDEN CURRANT (*Ribes aureum*) p. 230

A widely grown ornamental currant from the western U.S. and the hardiest and easiest grown of the 3 here included. **Size:** 4–6 feet high, about half as wide, its twigs not spiny. **Leaves:** Alternate, simple, nearly round. 1½–2½ inches long, 3-lobed, the lobes sparsely toothed. **Flowers:** Yellow, fragrant, in drooping clusters;

May. **Fruit:** A purplish-brown smooth berry. **Hardy:** Everywhere. **Varieties:** None. **Culture:** See above. It is the most desirable *Ribes* for the amateur.

KERRIA (*Kerria japonica*). Family Rosaceae p. 230
Like so many plants christened *japonica* this attractive shrub is Chinese and much grown, especially in its double-flowered form. **Size:** 4–6 feet high, nearly as wide, its green twigs holding their color all winter. **Leaves:** Alternate, simple, ovalish, 1½–4 inches long, double-toothed. **Flowers:** Solitary, yellow, of 5 petals (in the typical form), ¾–1½ inches wide; May. **Fruit:** Small, dry. **Hardy:** From Zone 3 southward. **Varieties:** The so-called Japanese Rose (it is neither a rose nor Japanese) is the variety *pleniflora,* which has double, ball-like flowers nearly 2 inches in diameter and is the form most widely grown as it is more showy and more vigorous than the typical form. **Culture:** Easy in any ordinary garden soil, in full sunshine.

Flowers Never Yellow or Greenish Yellow

Leaves not evergreen (but evergreen in 2 species of Photinia). See p. 238
Leaves evergreen. *Camellia*

CAMELLIA
(*Camellia*)
Family Theaceae

Magnificent evergreen shrubs (rarely treelike), comprising only a handful of species, but found in hundreds of horticultural varieties, most of them with double flowers, and all derived from Asiatic species. Leaves alternate, simple, evergreen, short-stalked, and rather finely toothed on the margin. Flowers waxy, usually solitary, with 5 petals in the wild species, but much doubled and with many petals in the hybrid forms, typically red, but in the many hybrids pink, rose-colored, or white. Fruit (rarely produced in cultivated plants, and never in the double-flowered kinds) a dry, somewhat woody capsule, rather sparsely seeded.

The culture of camellias is exacting. They want a well-drained but moist soil with a pH of 5.5 to 6.5, and most of them do better if shaded from early morning sun. *Camellia japonica,* which is the finest of the species, cannot be grown safely much beyond the middle of Zone 5 and only spring-flowering varieties should be tried there, as most winter-blooming kinds will have their blossoms blasted by frost. Farther south there is a much wider choice of varieties, especially where the mild winters permit some varieties to bloom in January, February, and March.

Camellia Sasanqua needs the same soil conditions as *Camellia japonica,* but is much more climatically tolerant. Blooming from October to Christmas it is perfectly hardy up to the northern limit of Zone 5, but a little hazardous above that. No camellia should be moved without a tightly packed ball of earth (in a can or in roped-up burlap) as they will not tolerate exposure of their roots to sun and wind. All of them are best mulched with pine needles or cottonseed meal. Many of the best varieties are grafted, slow-growing, and quite expensive. The only two species of general interest are:

Plant bushy; leaves 3–4 inches long; twigs smooth. *Camellia japonica*

Plant more loosely branched; leaves 1½–2½ inches long; twigs hairy. *Camellia Sasanqua*

COMMON CAMELLIA (*Camellia japonica*) p. 230

Nearly 600 named forms of this ever-popular Asiatic shrub are mute evidence of the plant's long cultivation and the ingenuity of many Japanese and other breeders of it. **Size:** Usually shrublike, 6–15 feet high and as wide, rarely (in the South) treelike and up to 30 feet high; twigs smooth. **Leaves:** Alternate, simple, thick, leathery, evergreen, ovalish, 3–4 inches long, finely toothed. **Flowers:** Solitary, waxy, 3–5 inches wide, typically of 5 rounded petals, but double-flowered in nearly all the varieties below; red, pink, or white, with little or no odor. **Fruit:** A small, woody capsule. **Hardy:** See above. **Varieties:** Within the range of this book only early spring-flowering forms can be selected, as the winter-blooming sorts will have their blossoms blasted by frost. Of those that bloom in early spring, suited to Zones 5 and 6, the following are worth trying, but they should not be tried above the middle of Zone 5:

Berenice Boddy, pink	Leucantha, white
Lady Clare, pink	Orton Pink, pink
Mathothiana, red	Pink Perfection, pink
Jarvis Red, red	Sarah Frost, red

Other varieties tried farther north than Zone 5, and some new hybrids said to be much more tolerant of cold, are growing at the Boyce Thompson Institute, at Yonkers, N.Y. Venturesome camellia growers north of Zone 5 may soon have, according to one optimist, camellias "hardy enough for cold regions." Most Southern growers are somewhat skeptical. **Culture:** See above.

SASANQUA CAMELLIA (*Camellia Sasanqua*) p. 230

An Asiatic shrub much more tolerant of cold than *Camellia japonica,* less showy than that species, but a boon to those who treasure its relative hardiness and late autumn bloom. **Size:** 4–7 feet high, about half as wide, its twigs hairy. **Leaves:** Alternate, simple, evergreen, not so leathery as in *Camellia japonica,* ovalish, 1½–2½ inches long. **Flowers:** Solitary, pink or white, 2½–3½ inches wide, of 5 petals (in the typical form), waxy, blooming from mid-Oct. to Christmas. **Hardy:** Perfectly safe from Zone 5 southward and, less safely, in protected parts of Zone 4. **Varieties:** There are dozens, but a selection might include:

Cleopatra, rose pink (semidouble)
Crimson Tide, red (single)
Hinode-gumo, white, flushed pink (single)
Jean May, pink (nearly double)
Mine-no-yuki, white (semidouble)
Pink Snow, pink (nearly double)
Showa-no-sakae, pink (double)

Culture: See above.

Leaves Not Evergreen

(but Evergreen in 1 species of Photinia and in Some Plants in the Ginseng Family)

Leaves with stipules. See p. 239
(For definition and picture of stipules see the Picture Glossary.)
Leaves without stipules.
Flowers small, in a terminal cluster. Virginia Willow
Flowers large, solitary. Franklinia

VIRGINIA WILLOW (*Itea virginica*)
Family Theaceae p. 241
A neat little native shrub, growing naturally in swamps and wet woods from N.J. to Fla. and westward to the Mississippi Valley. **Size:** 3–8 feet high, usually about 4 feet, about a third as wide as its hairy branches are erect. **Leaves:** Alternate, simple, oval, 2–3 inches long, narrowed at both ends, finely toothed, crimson in the fall. **Flowers:** Small, white, with 5 narrow petals, crowded in a terminal, spirelike cluster 2½–5 inches long; May–June. **Fruit:** A dry, small, hairy capsule. **Hardy:** From Zone 3 southward. **Varieties:** None. **Culture:** Choose a moist, fairly rich soil, preferably in a partially shaded place, although it will grow in the open.

FRANKLINIA (*Gordonia altamaha*)
Family Saxifragaceae p. 231
A remarkably interesting shrub or small tree, found in Georgia in 1790 by John Bartram, but not since known as a wild plant, all our current supply originating from Bartram's Garden in Philadelphia. **Size:** 6–20 feet high, about half as wide, its branches sparse and the foliage thin. **Leaves:** Alternate, simple, narrowly oblong, 5–7 inches long, remotely toothed, long-persistent, but ultimately turning bright crimson in the late fall. **Flowers:** Solitary at the leaf joints, nearly stalkless, white or pinkish, fragrant,

waxy, about 3½ inches wide; Sept.–Oct., earlier in the South. **Fruit:** A woody, globe-shaped capsule, about ¾ inch in diameter. **Hardy:** From Zone 5 southward, and in protected sites in Zone 4. **Varieties:** None. **Culture:** Exacting. It needs freedom from wind, a rich woods soil with an acid reaction of pH 5–6, plenty of moisture, and preferably some shade. An historically interesting, relatively rare plant, valued for its late bloom and brilliant autumnal coloring.

Related plant (not illustrated):

Loblolly Bay (*Gordonia Lasianthus*). A native evergreen tree, 30–60 feet high, with thick leathery leaves 5–6 inches long. Flowers white, fragrant, solitary, distinctly stalked, 2–3 inches wide. Hardy from Zone 6 southward.

Leaves with Stipules

(NOTE: For definition and picture of stipules see the Picture Glossary)

Leaves with 3 or more main veins arising near the base.

Juice mucilaginous; flowers large, usually solitary. Rose-of-Sharon

Juice not mucilaginous; flowers small and in clusters. *Ceanothus*

Leaves with generally a single midrib and many side veins arising from it. Rose Family

ROSE-OF-SHARON (*Hibiscus syriacus*)

Family Malvaceae p. 231

Many gardeners call this Chinese shrub *Althaea,* although that name is better restricted to the hollyhock. The Rose-of-Sharon is the only hardy shrub of the genus *Hibiscus* and is much valued for its late bloom. **Size:** 5–15 feet high, about three fourths as wide, its branches mostly erect, the sap mucilaginous. **Leaves:** Alternate, simple, ovalish, 2–5 inches long, sharply toothed, some 3-lobed and some unlobed, often on the same plant, the 3 or more main veins arising near the base; of no autumnal color. **Flowers:** Solitary, short-stalked, 3–5 inches wide, typically with 5 petals, but much more in double-flowered horticultural forms, variously colored, most showy on dark days; Aug.–Oct. **Fruit:** A dry, 5-valved capsule. **Hardy:** From Zone 3 southward. **Varieties:** At least 40, some of which comprise forms with magenta flowers that are anathema to some. A selected list of currently available varieties might include:

- Ardens, bluish-purple (double)
- Coelestis, violet-blue (single)
- Jeanne d' Arc, white (double)
- Lucy, rose-pink (semidouble)
- Rubis, pink (single)
- Snowdrift, white (single)
- Snowstorm, white (single)

Culture: Relatively easy in most garden soils, in full sun. In the North young plants may winter-kill, but come up from the roots the following spring. Once effective bark is ripened there is no trouble about growing it, but it should be watered if there is a prolonged drought.

CEANOTHUS
(*Ceanothus*)
Family Rhamnaceae

A large group of rather showy shrubs, chiefly from the Pacific Coast, and containing only 2 species hardy within our range and of quite different habit. Leaves alternate, simple, always toothed on the margins. Flowers rather small, but in profuse clusters. Petals 5. Fruit a dry pod, splitting into 3 parts.

One of our cultivated species is the common New Jersey Tea of eastern woodlands, scarcely over 3 feet high and with white flowers. The other, and much more desirable, is the very showy Blue-Blossom, which is evergreen in California and hardy here only as indicated below. Here it usually drops its leaves in the fall, and is a shrub 8–15 feet high, rarely treelike. Both species will grow in any good garden soil.

NEW JERSEY TEA (*Ceanothus americanus*) p. 241
A native shrub of little garden interest except that it will grow in poor soil. **Size:** 2–3 feet high, about half as wide. **Leaves:**

Alternate, simple, ovalish, 2–4 inches long, finely but irregularly toothed, its 3 principal veins arising at the base. **Flowers:** Small, white, in flat-topped or roundish small clusters that are long-stalked; midsummer. **Fruit:** A small dry pod, less than ½ inch wide. **Hardy:** Everywhere. **Varieties:** None. **Culture:** Easy in any garden soil, even a poor one. The least desirable of any *Ceanothus.*

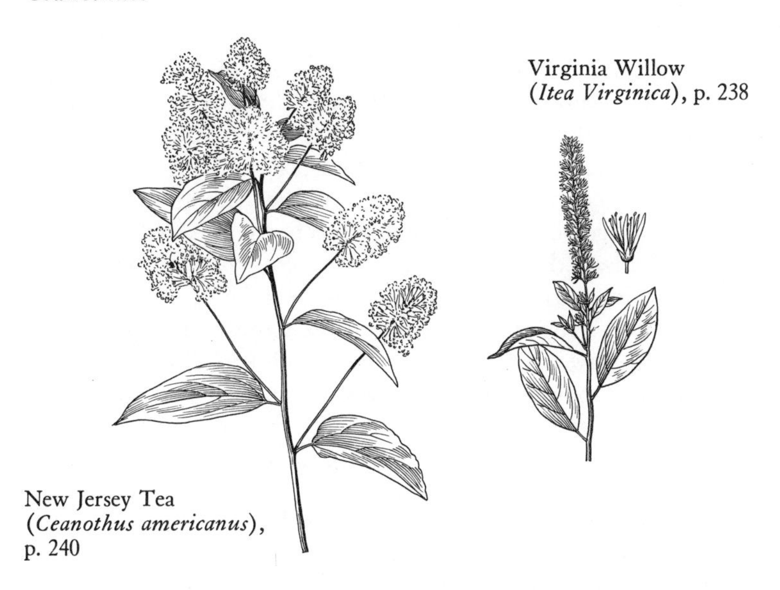

Virginia Willow (*Itea Virginica*), p. 238

New Jersey Tea (*Ceanothus americanus*), p. 240

BLUE-BLOSSOM (*Ceanothus thyrsiflorus*) p. 231

A showy evergreen or half-evergreen shrub of the Pacific Coast, unfortunately not hardy over most of the East. **Size:** 10–25 feet high, not stiffly erect, and (in England) often treated as a very showy espalier. **Leaves:** Alternate, simple, oblong, 1–2½ inches long, lustrous, evergreen in Calif. but half evergreen or losing its leaves in the East, its 3 main veins arising at the base. **Flowers:** Small, blue, in extremely profuse, ball-like clusters; June–July. **Fruit:** A dry capsule, splitting into 3 segments. **Hardy:** From Zone 5 southward. **Varieties:** About a dozen, but most of them not suited to the East. Those that are useful here might include Autumnal Blue, which has large trusses of blue flowers from

July to Sept; Cascade, blue and earlier-flowering (May–June); Delight, blooming in May. All are precariously hardy in the East. **Culture:** Put in open sunlight, in a light, porous soil; they often do well against a partially shaded, east-facing wall.

ROSE FAMILY
(*Rosaceae*)

A HUGE family of herbs, shrubs, and trees of first-rate garden importance, comprising about 100 genera and over 3000 species. In this book the family is divided into several sections for simplicity. The relatively tall trees (plums, cherries, and crabapples) will be found at pp. 194 to 213. Many of the medium-sized shrubs, all with alternate, simple, toothed or lobed leaves, will be found below. Still another group with leaves also alternate and simple, but without marginal teeth or lobes, will be found at p. 277. There is still a fourth group of the Rose Family that always have compound leaves, and these will be found at p. 301.

In all the four groups the leaves are alternate, mostly simple (but compound in a few genera like the rose itself and the mountain-ash). There may be marginal teeth to the leaves, as in those below, but in some as in the cherry laurel, *Cotoneaster,* and a few others there are no marginal teeth. Flowers nearly always showy, sometimes solitary, but mostly in clusters. Petals typically 5, but much more in some horticultural forms. Fruit quite various, fleshy, and with a single stone in plums and cherries; fleshy with several seeds in the apple and its relatives, but dry and inedible in many groups.

To recapitulate, all those below have alternate, simple leaves, all of which have teeth on the margins, and some also have lobed leaves. There is always a main midrib from which arise the secondary veins (but in the Ninebark they arise from the base of the leaf). Except for 2 species of *Photinia,* none is evergreen. The 8 genera below may be divided thus:

Flowers relatively small but borne in profuse clusters.
Leaves distinctly 3–5 lobed.
Bark usually peeling. Ninebark

Bark not peeling. *Stephanandra*

Leaves not lobed, merely toothed.

Fruit dry; leaves never evergreen.

Flower clusters long and drooping. Cream Bush

Flower clusters not drooping. *Spirea*

Fruit fleshy, red; leaves evergreen in two species. *Photinia*

Flowers larger, borne in sparse clusters, or solitary.

Flowers usually solitary, rarely in sparse clusters; prevailingly red. Flowering Quince

Flowers never solitary, white or pinkish.

Leaves distinctly narrowed toward the base; shrubs. Chokeberry

Leaves broad at the base; shrubs or small trees. Shadbush

NINEBARK (*Physocarpus opulifolius*) p. 231

A spirea-like native shrub, its bark peeling off in long strips, and covered in mid-spring with a profusion of small ball-like clusters of flowers. **Size:** 5–8 feet high, nearly or quite as wide from its arching habit. **Leaves:** Alternate, simple, toothed, ovalish or rounded, 2½–3½ inches long, sometimes slightly 3-lobed. **Flowers:** Scarcely ¼ inch wide, white, crowded in dense clusters; June. **Fruit:** A closely knit collection of 5 inflated podlike structures. **Hardy:** From Zone 2 southward. **Varieties:** *luteus* has the young leaves yellow, changing to bronzy yellow; *nanus* is a dwarf form with dark green foliage. **Culture:** Easy in any ordinary garden soil, in full sun. Not a particularly attractive member of the rose family.

Related plant (not illustrated):

Physocarpus monogynus. A western representative of the above, its leaves 3–5-lobed, with rather sparse clusters of white flowers. Hardy from Zone 3 southward.

STEPHANANDRA INCISA p. 231

A somewhat unimpressive Asiatic shrub, scarcely worth growing for its flowers, but its much-cut foliage rather handsome. **Size:** 5–8 feet high, quite as wide or wider from its arching habit.

Leaves: Alternate, simple, ovalish, long-pointed, 2–2½ inches long, cut almost to the middle, the lobes toothed; reddish in the fall. **Flowers:** Small, white or greenish white, in clusters that are about 2 inches long; June. **Fruit:** A dry, scarcely splitting pod. **Hardy:** From Zone 4 southward, and precariously in protected sites in Zone 3. **Varieties:** None. **Culture:** Simple in any ordinary garden soil, in full sun.

Related plant (not illustrated):

Stephanandra Tanakae. An Asiatic relative of the above with larger, more lobed, but not such finely dissected leaves. Flowers white; June–July. Hardy from Zone 4 southward.

CREAM BUSH (*Holodiscus discolor*) p. 231
A western U.S. shrub of secondary garden interest, although its trusses of spirea-like flowers are attractive. **Size:** 6–12 feet high, nearly or quite as wide from its arching habit. **Leaves:** Alternate, simple, oval, 2–4 inches long, toothed, white-felty beneath. **Flowers:** Very small, white, very numerous in a branching cluster; July. **Fruit:** A collection of 5 closely knit, dry structures. **Hardy:** From Zone 4 southward. **Varieties:** *ariaefolius* has the lower side of the leaves merely grayish green, and is the form usually cultivated. **Culture:** Prefers sandy loams in full sun, but not difficult to grow in any ordinary garden soil.

SPIREA
(*Spiraea*)

One of the most popular of all garden shrubs, comprising nearly 100 species, over 50 of which are known to be in cultivation in the U.S. Of these the 8 below are most likely to interest the amateur, and some of them, as the white-flowered, very early-blooming sorts, such as the Bridal Wreath, are so popular as to be perhaps a bit overplanted. All are extremely handsome, free-flowering shrubs with alternate, simple leaves, always toothed on the margin, but sometimes only faintly so. Flowers very small, but grouped in profuse clusters, the clusters sometimes flat-topped, or in others spirelike or ball-like. Fruit a dry, splitting pod, rather inconspicuous.

The spireas are of the easiest culture in any ordinary garden soil, and they will stand more abuse than most shrubs. If they have any preference it appears to be for a moist site, but most of them thrive anywhere. There are two distinct groups of garden spireas: (1) prevailingly white-flowered and early-blooming and (2) June- or July-blooming and often red or pink. The 8 garden species may thus be divided as below, but there are several wild species that cannot be included here.

April- and May-blooming; often collectively known as Bridal Wreath; flowers white, in dense ball-like or flat-topped clusters.
- *Flowers nearly always double.* — Bridal Wreath
- *Flowers scarcely ever double.*
  - *Leaves rather narrow, nearly linear.* — *Spiraea thunbergi*
  - *Leaves ovalish or obliquely squarish.*
    - *Leaves faintly 3-veined.* — *Spiraea vanhouttei*
    - *Leaves not 3-veined.* — *Spiraea cantoniensis*

Summer-blooming.
- *Flower cluster spirelike, not more or less flat-topped.* — *Spiraea billiardi*
- *Flower cluster not spirelike.*
  - *Leaves about ¾ inch long, puckered.* — *Spiraea bullata*
  - *Leaves 1½–3½ inches long, flat.* — *Spiraea Bumalda*

BRIDAL WREATH (*Spiraea prunifolia plena*) p. 231
An Asiatic shrub, its arching branches covered with a mass of white bloom, the flowers nearly always double. **Size:** 4–6 feet high, as wide or wider, with gracefully arching branches which are somewhat angled. **Leaves:** Alternate, simple, elliptic or elliptic-oblong, pointed both ends, ½–2 inches long, finely toothed, red or orange in the fall. **Flowers:** White, very profuse in nearly stalkless, roundish clusters that are about 2 inches in diameter; Apr.–May. **Fruit:** Negligible. **Hardy:** Everywhere. **Varieties:** None. **Culture:** See above.

ROSE FAMILY (*Rosaceae*), 2, 3, 5, 7
GINSENG FAMILY (*Araliaceae*), 1, 4
6, 8, 9 in various families

1. ***Fatshedera lizei*** p. 257
   A hybrid shrub, its large evergreen leaves somewhat resembling those of the English Ivy. Flowers small, green, in large clusters.

2. **Flowering Quince** (*Chaenomeles lagenaria*) p. 253
   A somewhat spiny Chinese shrub, now in several varieties, with red, pink, or white early-blooming flowers.

3. **Chokeberry** (*Aronia arbutifolia*) p. 254
   A low native shrub with early-blooming, apple-like flowers, followed by bright red fruit. Easily grown.

4. ***Fatsia japonica*** p. 257
   A small Japanese evergreen tree with large, lobed leaves and large clusters of small whitish flowers.

5. **Shadbush** (*Amelanchier canadensis*) p. 255
   A native tree or large shrub, covered in early spring with white flowers and small apple-like fruit.

6. **Sour Gum** (*Nyssa sylvatica*) p. 259
   A tall native tree with inconspicuous flowers, followed by smooth fruit. Autumn color dull red.

7. ***Photinia glabra*** p. 251
   An Asiatic evergreen shrub with rather showy clusters of small white flowers in summer. Fruit red.

8. **Papaw** (*Asimina triloba*) p. 259
   A native tree, rarely grown for the edible fruit, its flowers greenish yellow at first, ultimately purple.

9. ***Skimmia japonica*** p. 259
   A fragrant Japanese shrub, its crushed leaves highly aromatic. Flowers yellowish white, in Apr. or May.

1
2
3
4
6
7
8
9

1
2
3
4
5
6
7
8
9

MEZEREON FAMILY (*Thymelaeaceae*), 2, 6, 8, 9
LAUREL FAMILY (*Lauraceae*), 4, 5
1, 3, 7 in various families

1. **Smoke Tree** (*Cotinus Coggygria*) p. 264
   A shrub more showy in "fruit" than in flower, especially the plumy, hairy stalks of its sterile flowers.

2. ***Daphne odora*** p. 263
   A low Asiatic evergreen shrub with ball-like clusters of small purplish very fragrant flowers. Not hardy northward.

3. ***Sarcococca hookeriana*** p. 261
   An evergreen relative of the box from the Himalayas, its leaves longer than in the box. Not hardy northward.

4. **Laurel** (*Laurus nobilis*) p. 260
   The laurel of the poets and history, a fine evergreen foliage plant, easily sheared for hedges or other forms.

5. **Spicebush** (*Lindera Benzoin*) p. 265
   An aromatic native shrub, its yellow flowers small but numerous on bare twigs in Mar.–Apr. Fruit not persistent.

6. **Garland-Flower** (*Daphne Cneorum*) p. 262
   A practically prostrate evergreen shrub, forming mats. Flowers small, fragrant, and pink in Apr. or early May.

7. **Sea Buckthorn** (*Hippophae rhamnoides*) p. 266
   Male and female flowers on separate plants and negligible. Fruit profuse, yellow-orange, and persistent.

8. **Mezereon** (*Daphne Mezereum*) p. 264
   The easiest grown of the garden daphnes, its foliage not evergreen. Flowers fragrant, rosy-purple, in Mar.–Apr.

9. **Leatherwood** (*Dirca palustris*) p. 265
   A not very showy native shrub with small yellow flowers that bloom before the leaves unfold.

SPIRAEA THUNBERGI p. 249

Not very different from the above, but its flowers single; an Asiatic shrub of wide cultivation. **Size:** 3–5 feet high, about as wide and very twiggy. **Leaves:** Alternate, simple, narrowly lance-shaped or even linear, ¾–1¾ inches long, sharply toothed, bright green, becoming orange or red in the fall. **Flowers:** White, single, in nearly stalkless roundish clusters which are very profuse; Apr.–May before the leaves unfold. **Hardy:** From Zone 3 southward. **Varieties:** None. **Culture:** See above. Some of its numerous twigs have a tendency to die out and need pruning to keep the shrub in good shape.

Related plant (not illustrated):

Garland Spirea (*Spiraea arguta*). A hybrid spirea, a little taller than *Spiraea thunbergi* (one of its parents) and still more profuse in its bloom. Hardy from Zone 3 southward.

SPIRAEA VANHOUTTEI p. 231

By far the most commonly cultivated spirea in America and often (perhaps unfortunately) called Bridal Wreath, which properly is *Spiraea prunifolia plena.* **Size:** 4–6 feet high, as wide or wider, with beautifully arching branches. **Leaves:** Alternate, simple, somewhat obliquely squarish or angularly ovalish, ¾–1¾ inches long, green above, paler beneath, faintly 3-veined, often red or orange in the autumn. **Flowers:** White, very profuse, in dense, more or less flat-topped clusters; May or even early June. **Fruit:** Negligible. **Varieties:** None. **Culture:** See above. It stands city smoke and dust better than most shrubs.

Related plant (not illustrated):

Spiraea trilobata. An Asiatic shrub resembling *Spiraea vanhouttei,* but smaller in all its parts. Flowers white, in May. Hardy from Zone 3 southward.

SPIRAEA CANTONIENSIS p. 231

An Asiatic shrub closely related to *Spiraea vanhouttei,* but its leaves not faintly 3-veined. **Size:** 3–5 feet high, about as wide. **Leaves:** Alternate, simple, somewhat angularly oblongish, wedge-shaped at the base, 1–2½ inches long, deeply toothed, pale beneath. **Flowers:** Pure white, in profuse roundish clusters 1–2 inches wide; June. **Fruit:** Negligible. **Hardy:** From Zone 4 southward and doing well in the South. **Varieties:** None. **Culture:** See above. It is sometimes called Reeve's Spirea.

Left: *Spiraea thunbergi,* p. 248
Right: *Spiraea billiardi,* below.

SPIRAEA BILLIARDI above

An enormously popular, summer-flowering spirea of hybrid origin, and when established often forming large patches from the spreading of its underground stems. **Size:** 4–6 feet high, about as wide. **Leaves:** Alternate, simple, oblongish, 2–3½ inches long, doubly toothed, but without teeth toward the base, grayish beneath. **Flowers:** Small, bright rose-red, in a spirelike cluster that is 4–7 inches long and very showy; July–Aug. **Fruit:** Negligible. **Hardy:** From Zone 3 southward. **Varieties:** None, except a reputed white-flowered form. **Culture:** See above.

Related plants (not illustrated):

Hardhack (*Spiraea tomentosa*). A native pasture shrub, 2–3 feet high, its leaves white-felty beneath and with spirelike clusters of rose-pink flowers in midsummer. Hardy from Zone 2 southward.

Spiraea macrothyrsa. A hybrid shrub, closely related to *Spiraea billiardi,* the branches of its flower clusters being almost horizontal; pink. Hardy from Zone 3 southward.

Spiraea bullata, below.

SPIRAEA BULLATA above

A low Japanese shrub with rusty-hairy twigs, too small for the shrubbery, but useful in the rock garden. **Size:** 9–15 inches high, the plant very compact. **Leaves:** Alternate, simple, roundish, ½–1 inch long, strongly puckered. **Flowers:** Tiny, scarlet, so profuse in each cluster and the clusters so numerous that the bloom practically covers the plant; July. **Fruit:** Negligible. **Hardy:** From Zone 5 southward. **Varieties:** None. **Culture:** See above.

SPIRAEA BUMALDA p. 231

Perhaps the most widely cultivated of all the summer-blooming spireas and of hybrid origin. **Size:** 18–24 inches high, about as wide, the plants bushy. **Leaves:** Alternate, simple, ovalish or lance-shaped, doubly toothed, 1½–3½ inches long. **Flowers:** White or pink, but see below; July. **Fruit:** Negligible. **Hardy:** From Zone 3 southward. **Varieties:** The plant is scarcely grown in the typical form, for the varieties, of which there are several, are much superior. Two of the best are Anthony Waterer, a more compact form, with narrower leaves and bright crimson flowers, and *froebeli,* which is somewhat taller and is covered with a profusion of bright crimson flowers. There are several others, but they are scarcely different from Anthony Waterer, which is the plant most widely grown. **Culture:** See above.

PHOTINIA
(*Photinia*)

Handsome but little-grown shrubs, rarely trees, with striking evergreen leaves in the first 2 of the 3 cultivated species. Flowers very small, white, in profuse, more or less flat-topped clusters, followed by a red, fleshy fruit. Except for the hardiness restrictions, noted below, the photinias present no cultural difficulties.

Leaves evergreen. — *Photinia glabra*
Leaves not evergreen. — *Photinia villosa*

PHOTINIA GLABRA p. 246

An Asiatic evergreen shrub, little grown, but a valuable garden addition because of its lustrous foliage, summer bloom, and rather persistent red fruits. **Size:** 8–10 feet high, branchy and about as wide. **Leaves:** Alternate, simple, evergreen in the South, less so in the North, oblongish, 2–3½ inches long, finely toothed and wedge-shaped at the base. **Flowers:** Small, white, with 5 petals, arranged in a cluster 2½–6 inches wide; June–July. **Fruit:** A fleshy berry, pea-size, red, and persistent until late Nov. **Hardy:** From Zone 5 southward. **Varieties:** None. **Culture:** See above.

Related plant (not illustrated):

Photinia serrulata. A finer plant than the above, but not so widely planted. It is a Chinese evergreen shrub or small tree, sometimes up to 30 feet high, its lustrous leaves 4½–8 inches long, its small flowers in flat-topped clusters 4–6 inches wide, its red berries persistent into late Nov. Hardy from Zone 5 southward.

PHOTINIA VILLOSA

below

An Asiatic shrub or small tree, its foliage not evergreen. **Size:** 8–15 feet high. **Leaves:** Alternate, simple, oblongish, 1–3 inches long, pointed at the tip, finely toothed, hairy beneath, turning red in the fall. **Flowers:** Small, white, in clusters 1–2 inches wide, the flower stalks warty. **Fruit:** Red, pea-size, persistent through Dec. **Hardy:** From Zone 3 southward. **Varieties:** None. **Culture:** See above.

Photinia villosa, above.

FLOWERING QUINCE
(*Chaenomeles*)

Widely grown, often somewhat spiny shrubs, the twigs quite hairless. Leaves short-stalked, alternate, simple, always toothed on the margin. Flowers typically red, in very sparse clusters or even apparently solitary, blooming before the leaves unfold or with their unfolding. Petals 5, rather showy. Fruit fleshy, edible.

Only two species are much grown, the first one being nearly the most popular cultivated bush in the U.S.

Twigs smooth. *Chaenomeles lagenaria*
Twigs rough. *Chaenomeles japonica*

CHAENOMELES LAGENARIA p. 246

Commonly called Japanese Quince or Japanese Flowering Quince, this Chinese shrub has long been cultivated in Japan and here it is ubiquitous. **Size:** 4–6 feet high, the twigs smooth and without hairs; nearly as wide. **Leaves:** Alternate, simple, oblong-oval, 2–3 inches long, finely toothed. **Flowers:** Typically red (but see below), appearing before or with the unfolding of the leaves, about 1 inch wide, showy; Mar.–Apr. **Fruit:** Apple-like, but somewhat egg-shaped, about 2 inches long, greenish yellow, speckled, used in preserves. **Hardy:** From Zone 3 southward. **Varieties:** Over 40, many of them ill-defined and duplicates of each other.

A selection might include:

Incende, orange red (double)
Kermesina semiplena, scarlet (semidouble)
Rosea plena, pink (double)
Blood Red, red (single)
Marmorata, carmine (single)
Umbilicata, red (single)
Nivalis, white (single)

Culture: Easy anywhere. Much used for hedges as it stands clipping very well and its close spiny twigs make it all but impenetrable.

CHAENOMELES JAPONICA p. 254

A Japanese shrub related to the above, but usually lower and with rough twigs. **Size:** 2–3 feet high, about as wide. **Leaves:** Alternate, simple, broadly oval, 1½–2 inches long, coarsely toothed. **Flowers:** Brick red, about 1 inch wide; Mar.–Apr. **Fruit:** Yellow, about 1½ inches long. **Hardy:** From Zone 3 southward.

Chaenomeles japonica, p. 253.

Varieties: None. **Culture:** See above. It is a less desirable plant than *Chaenomeles lagenaria,* but it does stand city conditions.

CHOKEBERRY (*Aronia arbutifolia*) p. 246
A rather attractive early-blooming native shrub, of secondary garden interest, but the white flowers and red fruits relatively showy. **Size:** Usually 3–5 feet high, rarely taller. **Leaves:** Alternate, simple, oblongish, about 2 inches long, tapering toward the base, gray beneath, toothed, red in the fall. **Flowers:** White, with blackish anthers and 5 roundish petals, small, borne in terminal clusters; late Apr. and early May. **Fruit:** Red, berrylike, pea-size, persistent through the winter. **Hardy:** From Zone 2 southward. **Varieties:** Nurserymen offer a form with reputedly more brilliant-colored fruit under the name *Brilliantissima* (a name unknown to science). **Culture:** Easy in any ordinary garden soil, preferably in full sun, and if the site is moist that appears to be some advantage.

Related plant (not illustrated):

Aronia melanocarpa. A lower native relative of the above, with similar flowers but black fruit. Hardy from Zone 3 southward.

SHADBUSH
(*Amelanchier*)

Closely related to *Aronia* and of secondary garden importance, these shrubs (rarely trees) are natives of eastern North America. The two below (there are many native and ill-defined species) differ from *Aronia* in having the base of the leaf always rounded, not tapering to a narrow base. Flowers few in each cluster, white, the 5 petals narrow. Fruit like a miniature apple. Their culture is without difficulty in any ordinary garden soil.

Young foliage white-hairy. *Amelanchier canadensis*
Foliage always smooth. *Amelanchier laevis*

AMELANCHIER CANADENSIS p. 246

A native tree or large shrub of little garden interest, but covered with bloom in early spring. **Size:** Usually shrubby and 8–20 feet high, rarely a tree up to 45 feet, about half as wide. **Leaves:** Alternate, simple, ovalish, but rounded at the base, 2–3 inches long, white-hairy in youth, ultimately green. **Flowers:** White, the 5 petals strap-shaped, borne in rather profuse clusters about 2 inches long, before or with the unfolding of the leaves; Apr.–May. **Fruit:** Like a miniature apple, red purple, not persistent. **Hardy** From Zone 3 southward. **Varieties:** None. **Culture:** See above.

AMELANCHIER LAEVIS p. 256

A native tree, rarely shrubby, much less grown than *Amelanchier canadensis,* but a profuse bloomer in early spring. **Size:** 20–30 feet high, about as wide. **Leaves:** Alternate, simple, ovalish, but rounded or heart-shaped at the base, about 3½ inches long, bright green. **Flowers:** White, in drooping clusters, the 5 petals strap-shaped, blooming with or after the unfolding of the leaves; May. **Fruit:** Like a miniature apple, purplish black, not persistent. **Hardy:** From Zone 3 southward. **Varieties:** None. **Culture:** See above. An attractive but little-planted native tree.

Related plants (not illustrated):

There are several native shrubs that are not in the catalogs, but may easily be dug from the wild when dormant. One of the best is:

Amelanchier stolonifera. A shrub 2–3 feet high, often forming patches. It is covered with white bloom about the time the leaves unfold, followed by black-purple, sweet, edible fruit. Hardy from Zone 3 southward.

Amelanchier laevis, p. 255.

Leaves Large, Deeply Lobed or Divided

THE leaves in the group below are usually very large (7–8 inches wide) and deeply lobed. The only family is

GINSENG FAMILY
(*Araliaceae*)

SHRUBS or trees, often very prickly, comprising some striking garden plants grown mostly for their handsome foliage. In those below the leaves are alternate, simple, very large (7–18 inches wide) and deeply lobed, giving them a near-tropical aspect. While the plants are grown chiefly for their striking foliage, their flower clusters are immense, often compound, and much branched. Flowers small, usually greenish yellow. Petals 4 or 5. Fruit fleshy.

Only 2 genera are here considered:

Leaves 3–5-lobed.	*Fatshedera*
Leaves 7–9-lobed.	*Fatsia*

FATSHEDERA (*Fatshedera lizei*) p. 246
A hybrid shrub derived from crossing the English Ivy with the species following this one. **Size:** 4–6 feet high, about as wide, its twigs rusty-hairy when young. **Leaves:** Alternate, simple, evergreen, long-stalked, resembling the English Ivy, but 10–14 inches wide, 3–5-lobed. **Flowers:** Very small, light green, the petals 5, grouped in small dense clusters which are again grouped in a large branching cluster that may be 8–10 inches long and half as wide; Sept.–Oct. **Fruit:** Fleshy. **Hardy:** From Zone 6 southward, precariously in the southern part of Zone 5. **Varieties:** None. **Culture:** Needs a rich garden loam, a reasonably cool, moist site and freedom from wind.

FATSIA (*Fatsia japonica*) p. 246
A beautiful Japanese evergreen tree or shrub with tropical-looking immense leaves. **Size:** 10–20 feet high, as wide or wider, due to its branching habit. **Leaves:** Alternate, simple, stiff, 7–9-lobed, but nearly round in outline, 9–15 inches wide, the leafstalk 8–12 inches long. **Flowers:** Small, whitish, with 5 petals, grouped in globelike clusters that are themselves grouped in a larger, branched cluster 9–18 inches wide and showy; Sept.–Oct. **Fruit:** Black, berrylike, about ¼ inch in diameter. **Hardy:** Only from the southern part of Zone 6 southward. **Varieties:** One with variegated leaves. **Culture:** Prefers rich, sandy loam, but will grow in most ordinary garden soil, preferably in partial shade and out of the wind.

Shrubs and Trees with Alternate Simple Leaves which Have No Teeth or Lobes on the Margin

The following trees and shrubs differ from those in the preceding section in having alternate, simple leaves that, in all but 3 genera, have no teeth or lobes on the leaf margin. (In *Prinsepia, Pyracantha,* and *Exochorda* there may be minute teeth, especially toward the tip of the leaf. These 3 genera are included here because they do not have obvious teeth and all belong to the Rose Family in which, as here restricted, there are no marginal teeth at all.)

There are 21 genera in this group of plants that have no marginal teeth (3 exceptions noted above) and they are otherwise quite unrelated. They may be distinguished thus:

- *Leaves reduced to small scales that hug the twigs, the plants thus apparently leafless.* Tamarisk. See p. 269
- *Leaves with a normally expanded blade.*
  - *Leaves with stipules. (For definition and picture of stipules see the Picture Glossary).* See p. 271
  - *Leaves with no stipules.*
    - *Trees.*
      - *Petals 6; fruit 2–6 inches long.* Papaw
      - *Petals 5; fruit about ⅔ inches long.* Sour Gum
    - *Shrubs.*
      - *Leaves evergreen (2 exceptions in Daphne.)*
        - *Crushed foliage highly aromatic (sometimes unpleasantly so).*
          - *Flowers in terminal clusters.* *Skimmia*
          - *Flowers not in terminal clusters.* Laurel
        - *Crushed foliage not aromatic, if so only mildly.*
          - *Shrubs 4–6 feet high.* *Sarcococca*
          - *Low shrubs 12–30 inches high.* *Daphne*
      - *Leaves not evergreen.*
        - *Fruit a collection of long, plumed sterile flowers.* Smoke-Tree
        - *Fruit not so.*
          - *Low shrub, 2–4 feet high.* Leatherwood
          - *Taller shrubs, 6–12 feet high.*
            - *Flowers appearing before the leaves expand.* Spicebush
            - *Flowers appearing after the leaves expand.* Oleaster Family

PAPAW (*Asimina triloba*). Family Annonaceae p. 246
A medium-sized native tree grown for its somewhat showy flowers and its edible, banana-flavored, aromatic fruit. **Size:** 15–30 feet high, about three fourths as wide; foliage dense. **Leaves:** Alternate, simple, without marginal teeth, oval-oblong, wedge-shaped at the base, 8–12 inches long, drooping, fluttering in any breeze. **Flowers:** Solitary or few at the leaf joints, green at first, ultimately purple, about 2 inches wide, blooming as the leaves expand, the petals 6; May. **Fruit:** Fleshy, oblong, suggesting an old potato in color and shape, 2–6 inches long, the flesh yellow, aromatic, banana-flavored, and much prized by the early settlers. **Hardy:** From Zone 4 southward. **Varieties:** None. **Culture:** Not particularly easy as it inhabits rich, moist woods and needs a rich garden soil and at least partial shade. Not easy to transplant and a ball-and-burlapped specimen is the safest purchase.

SOUR GUM (*Nyssa sylvatica*). Family Cornaceae p. 246
Often called Pepperidge, this native shade tree is difficult to dig from the wild, and not too easy to grow in the garden, although its superb fall foliage tempts many to try it. **Size:** 60–90 feet high, about half as wide, its branches horizontal, but drooping very gradually and gracefully at the ends. **Leaves:** Alternate, simple, without marginal teeth, 3–5 inches long, somewhat broader toward the pointed tip, brilliant scarlet or brick-red in the fall. **Flowers:** Small, greenish, not showy, borne in dense heads, the petals 5; May–June. **Fruit:** Small, about ⅔ inch long, black purple, usually in clusters of 1–3. **Hardy:** From Zone 3 southward. **Varieties:** None. **Culture:** Naturally an inhabitant of low moist places, it needs such sites for best growth. Ball-and-burlapped trees will grow in most reasonably rich, moist garden soils, but even with careful planting its establishment is fickle.

Related plant (not illustrated):

Tupelo Gum (*Nyssa aquatica*). A taller native tree with larger leaves, hardy with safety only from Zone 5 southward.

SKIMMIA (*Skimmia japonica*). Family Rutaceae p. 246
A beautiful, densely branched, evergreen Japanese shrub, its foliage delightfully aromatic when crushed. **Size:** 3–5 feet high,

nearly as wide. **Leaves:** Alternate, simple, without marginal teeth, often more or less crowded at the ends of the twigs, elliptic or oblongish, 3–5 inches long. **Flowers:** In small clusters, the sexes on different plants, yellowish white, about ⅓ inch long, very fragrant; Apr.–May. **Fruit:** On female plants (and only if there is a male plant nearby), bright red, nearly round, about ⅓ inch thick, persistent into early winter. **Hardy:** From Zone 6 southward. **Varieties:** None. **Culture:** It needs a rich garden soil and must be grown under shade. Not suited to open windy places.

Related plant (not illustrated):

Skimmia reevesiana. A Chinese relative of the above, somewhat more difficult to acquire, lower, and with male and female flowers on the same plant, hence always fruit-bearing. Hardy from Zone 5 southward.

LAUREL (*Laurus nobilis*). Family Lauraceae p. 247

The laurel of history and the poets and a famous evergreen tree of the Mediterranean region, commonly called Bay or Bay Tree by gardeners, and much used as sheared, tubbed specimens for accents around pools, steps, and balustrades. **Sizes:** Treelike forms (rare in the U.S.) may be 40–60 feet high, but as universally grown here shrubby and sheared into globes, pyramids, and standards. **Leaves:** Evergreen, simple, alternate, oblongish, 3–4 inches long, without marginal teeth, aromatic when crushed. **Flowers:** Negligible. **Fruit:** A small, ultimately black berry. **Hardy:** From Zone 6 southward, precariously in the southern part of Zone 5. **Varieties:** None readily available. **Culture:** Mostly grown as tubbed, sheared (and expensive) specimens for formal gardens. Can be grown as a shrub or small tree outdoors only from Zone 6 southward, in rich, moist, garden soil, preferably in partial shade and in a wind-free site.

SARCOCOCCA
(*Sarcococca*)
Family Buxaceae

Evergreen Asiatic shrubs closely related to the box, but with longer leaves. Unlike box, which has opposite leaves, these have alternate ones, quite without marginal teeth. Flowers small, whitish, without petals. Fruit berrylike. These leathery-leaved shrubs are grown only for their foliage. Their cultivation is the same as for box (see p. 147).

Fruit black. — *Sarcococca hookeriana*
Fruit maroon. — *Sarcococca ruscifolia*

SARCOCOCCA HOOKERIANA p. 247

A Himalayan evergreen relative of the box, grown mostly for its foliage as the flowers are inconspicuous. **Size:** 4–6 feet high, densely branched and hence nearly as wide. **Leaves:** Alternate, simple, without marginal teeth, evergreen, 2–3 inches long, generally oblongish or lance-shaped. **Flowers:** Negligible. **Fruit:** Nearly round, black, about ⅓ inch in diameter. **Hardy:** From Zone 5 southward. **Varieties:** *humilis* is half the height, grows into considerable patches, and makes a good, if rather high, ground cover, especially in shady places. **Culture:** See above.

Sarcococca ruscifolia, below.

SARCOCOCCA RUSCIFOLIA above

A Chinese relative of the above and a valuable ground cover for shady places from Zone 5 southward. **Size:** As grown here 2–3 feet high, but in the wild often 4–6 feet high, its underground stems spreading and making considerable patches. **Leaves:** Alternate, simple, without marginal teeth, ovalish or elliptic, 2–3 inches long. **Flowers:** Fragrant, white, small, in 4-flowered clusters; Sept.–Feb. **Fruit:** About ¼ inch in diameter, dark red. **Hardy:** From Zone 5 southward. **Varieties:** None. **Culture:** See above.

DAPHNE
(*Daphne*)
Family Thymelaeaceae

Delightful little shrubs, some of them evergreens and with alluringly fragrant flowers. Leaves alternate, simple, without marginal teeth and no stipules. Flowers small, but crowded in flat-topped or ball-like clusters, some of them blooming before the leaves unfold. Fruit leathery or fleshy. Foliage mostly evergreen, but the leaves falling in *Daphne burkwoodi* and in the Mezereon.

The daphnes are widely grown in low borders and in the rock garden, but some of them are not hardy over much of the country as noted below. Use a sandy loam, preferably a little alkaline (*i.e.,* a cupful of lime scattered over each square yard of soil). Of the 50 known species the following 4 are well worth growing. None of them is easy to establish so that it is best to start with small, potted specimens.

Leaves evergreen.
  Low, almost prostrate shrub; flowers very fragrant. — Garland-Flower
  Erect shrubs.
    Flowers in terminal clusters. — *Daphne odora*
    Flowers from sides of the twigs. — Spurge Laurel
Leaves not evergreen. — Mezereon

GARLAND-FLOWER (*Daphne Cneorum*) p. 247
A low-creeping European evergreen shrub, forming large mats and useful as a ground cover. **Size:** Practically prostrate. **Leaves:** Simple, alternate, without marginal teeth, oblong, about 1 inch long, blunt at the tip and with a minute point. **Flowers:** Fragrant, pink, about 1/3 inch long, in terminal clusters; Apr.–May. **Fruit:** Yellowish brown, 1-seeded, small. **Hardy:** From Zone 4 southward, and with a straw mulch precariously hardy in Zone 3. **Varieties:** *alba* has white flowers. **Culture:** See above.

Related plant (not illustrated):

Daphne burkwoodi. A hybrid evergreen shrub, 2–4 feet high, usually offered under the varietal name of Somerset. It has extremely fragrant flowers, white fading to pink. Hardy from Zone 5 southward.

DAPHNE ODORA p. 247

An upright evergreen shrub from China and Japan, much grown for its ball-like clusters of very fragrant flowers. **Size:** 3–5 feet high, nearly as wide. **Leaves:** Alternate, simple, without marginal teeth, oblongish or elliptic, 2–3 inches long, narrowed at both ends, but bluntly pointed. **Flowers:** Small, rosy purple, very fragrant, in dense, headlike, terminal clusters; Mar.–Apr. **Hardy:** From Zone 5 southward. **Varieties:** *marginata* has yellow-bordered leaves. **Culture:** See above.

Spurge Laurel
(*Daphne Laureola*), below.

SPURGE LAUREL (*Daphne Laureola*) above

An erect, bushy evergreen, native in the Pyrenees, unlike most daphnes in having essentially scentless flowers. **Size:** 18–30 inches high, about as wide. **Leaves:** Alternate, simple, without marginal teeth, oblongish, 2–3 inches long, evergreen. **Flowers:** Small, yellowish green, in nearly stalkless clusters; Mar.–Apr. **Fruit:** Small, bluish black. **Hardy:** From Zone 5 southward, precariously hardy in Zone 4 with protection. **Varieties:** *philippi* is a smaller plant, more hardy than the type, and has flowers that are violet on the outside. **Culture:** Not easy, and unlike its relatives it requires a slightly acid, sandy loam and partial shade.

MEZEREON (*Daphne Mezereum*) p. 247

The most popular of all the daphnes with nonevergreen foliage, perhaps because of its extremely fragrant flowers and its relative hardiness; native in Eurasia. **Size:** 18–36 inches high, about as wide. **Leaves:** Alternate, simple, without marginal teeth, oblongish, 2–3 inches long, wedge-shaped at the base. **Flowers:** Small, in stalkless clusters, blooming before the leaves unfold, lilac purple or rosy purple and very fragrant; Mar.–Apr. **Fruit:** Small, scarlet. **Hardy:** From Zone 3 southward. **Varieties:** *alba* has white flowers; *grandiflora* has larger, bright purple flowers that are fall- and winter-blooming. **Culture:** See above.

Related plant (not illustrated):

Daphne Genkwa. An Asiatic shrub, 1–3 feet high, its leaves opposite. Flowers lilac white, scarcely fragrant; Apr.–May. Hardy from Zone 5 southward.

SMOKE-TREE (*Cotinus Coggygria*).
Family Anacardiaceae p. 247

In Aug. and Sept. few shrubs are so attractive as the Smoke-Tree, because the fruiting clusters seem, in the distance, like a smoky haze. **Size:** 10–15 feet high, nearly as wide, its foliage rather thin. **Leaves:** Alternate, simple, without marginal teeth, ovalish, 2–3 inches long, abruptly narrowed at the base; yellow and red in the fall. **Flowers:** Small, yellowish, not showy, but in the cluster are numerous sterile flowers. **Fruit:** Negligible, but the numerous plumed stalks of the sterile flowers, 7–10 inches long, are covered with purplish-green hairs. **Hardy:** From Zone 4 southward. **Varieties:** *atropurpureus* has a purple fruiting cluster; *rubrifolius* has reddish leaves. **Culture:** Will grow in any ordinary garden soil, in sun or partial shade.

Related plant (not illustrated):

American Smoke-Tree (*Cotinus obovatus,* but usually offered as *Cotinus americanus*). A native tree, 15–20 feet high, its fruiting cluster not so showy as that of the one above. Hardy from Zone 5 southward.

LEATHERWOOD (*Dirca palustris*). Family Thymelaeaceae p. 247
A not very well known native shrub with tough but pliant wood and bark used to make rope, hence its other name of rope bark. **Size:** 2–4 feet high, nearly as wide. **Leaves:** Alternate, simple, without marginal teeth, elliptic, 3–5 inches long and short-stalked. **Flowers:** Small, in nearly stalkless clusters at the leaf joints, yellow, blooming before the leaves unfold; Mar.–Apr. **Fruit:** Fleshy, egg-shaped, reddish or green. **Hardy:** From Zone 3 southward. **Varieties:** None. **Culture:** Although an inhabitant of low, moist places in the wild, it will grow in any ordinary garden soil. Not a very showy bush.

SPICEBUSH (*Lindera Benzoin*). Family Lauraceae p. 247
A native aromatic shrub found wild in the woods and swamps of eastern North America and cultivated for its profuse early bloom. **Size:** 8–15 feet high, about three fourths as wide. **Leaves:** Alternate, simple, without marginal teeth, oblong but wedge-shaped at the base, 3–5 inches long, yellow in autumn. **Flowers:** Very small, yellow, profuse, crowded in small clusters on the bare twigs as it blooms weeks before the leaves unfold; Mar.–Apr. **Fruit:** Fleshy, nearly round, bright red, not persistent. **Hardy:** From Zone 2 southward. **Varieties:** None. **Culture:** Very easy, but it does better in low, moist places and in partial shade.

OLEASTER FAMILY
(*Elaeagnaceae*)

GENERALLY scurfy-leaved shrubs or small trees with alternate, simple leaves that have no marginal teeth, and the twigs often more or less spiny. Flowers very small, without petals, generally greenish yellow, always borne in small clusters at the sides of the twigs, never terminal. Fruit fleshy, often very showy, with a single stone. The family is rather small, with only 3 genera, the 2 below, and another, *Shepherdia,* which has opposite leaves and is not included here. Some of the flowers are extremely fragrant.

Our 2 genera may be distinguished thus:

Fruit orange yellow; male and female flowers on separate plants. Sea Buckthorn
Fruits silvery or brownish (red in one); male and female flowers on the same plant. *Elaeagnus*

SEA BUCKTHORN (*Hippophae rhamnoides*) p. 247
A Eurasian spiny shrub much prized for its showy, persistent fruit that is little bothered by birds. **Size:** 10–25 feet high, about three fourths as wide, sometimes treelike, its twigs spiny. **Leaves:** Alternate, simple, lance-shaped or narrower, 1–3 inches long, silvery in youth, ultimately greenish on the upper surface. **Flowers:** Negligible, but male and female flowers on separate plants, and it is necessary to have both to ensure a crop of **Fruit:** Fleshy, but somewhat hard, nearly egg-shaped, about ¼ inch in diameter, profuse, orange yellow, showy and persistent for most of the winter. **Hardy:** From Zone 2 southward. **Varieties:** None. **Culture:** Easy, as it tolerates most ordinary garden soils, either in full sun or partial shade.

ELAEAGNUS
(*Elaeagnus*)

Scurfy-leaved shrubs with alternate, simple leaves, without marginal teeth, the twigs sometimes spiny. Flowers small, whitish yellow, without petals, the male and female on the same plant, often extremely fragrant, especially in *Elaeagnus pungens.* Fruit fleshy or dry but nearly always scurfy, silvery, brownish, or red in the gumi, and edible in some. Of the 40 known species the 5 below are in common cultivation as they stand smoke, dust, wind, and seashore exposure better than most shrubs, hence are much planted in the prairie states and along the coast. All are easy to grow in almost any garden soil. A simple key:

Fruit silvery or brownish, scarcely red.
Spring-flowering.

Fruits dryish or mealy.
Twigs silvery; fruit yellow, mealy. Oleaster
Twigs brownish; fruit silvery, dryish. Silverberry
Fruits fleshy and edible. *Elaeagnus umbellata*
Blooming in Oct.–Nov. *Elaeagnus pungens*
Fruit bright red. Gumi

OLEASTER (*Elaeagnus angustifolia*) p. 278
Often called the Russian Olive, this edible-fruited, spiny Eurasian shrub or small tree is most useful in dry, rather poor soils. **Size:** 10–20 feet high, about half as wide, its bark shreddy. **Leaves:** Alternate, simple, without marginal teeth, decidedly scurfy, silvery beneath, oblongish, 2–3½ inches long, and more decorative than the flowers or fruit. **Flowers:** Negligible, but fragrant. **Fruit:** Egg-shaped, about ½ inch long, yellow but silvery-scaled, the edible flesh sweet but mealy, not persistent. **Hardy:** From Zone 2 southward. **Varieties:** None any better than the typical form. **Culture:** See above.

SILVERBERRY (*Elaeagnus commutata*) p. 268
A native shrub with silvery foliage and apt to spread by some of its stems being close to the ground and rooting there. **Size:** 8–12 feet high, as wide or wider when mature from its rooting stems (stolons). **Leaves:** Alternate, simple, without marginal teeth, ovalish to oblong, 1–4½ inches long, short-stalked and silvery. **Flowers:** Silvery yellow, about 1 inch long, in clusters of 1–3, fragrant; May–July. **Fruit:** Silvery, short-stalked, about ½ inch long, its flesh mealy. **Hardy:** Everywhere. **Varieties:** None. **Culture:** See above.

ELAEAGNUS UMBELLATA p. 268
An Asiatic spiny shrub with silvery foliage and small but fragrant flowers. **Size:** 10–15 feet high, about three fourths as wide. **Leaves:** Alternate, simple, without marginal teeth, but the margins sometimes crisped, elliptic or oval-oblong, 1–3½ inches long, silvery. **Flowers:** Yellowish white, scaly on the outside, about ¾

inch long, fragrant; May–June. **Fruit:** Brown at first, ultimately reddish, scaly, nearly round, about ½ inch in diameter. **Hardy:** From Zone 3 southward. **Varieties:** *parvifolia* has silvery twigs. **Culture:** See above.

Left: *Elaeagnus pungens,* below. Center: Silverberry (*Elaeagnus commutata*), p. 267. Right: *Elaeagnus umbellata,* p. 267.

ELAEAGNUS PUNGENS above

A superlatively fragrant, fall-flowering Japanese shrub much grown in the South and hardy here only as indicated below. **Size:** 10–15 feet high, nearly or quite as wide, the twigs usually spiny. **Leaves:** Alternate, simple, without marginal teeth, but wavy-margined, oblongish, 2½–5 inches long, silvery beneath, evergreen in the South, scarcely so northward. **Flowers:** 1–3 hanging, silvery white, about ½ inch long, gardenia-scented, and so profuse as to make the shrub an alluring enticement; Oct.–Nov. **Fruit:** Small, brown and scaly at first, ultimately reddish. **Hardy:** Only from Zone 5 southward. **Varieties:** Many, mostly variations in the color of the foliage (blotched, margined, etc.), none of which is superior to the typical form. **Culture:** See above.

GUMI (*Elaeagnus multiflora*) p. 278

A smoke-resistant Asiatic shrub tolerating city gardens very well, with pleasantly acid, edible fruit. **Size:** 4–9 feet high, about three fourths as wide. **Leaves:** Alternate, simple, without marginal teeth, more or less elliptic, 1–3 inches long, silvery beneath. **Flowers:** Yellowish white, 1–2 in each cluster, with silvery and brownish scales on the outside; May. **Fruit:** Bright red, but scaly, about ½ inch long, its flesh pleasantly acid. **Hardy:** From Zone 3 southward. **Varieties:** None that are superior to the typical form. **Culture:** See above.

TAMARISK
(*Tamarix*)
Family Tamaricaceae

Apparently leafless but actually leafy shrubs or trees, the leaves small, scalelike, hugging the twigs, which are usually shed in the fall with the tiny leaves, which do not fall separately. The leaves are scarcely 1/16 inch long. Stems and twigs extremely pliant and swaying with the wind. The plants are useful along the seashore, where stiffer-branched shrubs might not survive. Flowers pink, very small, but gathered in large, showy, and rather fluffy clusters. Fruit a minute dry capsule, so small as to furnish no features that help in separating the species.

The genus *Tamarix,* which is essentially a desert group from the Old World, is best grown in a sandy soil and they survive in badly wind-swept places. Their culture is thus easy, but some of them are frost-sensitive, as indicated below. Of perhaps 60 known species the following 5 are most likely to be grown. Their identity is apt to be puzzling, and their leaf differences are too small to be noted here.

Flower clusters from the sides of the twigs. — *Tamarix africana*
Flower clusters terminal.
  Flowering in June–July. — Salt Cedar
  Flowering in Aug.–Sept.
    Twigs pinkish. — *Tamarix pentandra*
    Twigs not pinkish. — *Tamarix odessana*

TAMARIX AFRICANA below

A graceful shrub from the dry regions of the eastern Mediterranean, not so much cultivated as some of the other species. **Size:** 6–10 feet high, about half as wide. **Flowers:** Pink, the clusters about 3 inches long and borne along the sides of last season's twigs. **Hardy:** Only from Zone 6 southward. **Varieties:** None. **Culture:** See above.

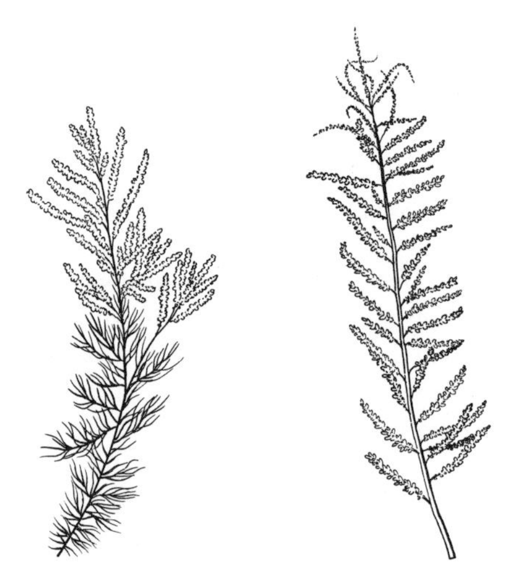

Left: *Tamarix odessana,* p. 271. Right: *Tamarix africana,* above.

SALT CEDAR (*Tamarix gallica*) p. 278

Often called French Tamarisk, this is the best and most widely grown of the group and is usually a shrub or small tree. **Size:** 15–25 feet high, about three fourths as wide, its twigs bluish green and not hairy. **Flowers:** White or pinkish, in terminal clusters; June–July. **Hardy:** From Zone 3 southward. **Varieties:** None. **Culture:** See above, but it will also grow in most ordinary garden soils.

TAMARIX PENTANDRA p. 278

A gracefully swaying shrub from Eurasian deserts, valuable because, with the Salt Cedar, it is the hardiest of the group. **Size:**

10–15 feet high, about half as wide, its twigs pinkish. **Flowers:** Pink or rose-pink, mostly in dense clusters which are grouped in a larger terminal cluster; Aug.–Sept. **Hardy:** From Zone 3 southward. **Varieties:** None. **Culture:** See above.

TAMARIX ODESSANA p. 270

A low shrub from the Caspian region, its branches upright. **Size:** 4–6 feet high, its branches not pinkish, less slender than most of the group and not hairy. **Flowers:** Pink, the slender clusters about 1½ inches long, borne on the twigs of the season; Aug.–Sept. **Hardy:** From Zone 4 southward. **Varieties:** None. **Culture:** See above.

Leaves with Stipules

(For definition and illustration of stipules, see the Picture Glossary.)

Petals 6 or more.	*Magnolia*
Petals less than 6.	
Petals 4.	*Loropetalum*
Petals typically 5.	Rose Family

MAGNOLIA
(*Magnolia*)
Family Magnoliaceae

MAGNIFICENT flowering shrubs or trees, one of them a stately evergreen with huge, fragrant blossoms. Leaves large, alternate, without marginal teeth, the leafstalk with stipules, which are sometimes conspicuous, especially in the spring. Flowers always solitary, never in clusters, terminal, large, extremely fragrant in some. Petals 6–8. Fruit fleshy, conelike, its scarlet seeds ultimately, but briefly, hanging by a threadlike stalk. Over 50 species are known of which perhaps 20 are grown in the U.S. Of these at least 9 are so much cultivated that they are included below.

The cultivation of magnolia is not easy. All have soft but brittle roots that must never be exposed to the sun and wind. Hence plants should be moved with a tight ball of soil, firmly tied up in burlap, and planted preferably in the spring, in good rich, moist, but not wet soil. They repay an application of manure every second year, spread as a mulch and not over 2 inches thick. Avoid dry and windy sites, as most of them, in their native habitat, are forest dwellers, and consequently resent too much exposure to winds. They are useless in big cities. Some species bloom before the leaves unfold, while others bloom in spring or early summer, so that magnolia bloom may be had from March until July by a selection from the 9 below.

Blooming before the leaves unfold.
- *Flowers white.*
  - *Flowers with a purple streak at the base of the petals.* — *Magnolia Kobus*
  - *Flowers without a purple streak.*
    - *Petals strap-shaped; leaves 1½–4 inches long.* — *Magnolia stellata*
    - *Petals oblongish, concave; leaves 4–7 inches long.* — Yulan
- *Flowers purple.*
  - *Shrub 4–10 feet high; leaves 4–7 inches long.* — *Magnolia liliflora*
  - *Treelike, often 15–25 feet high; leaves 3–6 inches long.* — *Magnolia soulangeana*

Flowers blooming after the leaves are expanded.
- *Leaves evergreen, rusty-hairy beneath.* — Evergreen Magnolia
- *Leaves not evergreen.*
  - *Leaves heart-shaped at the base, 12–36 inches long.* — Large-leaved Cucumber Tree
  - *Leaves not heart-shaped at the base, never more than 6 inches long.*
    - *Petals 6.* — *Magnolia sieboldi*
    - *Petals 9–12.* — Sweet Bay

MAGNOLIA KOBUS p. 278

A Japanese tree, but usually shrubby in cultivation, its lily-like white flowers blooming on the bare twigs before the leaves unfold. **Size:** 20–30 feet as a tree, usually much less as grown here. **Leaves:** Alternate, simple, without marginal teeth, oval, 2½–4 inches long, abruptly pointed, pale and nearly smooth beneath. **Flowers:** White, 4–5 inches wide, lily-like, with 6–9 thin petals which are purple-streaked at the base, and blooming before the leaves expand; Apr.–May. **Fruit:** 4–5 inches long. **Hardy:** From Zone 3 southward. **Varieties:** *borealis* is a hardier form but apt to delay blooming until quite large. **Culture:** See above.

Related plant (not illustrated):

Magnolia salicifolia. A Japanese tree, 20–30 feet high, its leaves 3–5 inches long. Flowers nearly 5 inches wide, white but often purple at the base, blooming before the leaves expand. Hardy from Zone 3 southward.

MAGNOLIA STELLATA p. 278

The earliest-blooming of the group, this Japanese shrub may have its flowers blasted by a late frost. It is a superb plant with starlike flowers on the bare twigs. **Size:** 4–15 feet high and quite as wide because of its spreading branches, the young growth of which is densely soft-hairy. **Leaves:** Alternate, simple, without marginal teeth, broadly oval to oblong, 1½–4 inches long, pale beneath. **Flowers:** White, fragrant, the 12–18 petals narrow and strap-shaped, spreading and at length turned back; Mar.–Apr. **Fruit:** 2 inches long, red. **Hardy:** From Zone 4 southward, but not suited to the Deep South. **Varieties:** *rosea* has pale pink flowers. **Culture:** See above.

Related plant (not illustrated):

Magnolia loebneri. A hybrid shrub, resembling *Magnolia stellata* but with larger leaves and flowers. Hardy from Zone 4 southward.

YULAN (*Magnolia denudata*) p. 278

A Chinese tree with widely spreading branches upon which the cup-shaped flowers bloom before the leaves expand. **Size:** 20–50 feet high, about as wide, its young twigs soft-hairy. **Leaves:** Oval, 4–7 inches long, 3–4 inches wide, tapering toward the base, rounded at the tip and with a short point. **Flowers:** White, fragrant, cup-shaped, 5–6 inches wide, blooming before the leaves expand, the oblongish petals concave; Apr.–May. **Fruit:** Brownish, cylinder-shaped, 3–4 inches long. **Hardy:** From Zone 3 southward. **Varieties:** None. **Culture:** See above. It needs plenty of space as its branches are spreading.

MAGNOLIA LILIFLORA p. 276

A beautiful Chinese shrub with lily-shaped flowers that bloom before the leaves expand. **Size:** 4–10 feet high, about as wide, with wide-spreading branches. **Leaves:** Alternate, simple, without marginal teeth, nearly oval, 4–7 inches long, narrowing to a short point at the tip. **Flowers:** Lily-shaped, nearly 8 inches wide, slightly fragrant, blooming before the leaves expand, white inside, but purple outside and very handsome; May. **Fruit:** Brown, oblong. **Hardy:** From Zone 4 southward. **Varieties:** *nigra* has larger flowers, purple inside and outside, and appearing when the leaves are partly expanded. It is the darkest-flowered of all the magnolias. **Culture:** See above.

MAGNOLIA SOULANGEANA p. 278

Perhaps the most widely planted of all magnolias, this hybrid tree or large shrub bears its flowers while still quite young, always on bare twigs. **Size:** 10–25 feet high, often treelike, but also shrubby and wide-spreading. **Leaves:** Alternate, simple, without marginal teeth, broadly oblong, 3–6 inches long, slightly soft-hairy beneath. **Flowers:** Cup-shaped, nearly 6 inches wide, generally purple (rarely white), scentless or fragrant; Apr.–May, before the leaves expand. **Fruit:** Cucumber-like. **Hardy:** From Zone 2 southward. **Varieties:** Many, and much more likely to be cultivated than the type. Among the best and certainly the most widely grown of all magnolias are: *lennei,* which has broader leaves and flowers that are white inside, but rose-purple on the outside; *alba superba,* which has pure white flowers; *alexandrina,* which has white flowers tinged purple outside; *rubra,* with red flowers (often offered as *Magnolia rustica rubra*). **Culture:** See above. It is the only magnolia that is moderately safe for city gardens.

EVERGREEN MAGNOLIA (*Magnolia grandiflora*) p. 278
This aristocrat of all magnolias is a magnificent evergreen native tree, widely planted in the South but not hardy northward, its foliage dense and casting deep shade. **Size:** 60–100 feet high, about three fourths as wide. **Leaves:** Alternate, simple, without marginal teeth, oblongish, 5–8 inches long, tapering both ends, leathery and evergreen, shining green above, rusty-woolly beneath. **Flowers:** White, extremely fragrant, waxlike, coming out of silky-hairy buds, cup-shaped, 6–8 inches wide, June–July. **Fruit:** Rusty-woolly, about 4 inches long. **Hardy:** From Zone 5 southward. **Varieties:** *lanceolata* has narrower leaves and is a little more hardy than the typical form, and one of this variety over 50 feet high once grew in a yard in Brooklyn, N.Y. **Culture:** See above.

LARGE-LEAVED CUCUMBER TREE
(*Magnolia macrophylla*) p. 279
The largest-leaved of all the magnolias and a native tree of the forests from Ky. to Fla., west to Ark. and La. **Size:** 30–50 feet high, widely branching and hence nearly as wide. **Leaves:** Alternate, simple, without marginal teeth, but heart-shaped or eared at the base, oblongish, 1–3 feet long, soft-hairy and with a bloom; not evergreen. **Flowers:** Cup-shaped, white, fragrant, 10–12 inches wide, blooming after the leaves expand; May–June. **Fruit:** Nearly round, rose-colored when ripe. **Hardy:** From Zone 4 southward. **Varieties:** None. **Culture:** See above.

Related plant (not illustrated):

Cucumber Tree (*Magnolia acuminata*). Also a native tree, 70–100 feet high. Leaves ovalish, 6–9 inches long. Flowers cup-shaped, greenish yellow, 2–3 inches high. Hardy from Zone 4 southward.

MAGNOLIA SIEBOLDI p. 276
This medium-sized tree from Korea and Japan is showy in bloom because of the contrast between the red stamens and the white petals. **Size:** 20–30 feet high, about three fourths as wide, its young twigs covered with soft, flattened hairs. **Leaves:** Simple, alternate, without marginal teeth, elliptic or broadly oval, 2–6

inches long, bluntly pointed, soft-hairy and with a bloom beneath. **Flowers:** Cup-shaped, fragrant, white, 3–4 inches wide, its red stamens conspicuous; June–July. **Fruit:** Crimson, about 1½ inches long. **Hardy:** From Zone 4 southward. **Varieties:** None. **Culture:** See above. It is often called Oyama Magnolia.

Related plants (not illustrated):

Magnolia obovata. A Japanese tree, 60–100 feet high, its leaves ovalish and 12 inches long or more. Flowers white, fragrant, cup-shaped, 5–7 inches wide, opening as the leaves expand. Fruit scarlet, about 8 inches long. Hardy from Zone 3 southward.

Umbrella Tree (*Magnolia tripetala*). A native tree, 20–40 feet high, its leaves oblong, 1–2 feet long. Flowers appearing with the leaves, cup-shaped, 8–10 inches wide, white and of unpleasant odor. Fruit rose-pink, about 4 inches long. Hardy from Zone 3 southward.

SWEET BAY (*Magnolia virginiana*) below

Perhaps the most delightfully fragrant shrub or small tree in our

Left: *Magnolia liliflora,* p. 274. Center: *Magnolia sieboldi,* p. 275. Right: Sweet Bay (*Magnolia virginiana*), above.

native flora, evergreen in the South, but usually dropping its leaves within the range of this book. **Size:** Shrubby in the North but a tree 30–60 feet high in the South. **Leaves:** Alternate, simple, without marginal teeth, oblongish, but pointed at the tip, pale bluish green beneath, 3–5 inches long. **Flowers:** Cup-shaped, white, about 3 inches wide, enticingly aromatic and fragrant; late May–June, after the leaves expand or with their expansion. **Fruit:** About 2 inches long, red. **Hardy:** From Zone 3 southward. **Varieties:** None. **Culture:** Naturally a swamp or bog plant, it prefers moist, somewhat acid sites, but will grow fairly well in ordinary garden soils.

LOROPETALUM (*Loropetalum chinense*)
Family Hamamelidaceae p. 279
A Chinese evergreen shrub closely related to the witch-hazel, and blooming in very early spring. **Size:** 6–12 feet high, about three fourths as wide. **Leaves:** Alternate, simple, evergreen, without marginal teeth, ovalish, 1–2 inches long. **Flowers:** White, the 4 petals strap-shaped, about 1 inch long; Mar. **Fruit:** A small, dry, woody capsule. **Hardy:** From Zone 5 southward. **Varieties:** None. **Culture:** It needs rich, moist garden soil and is not to be tried in an open, wind-swept place. Not much grown but interesting as an evergreen relative of the witch-hazel.

ROSE FAMILY
(*Rosaceae*)

Here belong a group of 6 genera of the Rose Family which differ from most of the others in not having compound leaves, or simple leaves with obvious marginal teeth. In those below the leaves are alternate, simple, generally without marginal teeth, or if there are any they are very small, or the leaf is slightly toothed only toward the tip. The flowers are all white (rarely pinkish white) and there are always 5 petals. The group contains shrubs and trees, some of them evergreen. For many other plants in the Rose Family and a description of it see p. 242.

MAGNOLIA FAMILY (*Magnoliaceae*), 1, 2, 5, 6, 8
OLEASTER FAMILY (*Thymelaeaceae*), 4, 7
TAMARISK FAMILY (*Tamaricaceae*), 3, 9

1. **Evergreen Magnolia** (*Magnolia grandiflora*) p. 275
   The aristocrat of all cultivated magnolias, with handsome leaves and immense white, fragrant flowers. Not hardy northward.

2. ***Magnolia Kobus*** p. 273
   Inclined to be a shrubby tree, its white, lily-like flowers blooming before the leaves expand.

3. **Salt Cedar** (*Tamarix gallica*) p. 270
   Has pinkish-white flowers, very profuse in early summer, and tiny scalelike leaves on very slender twigs.

4. **Oleaster** (*Elaeagnus angustifolia*) p. 267
   Will grow in dry, rather indifferent soils. Flowers negligible. Fruit yellow but silvery-scaled, edible. Hardy everywhere.

5. **Yulan** (*Magnolia denudata*) p. 273
   A Chinese tree whose cup-shaped, fragrant white flowers bloom before the leaves expand in late Apr.

6. ***Magnolia soulangeana*** p. 274
   The leaves shown in the drawing do not expand until the flowers have bloomed. A fine, hardy hybrid tree.

7. **Gumi** (*Elaeagnus multiflora*) p. 268
   A hardy Asiatic shrub, its leaves silvery beneath. Flowers white or yellowish white. Fruit red. Will stand city conditions.

8. ***Magnolia stellata*** p. 273
   A stunning shrub, with starlike white flowers in a great profusion, blooming before the leaves expand.

9. ***Tamarix pentandra*** p. 270
   Resembling No. 3, but later-blooming, so that if they are planted together, tamarisk bloom may last from early summer to Sept.

1
2
3
5
6
7
8
9

1
3
4
5
7
8
9

WITCH-HAZEL FAMILY (*Hamamelidaceae*), 5
ROSE FAMILY (*Rosaceae*), 2–4 and 6–9
MAGNOLIACEAE, 1

1. **Large-leaved Cucumber Tree** (*Magnolia macrophylla*) p. 275
   The largest-leaved of the magnolias, with immense cup-shaped flowers in May and June.

2. ***Cotoneaster horizontalis*** p. 286
   A half-evergreen, nearly prostrate, very vigorous shrub, with numerous white flowers and small apple-like fruit.

3. ***Cotoneaster acutifolia*** p. 288
   An erect Chinese shrub with white flowers followed by small, apple-like fruit.

4. **Pearl Bush** (*Exochorda racemosa*) p. 284
   An easily grown Chinese shrub with white, early-blooming showy flowers and dry fruits.

5. ***Loropetalum chinense*** p. 277
   Evergreen shrub related to the witch-hazel, its white flowers with only 4 petals.

6. **Rock Spray** (*Cotoneaster microphylla*) p. 290
   A beautiful, evergreen, sprawling Himalayan shrub, suitable for draping over walls.

7. ***Stransvaesia davidiana*** p. 280
   An erect Chinese evergreen shrub with white flowers and reddish fruit. Not hardy northward.

8. **Everlasting Thorn** (*Pyracantha coccinea*) p. 281
   Grown for its spectacular showy fruit that persists into early winter. Easily grown.

9. **Cherry Laurel** (*Laurocerasus officinalis*) p. 284
   A splendid evergreen shrub with small white flowers. It can be easily sheared into any shape.

ROSE FAMILY cont'd:

The 6 genera are divided thus:

Leaves evergreen.
- *Anthers red.* — *Stransvaesia*
- *Anthers yellow.*
  - *Fruit red or orange.* — Firethorn
  - *Fruit dark purple or blackish.* — Cherry Laurel

Leaves not evergreen.
- *Fruit dry.* — Pearl Bush
- *Fruit fleshy.* — *Cotoneaster*

STRANSVAESIA (*Stransvaesia davidiana*) p. 279
A rather straggling Chinese evergreen shrub with white flowers in a profuse terminal cluster. **Size:** 10–15 feet high or more, about three fourths as wide, the young twigs silky. **Leaves:** Evergreen, alternate, simple, without marginal teeth, oblongish, 3–4 inches long, pointed at the tip, turning bronzy red in the winter. **Flowers:** Small, white (with rather striking red anthers), grouped in a profuse terminal cluster that is nearly 4 inches wide; June. **Fruit:** Berrylike, red or pinkish, pea-size, persistent into Nov. **Hardy:** From Zone 5 southward. **Varieties:** *undulata* has wavy-margined leaves, is a little hardier and a smaller plant; *salicifolia* has narrower leaves and a densely hairy flower cluster. **Culture:** Within its hardiness restrictions, the plant is easily grown in most garden soils. As a broad-leaved evergreen it should be moved only as a ball-and-burlapped specimen. Not a very striking shrub.

FIRETHORN
(*Pyracantha*)

Showy evergreen shrubs, cultivated for their fine foliage and profuse red or orange fruit that in some species persists through the winter. Twigs often thorny. Leaves alternate, simple, the margins without teeth, or with minute ones, mostly near the tip. Flowers white, in a profuse branched cluster. Petals 5, nearly round. Fruit small, fleshy, brightly colored, with 5 small seeds.

A small group of 6 species, 3 of which are widely cultivated for their showy fruit. Not all of them are hardy everywhere so that hardiness notes below should be consulted before planting. Their culture is otherwise easy in any ordinary garden soil.

Flower cluster hairy. Everlasting Thorn
Flower cluster not hairy.
Leaves broadest above the middle.
Pyracantha crenato-serrata
Leaves broadest near or below the middle.
Pyracantha atalantioides

EVERLASTING THORN (*Pyracantha coccinea*) p. 279
A spiny evergreen Eurasian shrub, much grown for its spectacularly showy fruit; easily trained against a wall. **Size:** 12–20 feet high, rather narrow, its twigs thorny. **Leaves:** Alternate, simple, very finely wavy-toothed on the margin, ovalish, 1–1½ inches long, turning bronzy in the winter. **Flowers:** Small, white, the cluster hairy; May. **Fruit:** Berrylike, about ⅓ inch in diameter, brilliant scarlet, very profuse and persistent into late winter. **Hardy:** From Zone 4 southward. **Varieties:** *lalandi* is far more grown than the type because it is a little more hardy and its orange-red fruit is equally persistent. There are several other varieties offered, but they have little advantage over *Pyracantha coccinea* and its variety *lalandi*. **Culture:** See above.

PYRACANTHA CRENATO-SERRATA p. 282
A handsome Chinese evergreen shrub grown mostly for its persistent, brick-red fruit. **Size:** 5–9 feet high, rather narrow, the young twigs rusty-hairy. **Leaves:** Alternate, simple, the margins wavy but not toothed, broadest toward the tip, 1–3 inches long, evergreen. **Flowers:** Small, white, in smooth clusters that are 1½–2½ inches wide; May–June. **Fruit:** Nearly round, about ½ inch thick, brick-red, persistent into mid-Jan. **Hardy:** From Zone 5 southward. **Varieties:** Rather confused, some offered as *gibbsi* being usually *crenato-serrata*. **Culture:** See above.

Related plant (not illustrated):

Pyracantha crenulata. A Himalayan relative of the above, but less fruitful and of chief interest because its variety *rogersiana* is hardy up to the coastal region of Zone 4.

Left: *Pyracantha atalantioides,* below.
Right: *Pyracantha crenato-serrata,* p. 281.

PYRACANTHA ATALANTIOIDES above

A handsome evergreen Chinese shrub valued for its winter-persistent fruit. **Size:** 10–15 feet high, rather narrow. **Leaves:** Simple, alternate, oblongish or elliptic, broadest near or below the middle, 1–4 inches long, evergreen, its margin wavy and very finely toothed. **Flowers:** Small, white, but in very profuse clusters; May–June. **Fruit:** Crimson or scarlet, about ½ inch thick, winter-persistent. **Hardy:** From Zone 5 southward. **Varieties:** None. **Culture:** See above.

Related plants (not illustrated):

Pyracantha angustifolia. A Chinese shrub 8–12 feet high, its fruit orange or brick-red. Hardy from Zone 6 southward.

Cherry Laurel
(*Laurocerasus*)

Beautiful evergreen shrubs or trees grown mostly for their striking laurel-like leaves and for their white flowers. Leaves alternate, simple, without marginal teeth or very shallowly and minutely toothed. Flowers small, white, in finger-shaped clusters. Petals 5. Fruit fleshy, dark purple or blackish.

The cherry laurels require a rich, moist soil and freedom from wind, dust, and smoke. They should not be moved without a good ball of soil tightly roped in burlap, for the roots must not be exposed to the sun and wind. Only 2 species are commonly cultivated and neither is hardy northward:

Flower cluster longer than the leaves; leafstalk about 1 inch long. Portuguese Cherry Laurel
Flower cluster shorter than the leaves; leafstalk scarcely ½ inch long. Cherry Laurel

PORTUGUESE CHERRY LAUREL
(*Laurocerasus lusitanica*) below
A splendid evergreen shrub (or tree in the South) from Spain and Portugal, much resembling the next species, but with longer flower clusters. **Size:** 6–10 feet as a shrub, but treelike and much taller in the South. **Leaves:** Alternate, simple, oblongish, 3–5 inches long, long-pointed, very shallowly toothed on the margin, evergreen and leathery. **Flowers:** Small, white, in finger-shaped

Portuguese Cherry Laurel (*Laurocerasus lusitanica*), above.

clusters that are more or less upright, 7–9 inches long; June. **Fruit:** Egg-shaped, about ½ inch long, dark purple. **Hardy:** From Zone 5 southward. **Varieties:** None. **Culture:** See p. 283.

CHERRY LAUREL (*Laurocerasus officinalis*) p. 279
This magnificent evergreen from southeastern Europe to Persia has been cultivated for centuries, mostly for its handsome foliage and the fact that it can be sheared into any desired shape and hence is fine for an evergreen hedge. **Size:** 5–10 feet as a shrub, but often treelike (unless sheared) and twice this height; widely spreading if not sheared. **Leaves:** Alternate, simple, oblongish, 2½–6 inches long, remotely or not at all toothed, evergreen and short-stalked. **Flowers:** Small, white, the finger-shaped cluster shorter than the leaves; May–June. **Fruit:** Dark purple, about ½ inch long, not persistent. **Hardy:** From Zone 5 southward. **Varieties:** Several, but ill-defined and scarcely worth growing except for the variety *zabeliana,* which is considerably more hardy than the type and should prove safe up to the southern and coastal part of Zone 4. **Culture:** See above.

Related plant (not illustrated):

Laurocerasus caroliniana. A handsome native evergreen shrub or small tree, 15–40 feet high, with oblong leaves nearly without marginal teeth. Flowers small, white, in slender clusters. Hardy from Zone 5 southward.

PEARL BUSH (*Exochorda racemosa*) p. 279
A wide-spreading, not evergreen, Chinese shrub, much grown for its white, early-blooming flowers. **Size:** 5–10 feet high and at least as wide. **Leaves:** Alternate, simple, without marginal teeth, elliptic-oblong, 1–3 inches long. **Flowers:** White, nearly 2 inches wide, in a showy, terminal, but not very profuse cluster; Apr.–May. **Fruit:** A dry 5-angled capsule. **Hardy:** From Zone 4 southward. **Varieties:** None. **Culture:** Extremely easy in any ordinary garden soil, in full sun or partial shade.

Related plant (not illustrated):

Exochorda giraldi, variety *wilsoni.* A recently introduced Chinese shrub which is more upright and more freely flowering than *Exochorda racemosa.*

COTONEASTER
(*Cotoneaster*)

Extremely popular shrubs, rarely trees, grown largely for their decorative fruit, which resembles a miniature apple. Leaves alternate, simple, without marginal teeth, evergreen in some tender species, rarely so in those below, but sometimes persistent or half evergreen. Flowers small, white or pinkish, mostly in rather small, rarely profuse, clusters, or sometimes solitary. Petals 5, roundish and flat in some but standing upright in others. Fruit small, fleshy, containing 2–5 seeds. *Cotoneaster adpressa* is essentially a vine and will be found at p. 51.

Over 50 species are known, and of these over 30 are in cultivation in the U.S., but only 8 can be included here and their identity is somewhat difficult. The selection of only 8 species from the 29 rightly included in the 4th edition of *Taylor's Encyclopedia of Gardening* may seem a meager list. But their identity is confusing to the amateur and it seems better to concentrate on the reasonably well-known species and those most easily grown, than attempt a more comprehensive and far more complex treatment. The plants resent moving and hence should be planted only with a good-sized ball of earth, tightly wrapped in burlap; better still are young plants in pots. Spring planting is best, in a sunny site, in well-drained soil that is more alkaline than acid. The 8 below lose their leaves in the fall unless marked evergreen. They may be divided thus:

Petals standing erect, mostly pinkish; fruit red or black.
With red fruits.
Leaves smooth or nearly so beneath.
Branches horizontal; habit half prostrate.
Cotoneaster horizontalis
Branches erect. *Cotoneaster apiculata*
Leaves densely white-felty beneath.
Cotoneaster francheti
With black fruits. *Cotoneaster acutifolia*
Petals not erect, more or less roundish and flat.
Erect shrubs; flowers in clusters.

Leaves not evergreen or half evergreen.
Cotoneaster multiflora
Leaves nearly evergreen, especially in the South.
Cotoneaster salicifolia
Sprawling shrub; flowers only 1–3 in the cluster.
Rock Spray

COTONEASTER HORIZONTALIS p. 279

The best-known and most widely planted cotoneaster, this half-prostrate Chinese shrub needs plenty of space in which to spread or climb if attached to a wall. **Size:** 18–24 inches high, ultimately spreading into a large patch that may be 4–6 feet wide, its branches forked and horizontal. **Leaves:** Half evergreen in the North or even shedding in severe winters, nearly round, about ½ inch long, very profuse, turning orange-red in the fall. **Flowers:** Tiny, pinkish, extremely numerous, but solitary or in twos, the petals erect; May. **Fruit:** Red, pea-sized, showy from its abundance, persistent into Dec. **Hardy:** From Zone 4 southward. **Varieties:** *perpusilla* has smaller, but more abundant fruit. **Culture:** See above.

Related plant (not illustrated):

Cotoneaster simonsi. A Himalayan upright shrub, 7–12 feet high, the leaves half evergreen. Flowers white, the petals upright. Fruit scarlet. Hardy from Zone 5 southward.

COTONEASTER APICULATA p. 287

A half-evergreen Chinese shrub grown mostly for its bright scarlet fruit. **Size:** 3–5 feet high, about three fourths as wide, its branches erect. **Leaves:** Nearly round but with a sharp tip, dull on the upper surface, about ¾ inch wide. **Flowers:** Pink, mostly solitary, but numerous, the petals erect; June. **Fruit:** Scarlet, stalkless, pea-size, persistent into Nov. **Hardy:** From Zone 4 southward, and usually dropping its leaves there, but half evergreen southward. **Varieties:** None. **Culture:** See above.

Left: *Cotoneaster francheti,* below.
Right: *Cotoneaster apiculata,* p. 286.

COTONEASTER FRANCHETI above

A handsome open-branched, half-evergreen Chinese shrub, conspicuous for its relatively large leaves that are white-felty beneath. **Size:** 4–8 feet high, nearly or quite as wide, with spreading branches. **Leaves:** 1–1½ inches long, ovalish, densely white-felty beneath. **Flowers:** Pinkish, 8–12 in each cluster, the petals erect, not so profuse in bloom as the other species; June. **Fruit:** Egg-shaped, about ⅓ inch thick, orange-red, persistent only into Oct. and not profuse. **Hardy:** From Zone 5 southward. **Varieties:** None. **Culture:** See above. The farther south this is grown, the more likely it is to be completely evergreen.

Related plants (not illustrated):

Cotoneaster dielsiana. A Chinese shrub 2–7 feet high and differing from the above chiefly in having only 3–5 flowers in each cluster, and in having bright red fruit. Hardy from Zone 5 southward.

Cotoneaster zabeli. A Chinese shrub 4–6 feet high, its

leaves ½–1 inch long, yellow-felty beneath. Flowers pink, 3–9 in the cluster. Fruit red, about ¾ inch long. Hardy from Zone 3 southward.

COTONEASTER ACUTIFOLIA p. 279

A Chinese shrub, not evergreen, and one of the hardiest of the group. **Size:** 8–12 feet high, nearly as wide, with spreading branches. **Leaves:** Short-stalked, oblongish, 1½–4½ inches long, dull green above, paler beneath. **Flowers:** Pink or whitish in short-stalked clusters of 2–5, the petals erect; May–June. **Fruit:** Black, elliptic, about ½ inch long, not showy. **Hardy:** From Zone 2 southward. **Varieties:** None. **Culture:** See above.

Related plants (not illustrated):

Cotoneaster bullata. A hairy-twigged Chinese shrub, not evergreen, 4–6 feet high, its oblongish leaves 2–3 inches long and puckered. Flowers white or pink. Fruit scarlet, and winter-persistent. Hardy from Zone 4 southward.

Cotoneaster foveolata. A Chinese shrub 4–8 feet high with spreading branches and hairy leaves, 2½–4 inches long. Flowers pinkish, the petals erect. Fruit black. Hardy from Zone 4 southward.

COTONEASTER MULTIFLORA p. 289

A Chinese shrub, not evergreen, grown for its profuse white flowers and bright red fruit. **Size:** 8–10 feet high and as wide, or wider, with spreading branches. **Leaves:** Thin, broadly oval, 1–2 inches long, obviously stalked. **Flowers:** White, about ½ inch wide, profuse in a many-flowered, loose, smooth cluster, the petals flat; May–June. **Fruit:** Nearly round, about ⅜ inch in diameter, red, persistent until Christmas. **Hardy:** From Zone 5 southward, possibly in protected sites in Zone 4. **Varieties:** *calocarpa* has larger leaves and larger, light red or pink fruit. **Culture:** See above.

COTONEASTER SALICIFOLIA p. 289

A Chinese shrub, half evergreen in the North, but a true evergreen in the South, taller than most of the group and with an upright habit. **Size:** 7–12 feet high, its branches more or less erect. **Leaves:** 1½–3 inches long, red-veined and red-stalked. **Flowers:** White, the petals flat, the clusters densely woolly; June. **Fruit:** Red, nearly round, about ¼ inch thick, not persistent. **Hardy:** From Zone 5 southward. **Varieties:** *floccosa* has the leaves densely woolly on the under side in youth. **Culture:** See above.

Related plants (not illustrated):

Cotoneaster frigida. A Himalayan shrub or small tree, 7–20 feet high, not evergreen. Leaves ovalish, 3–5 inches long. Flowers white. Fruit brilliantly red, winter-persistent. Hardy from Zone 6 southward.

Cotoneaster glaucophylla. A Chinese shrub, closely related to *Cotoneaster salicifolia,* but its red fruit football-shaped. Hardy from Zone 5 southward.

Cotoneaster pannosa. A Chinese evergreen shrub, 3–6 feet

Left: *Cotoneaster multiflora,* p. 288.
Right: *Cotoneaster salicifolia,* p. 288.

high, its branches arching. Leaves about 1 inch long, white-woolly beneath. Flowers white, in profuse clusters. Fruit dull red. Hardy from Zone 6 southward.

ROCK SPRAY (*Cotoneaster microphylla*) p. 279
A low, evergreen Himalayan shrub, its branches somewhat upright, but the plant forming dense masses. **Size:** 2–3 feet high, many times as wide. **Leaves:** About 1/4 inch long, shining above, pale beneath, very numerous. **Flowers:** White, tiny, very profuse, the petals spreading; May–June. **Fruit:** Scarlet, nearly round, pea-size, persistent until Mar. **Hardy:** From Zone 4 southward, and much planted to drape over banks and walls. **Varieties:** *thymifolia* has even smaller flowers and leaves. **Culture:** See above.

Related plants (not illustrated):

Cotoneaster congesta. A low Himalayan shrub 1–3 feet high, with solitary white flowers and bright red fruit. Hardy from Zone 5 southward.

Cotoneaster conspicua. An evergreen Chinese shrub 4–6 feet high, the oblongish leaves 3–5 inches long and softly prickle-tipped. Flowers white, solitary, nearly 1/2 inch wide. Fruit pea-size, bright red, and winter-persistent. Hardy from Zone 5 southward.

Trees and Shrubs with Compound Leaves

(For a definition and illustration of a compound leaf see the Picture Glossary.)

THERE are a few plants with compound leaves that will be found elsewhere because their affinities are obviously with their simple-leaved relatives. The Box-Elder is such an exception and will be found at MAPLE (see p. 168). Disregarding such exceptions, which have always been noted wherever they occur, the shrubs and trees below all have compound leaves and may be separated thus:

Petals usually 6. Barberry Family (See p. 293)
Petals 4. See p. 298
Petals prevailingly 5 (often much more in double-flowered roses).
Plants with prickles, bristles, spines, or thorns on the stem or leaves or on both. (Exceptions are the Mexican Orange and the Cork-Tree.) See p. 300
Plants unarmed.
Leaves opposite. See p. 314
Leaves alternate.
Leaves twice-compound. (For definition and illustration of twice-compound see the Picture Glossary.) Chinaberry
Leaves only once-compound.
Leaflets more than 5, usually much more. See p. 314
Leaflets 3 or 5.
Leaflets 3. Hop-Tree
(NOTE: *There are 2 species of* Rhus *that have only 3 leaflets. For these exceptions see* SUMAC *at p. 318.*)
Leaflets usually 5.
Leaflets arranged finger-fashion. *Acanthopanax*
Leaflets arranged feather-fashion.
Flowers brownish purple. Yellowroot
Flowers yellow. Hardhack

CHINABERRY (*Melia Azedarach*)
Family Meliaceae p. 294
A showy Chinese and Himalayan tree much grown in the South for its spectacular bloom. **Size:** 20–50 feet high, three fourths as

wide, its bark furrowed. **Leaves:** Twice-compound, 15–30 inches long, composed of many ovalish leaflets 1–2½ inches long, sharply toothed or even lobed. **Flowers:** In compound clusters 5–8 inches long, profuse and showy, the 5 petals lilac or purplish (rarely white); Apr.–May. **Fruit:** Fleshy, yellow, about ¾ inch thick, persistent after leaf-fall and a favorite of birds. **Hardy:** From Zone 6 southward. **Varieties:** Texas Umbrella Tree (variety *umbraculifera*) has drooping foliage on erect, crowded branches which spread from the trunk like spokes, giving an umbrella-like outline. It is hardy up to Zone 5. **Culture:** Easy in any dry, sandy soil in a region of much summer heat. Not long-lived.

HOP-TREE (*Ptelea trifoliata*). Family Rutaceae p. 294

A medium-sized native shrub or small tree, grown mostly for its moderately decorative fruit as its flowers are inconspicuous. **Size:** 10–20 feet high, about half as wide. **Leaves:** Compound, composed of 3 elliptic or oblong leaflets, 2–4½ inches long, essentially stalkless, faintly dotted (strong-smelling when crushed), toothless or obscurely toothed. **Flowers:** Greenish and inconspicuous. **Fruit:** Dry, veiny, papery, greenish, nearly round and broadly winged, about ¾ inch long, its 2 seeds plump. **Hardy:** From Zone 2 southward. **Varieties:** None. **Culture:** Prefers a reasonably good garden soil, preferably a moist one, and partial shade. Not an important ornamental.

ACANTHOPANAX (*Acanthopanax sieboldianus*)
Family Araliaceae p. 294

An essentially unarmed Japanese shrub (rarely with a few prickles) and a very handsome shrub for the lawn. **Size:** 5–9 feet high, about as wide. **Leaves:** Compound, composed of 5–7 leaflets that are arranged fan-fashion, and are wedge-shaped, nearly stalkless, 1–2½ inches long, and so long-persistent in the late fall that the shrub is a valuable foliage plant. **Flowers:** Greenish, inconspicuous, often wanting; July–Aug. **Fruit:** Lacking in all cultivated plants. **Hardy:** From Zone 4 southward. **Culture:** Easy in any ordinary garden soil, and as it stands city smoke, dust, and wind, it is a valuable foliage plant for such places and in better gardens.

YELLOWROOT (*Xanthorhiza simplicissima*)
Family Ranunculaceae p. 294
A medium-high ground cover, native in the eastern U.S., grown mostly for its foliage as the flowers and fruit are negligible. **Size:** Nearly uniformly 24 inches high, the bark and root yellow. **Leaves:** Compound, composed of 5 ovalish, deeply cut or even lobed leaflets, 2–4 inches long. **Flowers:** Brownish, small, inconspicuous; Apr.–May. **Fruit:** Dry, pod-like, 1-seeded. **Hardy:** From Zone 3 southward. **Varieties:** None. **Culture:** Requires a moist, shady place, free of wind; it spreads readily and makes large patches.

HARDHACK (*Potentilla fruticosa*). Family Rosaceae p. 294
A low, long-blooming native shrub, suitable for limestone areas, but easily grown in most gardens. **Size:** 1–4 feet high, much branched, and nearly as wide. **Leaves:** Compound, composed of 5 (rarely 3 or 7) lance-shaped leaflets about 1 inch long, covered with short, silky hairs, the margins slightly rolled. **Flowers:** Many, yellow, in small clusters, the petals 5; May–Aug. or even later. **Hardy:** From Zone 2 southward. **Varieties:** Over 30 of which the best are: *purdomi,* with pale yellow flowers; *veitchi,* with cream-white flowers; *mandshurica,* having leaves with dense white hairs on both sides. **Culture:** Grows naturally in limestone pastures in the North, but will grow in any ordinary garden soil, in full sun.

BARBERRY FAMILY
(*Berberidaceae*)

Most of the plants of this family are the barberries themselves, which have simple leaves and are hence found at p. 225. But the 2 genera below have alternate, compound leaves, the leaflets either spiny-toothed or wholly without teeth. Flowers with 3 or 6 petals. Fruit fleshy. The 2 genera are:

Flowers white; leaflets without teeth. Nandin
Flowers yellow; leaflets spiny-toothed. *Mahonia*

BARBERRY FAMILY (*Berberidaceae*), 1, 6
RUE FAMILY (*Rutaceae*), 2, 4
3, 5, 7, 8, 9 in various families

1. ***Mahonia bealei*** p. 297
A tall evergreen Chinese shrub much grown for its handsome glistening foliage and yellow flowers.

2. **Hop-Tree** (*Ptelea trifoliata*) p. 292
A native shrub or small tree grown chiefly for its papery, greenish fruit. Quite hardy.

3. **Chinaberry** (*Melia Azedarach*) p. 291
Not hardy northward but much grown in the South for its superb flower clusters.

4. ***Evodia danielli*** p. 299
An aromatic, medium-sized Asiatic tree, its midsummer flowers liked by bees.

5. ***Acanthopanax sieboldianus*** p. 292
A Japanese shrub, much used as a lawn foliage plant, since it frequently lacks both flowers and fruit.

6. **Nandin** (*Nandina domestica*) p. 296
An Asiatic shrub with compound leaves, white flowers, and persistent red fruit. Not hardy northward.

7. **Golden-Rain Tree** (*Koelreuteria paniculata*) p. 299
A summer-blooming, medium-sized Asiatic tree, its yellow flowers followed by persistent fruit.

8. **Yellowroot** (*Xanthorhiza simplicissima*) p. 293
A small native shrub grown mostly for its attractive foliage because its flowers are not showy.

9. **Hardhack** (*Potentilla fruticosa*) p. 293
A low, long-blooming native shrub, hardy everywhere but partial to limestone regions.

1
2
3
4
5
6
7
8
9

1
2
5
4
3
6
7
9
8

GINSENG FAMILY (*Araliaceae*), 1
ROSE FAMILY (*Rosaceae*), 2–9

1. **Hercules-Club** (*Aralia spinosa*) p. 300
   A medium-sized native tree, so prickly that it repels intruders. Leaves large and showy.

2. **Father Hugo's Rose** (*Rosa hugonis*) p. 303
   A beautiful single-flowered yellow rose from China, easily grown in poorish soils.

3. **Damask Rose** (*Rosa damascena*) p. 305
   One of the old-fashioned "species" roses, with extremely fragrant flowers.

4. ***Rosa rugosa*** p. 307
   A rough-leaved, very sturdy Asiatic rose that stands wind and exposure.

5. **Provence Rose** (*Rosa gallica*) p. 304
   Extremely fragrant "species" rose, cultivated for centuries for its persistent scent.

6. **Hybrid Rose Crimson Glory** p. 311
   A shrub, 3–4 feet high, its bloom often tending to repeat after first flowering.

7. **Hybrid Rose Golden Dawn** p. 311
   A yellow-flowered hybrid rose, usually 3–4 feet high. Not apt to repeat its bloom.

8. **Hybrid Rose Dr. Walter van Fleet** p. 311
   A splendid climbing rose, often 15–20 feet high, its flowers pink or blush.

9. **Hybrid Rose Souvenir de la Malmaison** p. 311
   A somewhat repeating hybrid rose, 4–5 feet high, its flowers flesh-pink.

NANDIN (*Nandina domestica*) p. 294
An upright, stiffish, Chinese and Japanese evergreen shrub, commonly called Sacred Bamboo although it has nothing to do with any bamboo. **Size:** 6–8 feet high, often more in the South, rather narrow and with upright stems. **Leaves:** Twice- or thrice-compound, very graceful, the ultimate leaflets narrow, 1–2 inches long, toothless, and very handsome in their red color in the fall. **Flowers:** Small, white, the petals not showy, but the branched cluster nearly a foot long and handsome; June–July. **Fruit:** A small 2-seeded red berry, persistent until Christmas, and the chief attraction of the plant. **Hardy:** From Zone 5 southward, often in protected places in Zone 4 if mulched over the winter with a 2-inch layer of manure. If killed to the ground such mulched plants usually send up new shoots in the spring. **Culture:** Easy in most garden soils, but it does better if the site is not too windy, and if in Zone 4 it is protected from north and west winds.

MAHONIA
(*Mahonia*)

Beautiful, evergreen shrubs with alternate, compound leaves, the leaflets generally spiny-toothed on the margin. Flowers yellow, in profuse clusters that arise at the sides of the stem, rarely terminal. Petals 6. Fruit berrylike, usually with a bloom, prevailingly blue.

The mahonias, which include the Oregon Grape, are sometimes low shrubs suitable as ground covers in sheltered, shady places, but 3 of them are rather tall. They are not particular as to soil, but resent exposure to the wind and sun in winter, and thus do better if there is a good snow blanket. Over 45 species are known, of which the following are most popular:

Shrubs tall, at least 4–10 feet high.
  Flower clusters 3–4 inches long. — California Barberry
  Flower clusters 5–10 inches long. — *Mahonia bealei*
Low shrubs, 1–3 feet high. — Oregon Grape

CALIFORNIA BARBERRY (*Mahonia pinnata*) p. 298
A Pacific Coast evergreen shrub, thought by some to be only a variety of the Oregon Grape, but consistently higher. **Size:** 8–12 feet high, the stems erect and often leafless below. **Leaves:** Compound, usually in tufts toward the top, the 7–13 leaflets ovalish, spiny-toothed, evergreen, handsome, shiny above, paler beneath. **Flowers:** Pale yellow, the clusters at the leaf joints, 3–4 inches long; June. **Fruit:** Nearly round, purplish black, not persistent. **Hardy:** From Zone 6 and probably from Zone 5 southward. **Varieties:** None. **Culture:** See above. Best suited to the Pacific Coast.

MAHONIA BEALEI p. 294
A beautiful evergreen Chinese shrub much grown for its splendid glistening foliage. **Size:** 8–12 feet high, rather narrow because of its stiff erect branches. **Leaves:** Compound, leathery, composed of 9–15 round-oval leaflets, with a few large teeth on the margin, the end leaflet larger than the lateral ones, bluish green above and with a slight bloom beneath. **Flowers:** Yellow, fragrant, in erect clusters about 6 inches long; Apr.–May. **Fruit:** Bluish black, not persistent. **Varieties:** None. **Culture:** See above.

Related plants (not illustrated):

Mahonia fortunei. A Chinese shrub, 3–5 feet high, its leaflets 6–18, the margins with 5–10 teeth. Flowers yellow. Hardy from Zone 5 southward.

Mahonia nervosa. Also known as Oregon Grape; a shrub 1–2 feet high, the 11–19 ovalish leaflets with spiny teeth on the margin. Flowers yellow, the clusters 8 inches long. Hardy from Zone 4 southward.

Mahonia japonica. Quite similar to *Mahonia bealei,* but the flower clusters drooping. Hardy from Zone 4 southward.

OREGON GRAPE (*Mahonia Aquifolium*) p. 298
One of the best tallish ground covers for shady places and a native of the Pacific Coast. **Size:** 1–3 feet high, sometimes more if not clipped, and spreading by underground stems to make large patches. **Leaves:** Evergreen, compound, holly-like, the 5–9 ovalish

leaflets shiny, stiff, the marginal teeth spiny. **Flowers:** Yellow, fragrant, in dense, erect terminal clusters, about 3 inches high; Apr.–May. **Fruit:** A small, bluish, edible berry, not persistent. **Hardy:** From Zone 3 southward. **Varieties:** Several, but scarcely superior to the typical form. **Culture:** See above. It makes splendid, relatively low-growing patches of evergreen foliage and prefers shady or partly shady places.

Related plant (not illustrated):

Creeping Barberry (*Mahonia repens*). Closely related to the above but only about 1 foot high, the leaflets 3–7. Fruit black, with a bloom. Hardy from Zone 3 southward. Not so handsome as *Mahonia Aquifolium*.

Left: California Barberry (*Mahonia pinnata*), p. 297.
Right: Oregon Grape (*Mahonia Aquifolium*), p. 297.

Petals 4

WHILE the shrubs of the Barberry Family have 3 or 6 petals in each flower, the 2 plants below have only 4 petals in each flower.

The 2 may be distinguished thus:

Flowers white; leaflets finely toothed. Evodia danielli
Flowers yellow; leaflets distinctly lobed toward the base. Golden-Rain Tree

EVODIA (*Evodia danielli*). Family Rutaceae p. 294
A welcome midsummer-blooming tree from China and Korea with aromatic foliage, its flowers much favored by bees. **Size:** 15–25 feet high, about three fourths as wide, often shrublike, its foliage thin. **Leaves:** Compound, composed of 7–11 more or less ovalish leaflets that are 2–4½ inches long with very finely toothed margins. **Flowers:** Scarcely ⅛ inch wide but grouped in a profuse cluster nearly 8 inches wide, whitish; July–Aug. **Fruit:** A collection of small, hooked pods. **Hardy:** From Zone 3 southward. **Culture:** Easy in any ordinary garden soil, preferably in full sun.

Related plant (not illustrated):

Evodia hupehensis. Closely related to the above, but a tree up to 60 feet high. Hardy from Zone 3 southward.

GOLDEN-RAIN TREE (*Koelreuteria paniculata*)
Family Sapindaceae p. 294
A beautiful summer-blooming Asiatic tree, often called Varnish-Tree and China-Tree. **Size:** 20–30 feet high, nearly as wide, with a domelike outline. **Leaves:** Compound (rarely twice-compound), 9–14 inches long, composed of 7–15 ovalish-oblong leaflets that are coarsely toothed and even deeply cut on the margin and without autumnal color. **Flowers:** Small, yellow, with 4 petals, arranged in a showy cluster 12–18 inches long; July. **Fruit:** A papery pod about 2 inches long, yellow, in long, loose clusters and showy in late summer. **Hardy:** From Zone 3 southward. **Varieties:** *fastigiata* has a narrower crown. **Culture:** Easy in any ordinary garden soil. Valuable as one of the few yellow-flowered trees.

Petals Prevailingly 5 (Often More in Double-flowered Roses). Plants with Prickles, Bristles, Spines or Thorns on the Stem or Leaves or on Both. (Exceptions Are the Mexican Orange and the Cork-Tree.)

Leaves twice-compound. — *Hercules'-Club*
  (For definition and illustration of twice-compound see the Picture Glossary.)
Leaves only once-compound.
  Leaves with obvious stipules. — Rose
    (For definition and illustration of stipules see the Picture Glossary.)
  Leaves without stipules. — Rue Family

HERCULES'-CLUB (*Aralia spinosa*)
Family Araliaceae p. 295

To form an impenetrable prickly barrier, few plants exceed the Hercules'-Club, a medium-sized native tree, often shrubby. **Size:** 8–40 feet high, nearly as wide due to its spreading branches, its twigs and trunk armed with stout spines. **Leaves:** Twice-compound, very long-stalked, 15–30 inches long, its ultimate leaflets ovalish, about 4 inches long and very prickly. **Flowers:** Small, greenish white, in an immense compound cluster that may be 15–25 inches wide; Aug.–Sept. **Fruit:** Black, berrylike, pea-size, borne in large compound clusters, rather showy but not persistent. **Hardy:** From Zone 3 southward. **Varieties:** None. **Culture:** Growing naturally in moist or swampy woods, it prefers a moist site in the garden, but thrives also in ordinary soil and will even stand city conditions.

Related plant (not illustrated):

Angelica Tree (*Aralia elata*). A Japanese shrub or small tree, 10–25 feet high, very prickly. Leaves twice-compound, the leaflets narrowly elliptic, about 9 inches long. Flowers whitish. Fruit black. Hardy from Zone 4 southward.

Rose
(*Rosa*)
Family Rosaceae

An enormous group of prickly or bristly shrubs, stretching from the wild, single-flowered species of our roadsides to the much-hybridized horticultural forms. These, for most amateurs, are *the* rose. Such a concept is completely incorrect, for the basis of our much-doubled rose is, or was, always the wild species from which, by selection and hybridization, all our modern horticultural forms (perhaps 3000) have been derived. The horticultural forms of the rose, the main classes into which they are divided, and their culture, will be found at the end of the enumeration of the *species* of *Rosa* below.

Rosa, as a genus, comprises perhaps 200 wild species of which over 75 are known to be in cultivation in this country. Of these only a few basic types will be found below—all of them being what are known as "species roses," which means that most of them are not double-flowered, and some of them are extremely fragrant, a quality that they have handed down to some of their modern descendants.

Leaves always compound, the leaflets arranged feather-fashion, with an odd leaflet at the tip, the stems always prickly or bristly. The base of the leafstalk is provided with a pair of conspicuous stipules, which often hang on for a considerable time. (Many stipules are apt to be fleeting. See the Picture Glossary.) Petals normally 5, followed by the rose "hip," a fleshy covering of the true fruit, which is found inside and is small, seedlike, and bony. From the species roses below we have had to exclude the wild roses of eastern North America, for they are little cultivated and the seeker should consult Dr. H. A. Gleason's treatment of them in the new edition of Britton & Brown's *Illustrated Flora,* issued by the New York Botanical Garden in 1952. All the plants below are in cultivation for their beauty, or are included here because they are the ancestors of many modern horticultural forms of the rose.

These species roses are much less demanding in their care than the horticultural forms at the end of this enumeration. The species roses will grow in any ordinary, preferably rich, garden soil, need little or no pruning, and do not need to be cut back to nearly ground level each fall.

The species are not easy to identify, but the following key may help:

Flowers solitary (in small clusters in the damask rose).
- *Flowers generally white or yellow; leaflets 5–11 or even more.*
  - *Twigs prickly and very bristly.* Scotch Rose
  - *Twigs prickly, but only slightly bristly.*
    - *Leaflets simply toothed.* Father Hugo's Rose
    - *Leaflets doubly toothed.* Harison's Yellow Rose
- *Flowers pink or red; leaflets 3–5.*
  - *Prickles of unequal size: leaflets doubly-toothed.*
    - *Flower stalk erect.* Provence Rose
    - *Flower stalk nodding.* Cabbage Rose
  - *Prickles essentially uniform.* Damask Rose

Flowers not solitary, usually in small clusters.
- *Plants climbing, clambering, or nearly prostrate.*
  - *Plant nearly prostrate, almost evergreen; leaves smooth and shining.* Memorial Rose
  - *Plants climbing or clambering, not prostrate; leaves hairy.* Multiflora Rose
- *Plants erect, never climbing or prostrate.*
  - *Leaves with deeply impressed veins on the upper side.* *Rosa rugosa*
  - *Leaves not so.*
    - *Leaflets 3–5, roundish, smooth.* Sweetbrier
    - *Leaflets 5–7, oblongish, hairy beneath.* China Rose

SCOTCH ROSE (*Rosa spinosissima*) below

A very prickly and bristly Eurasian rose with solitary but very numerous flowers. **Size:** 2–3 feet high, about as wide. **Leaves:** Compound, composed of 7–9 (rarely 5 or 11) roundish leaflets ½–¾ inch wide. **Flowers:** Solitary, but profuse, about 2 inches wide, typically single; white, yellow, or more rarely pink; June. **Fruit:** (hip): Nearly round, dark brown or black, about ½ inch thick, not showy. **Hardy:** From Zone 3 southward. **Varieties:** Many, and the most desirable are: *altaica,* pale yellow flowers 3 inches wide; *alba-plena,* double white flowers; *lutea,* bright yellow flowers and larger leaflets than the type; *hispida,* twice as high as the type, with yellow flowers. **Culture:** See above.

Left: A double-flowered form of Harison's Yellow Rose (*Rosa harisoni*), p. 304.
Right: Scotch Rose (*Rosa spinosissima*), above.

FATHER HUGO'S ROSE (*Rosa hugonis*) p. 295

One of the finest single-flowered yellow roses, coming originally from Central China and first collected by Father Hugo. **Size:** 6–8 feet high, as wide or wider due to its spreading branches, its twigs prickly and a little bristly. **Leaves:** Compound, composed of 5–13 ovalish or elliptic leaflets ½–¾ inch long, simply toothed on the margin. **Flowers:** Solitary but very profuse, single, yellow, about 2 inches wide; May. **Fruit** (hip): Dark red or scarlet, nearly round, about ¾ inch thick, not persistent. **Hardy:** From Zone 3 southward. **Varieties:** None. **Culture:** See above. It will stand fairly poor soils and should not be manured.

HARISON'S YELLOW ROSE (*Rosa harisoni*) p. 303
A hybrid shrub derived from crossing the Scotch Rose and *Rosa foetida.* **Size:** 3–6 feet high, nearly as wide. **Leaves:** Compound, composed of 5–9 broadly oval leaflets ¾–1½ inches long, doubly toothed. **Flowers:** Solitary, pale yellow, often somewhat doubled, about 2 inches wide; June. **Fruit:** Blackish red, globular, not showy. **Hardy:** From Zone 3 southward and it will stand exposure to bitter winds. **Varieties:** None. **Culture:** See above.

Related plant (not illustrated):

Austrian Brier (*Rosa foetida*). An Asiatic shrub, one of the parents of Harison's Yellow Rose, 7–10 feet high. Leaflets 5–9. Flowers yellow, nearly 3 inches wide, unpleasantly scented. Hardy from Zone 3 southward, but not an easy rose to grow. A double-flowered form, the variety *persiana,* is the Persian Yellow Rose.

PROVENCE ROSE (*Rosa gallica*) p. 295
A very fragrant treasure from southwestern Europe and adjacent Asia, and the source of many varieties used in France for making attar of roses. **Size:** 3–4 feet high, about three fourths as wide, the stems densely prickly, the prickles of unequal size. **Leaves:** Compound, composed of 3–5 elliptic leaflets 2–2½ inches long, doubly toothed on the margin. **Flowers:** Erect, solitary, nearly 3 inches wide, pink or crimson, single, very fragrant, the petals retaining their odor when dry better than most roses; June. **Fruit** (hip): Nearly round, brick-red, persistent into Nov. **Hardy:** From Zone 3 southward. **Varieties:** *officinalis,* the Apothecary's Rose, is double-flowered and has been cultivated in Europe for centuries. **Culture:** See above.

CABBAGE ROSE (*Rosa centifolia*) p. 306
Often called the Hundred-leaved Rose, this prickly and bristly shrub from the Caucasus was known in cultivation before the

Christian era. **Size:** 4–6 feet high, its creeping rootstocks often resulting in considerable patches. **Leaves:** Compound, composed of usually 5 leaflets 1½–2 inches long and hairy both sides. **Flowers:** Nodding, solitary, very fragrant, greatly doubled, about 3 inches wide; June–July. **Fruit** (hip): Football-shaped, red. **Hardy:** From Zone 3 southward. **Varieties:** The Moss Rose (variety *muscosa*) has the flower stalks and the leafy part of the flower covered with a sticky-mossy coating. **Culture:** See above.

DAMASK ROSE (*Rosa damascena*) p. 295

An extremely fragrant rose cultivated in Europe as a source of attar of roses, and probably a native of Asia Minor. **Size:** 5–7 feet high, the stems with hooked prickles, and often also bristly, the prickles essentially uniform. **Leaves:** Compound, composed of 5 (rarely 7) ovalish or oblong leaflets 2–2½ inches long, simply toothed on the margin. **Flowers:** Solitary or in small clusters, about 2½ inches wide, pink or red, double, very fragrant; June–July. **Fruit:** Red, bristly, about 1 inch long, not persistent. **Hardy:** From Zone 4 southward. **Varieties:** York and Lancaster Rose (variety *versicolor*) has white-striped, semidouble flowers, or some white and some pink flowers on the same plant. **Culture:** See above.

Related plant (not illustrated):

Rosa alba. A probable hybrid of chief interest as one of the parents of many long-cultivated roses. It is a shrub 4–6 feet high, with usually 5 leaflets. Flowers double or semidouble, white or pink, about 3 inches wide, fragrant. Hardy from Zone 4 southward.

MEMORIAL ROSE (*Rosa wichuraiana*) p. 306

A nearly prostrate, almost evergreen rose from Asia and the parent of such old favorites as Dorothy Perkins, Dr. Walter Van Fleet, and many fine modern varieties. **Size:** Trailing or nearly prostrate, the strong prickles hooked. **Leaves:** Almost evergreen, compound, composed of 7–9 roundish, blunt, shining green, leathery leaflets ¾–1 inch long. **Flowers:** Nearly 2 inches wide, white, fragrant, in clusters; July–Oct. **Fruit:** Egg-shaped, about ½ inch long, reddish, persistent until Mar. **Hardy:** From Zone 3 southward. **Varieties:** None except the hybrid forms derived from it, which are many. **Culture:** See above.

Related plant (not illustrated):

Tea Rose (*Rosa odorata*). A half-evergreen, erect Chinese shrub with tea-scented foliage. Leaflets 5–7, ovalish, 2–3 inches long, shining green above. Flowers solitary or in small clusters, white, pink or yellow, usually double, about 3 inches wide. Hardy from Zone 6 southward and not much grown, but the origin of many fine hybrids.

Left: Memorial Rose (*Rosa wichuraiana*), p. 305.
Center: Multiflora Rose (*Rosa multiflora*), below.
Right: Cabbage Rose (*Rosa centifolia*), p. 304.

MULTIFLORA ROSE (*Rosa multiflora*) above

An immensely vigorous and very prickly rose from Japan and Korea, now widely used to make a practically impenetrable fence or loose hedge, its stems scrambling and needing support while young. **Size:** 5–8 feet high, if supported, otherwise trailing, its vicious prickles hooked. **Leaves:** Compound, composed of 5–9 oblongish leaflets 1–1½ inches long. **Flowers:** Prevailingly white (but see below), not over ¾ inch wide, fragrant and arranged in many-flowered, very profuse clusters; July–Aug. **Fruit** (hip): Small, nearly globular, red, persistent through the winter, and much favored by birds. **Hardy:** Everywhere. **Varieties:** Many named forms, often pink or red, one of them being the old favorite Crimson Rambler. **Culture:** See above. It is perhaps the easiest of all roses to grow, and planted about 2–3 feet apart it will, in a few years, make a farm fence which is hogproof.

ROSA RUGOSA p. 295

An upright, stiffish Asiatic shrub which if left alone will make patches 10–20 feet wide in a few years. **Size:** 4–6 feet high, its stems densely prickly and bristly; about as wide. **Leaves:** Compound, composed of 5–9 elliptic leaflets 1½–2 inches long, the veins conspicuously impressed on the upper side, the surface rough. **Flowers:** Nearly 3½ inches wide, red or white, in few-flowered clusters, or solitary; July–Aug. **Fruit:** Red, smooth, not persistent. **Hardy:** From Zone 3 southward. **Varieties:** Many named forms, few of them any better than the typical plant. **Culture:** See above. It is the best of all roses for exposed windy sites, especially along the coast, where it often escapes into sandy or indifferent soils.

SWEETBRIER (*Rosa Eglanteria*) p. 308

Often called the Eglantine, this much-branched European shrub has pleasantly aromatic foliage and its twigs were once used (in England) to make sweetmeats. **Size:** 5–8 feet high, about as wide, its stems beset with hooked prickles and bristles. **Leaves:** Compound, composed of 5–7 nearly round, sticky leaflets about 1 inch long. **Flowers:** About 2 inches wide, in small clusters or sometimes solitary; June. **Fruit** (hip): Orange or scarlet, persistent through Nov. **Hardy:** Everywhere. **Varieties:** None readily available, but a double-flowered form is sometimes listed.

Culture: See above. Commonly grown for its aromatic foliage, and an occasional escape in the eastern U.S.

Left: China Rose (*Rosa chinensis*), below.
Right: Sweetbrier (*Rosa Eglanteria*), p. 307.

CHINA ROSE (*Rosa chinensis*) above

A protean source of many garden roses, this long-flowering, half-evergreen Chinese shrub, which is unsuited to the North, has been one of the parents of innumerable varieties. **Size:** 2–3 feet high, nearly as wide, its stems with hooked prickles or sometimes practically unarmed. **Leaves:** Compound, composed of 3–5 oval-ish, shining green leaflets 2–2½ inches long. **Flowers:** About 2 inches wide, solitary or in long-stalked, few-flowered clusters, crimson to pink or even white; July–Oct. **Fruit** (hip): Scarlet, about ¾ inch long. **Hardy:** From Zone 6 southward. **Varieties:** Many, of which the most important are: *semperflorens,* the Chinese Monthly Rose, which has usually solitary, long-stalked crimson or dark pink flowers; *minima,* the Fairy or Miniature Rose, a low shrub with very small, mostly single flowers. Related to it is the so-called Baby Rambler; *viridiflora,* the Green Rose, has all its petals like small green leaves. **Culture:** See above.

Related plant (not illustrated):

Noisette Rose (*Rosa noisettiana*). An erect hybrid shrub, 7–10 feet high. Leaflets 5–7. Flowers white, pink, red, or yellow in many-flowered clusters. Hardy from Zone 6 southward.

Hybrid Roses

While the species roses enumerated above are of comparatively easy culture, require little or no pruning, and usually flower upon stems that persist above ground over the winter, scarcely any of this is true of the hybrid roses, of which nearly 3000 varieties are known, divided into several classes.

It is obviously impossible to deal with such a mass of horticultural forms of the rose in a book like this, and for an extensive review of them the seeker is referred to the article Rose, by Roy E. Shepherd, in the fourth edition of *Taylor's Encyclopedia of Gardening,* issued in 1961.

These hybrid roses, which, for most gardeners, are the only ones they know, have been developed over the years by many breeders in France, Holland, England, and here. They are far finer than the species roses noted above, but their culture is more demanding. The suggestions of Mr. Shepherd regarding their culture are summarized below:

(1) Choose a site with good rich garden soil. If this is impossible, dig out all poor soil and fill the hole with good soil and let it settle. See that the hole is big enough to allow the roots to be spread without coiling or injury and put the plant at about the same level it was when growing in the nursery, or slightly lower if the climate is severe. In planting fill in a little soil over the roots (about 4 inches) and pour half a bucket of water in the hole before filling the hole with the balance of the soil. This water will help obviate air pockets around the roots, and when the water has seeped away, the rest of the soil can be filled in up to ground level. Sprinkle a little fertilizer (a 5–10–5 fertilizer is fine) over the ground, but do not let it touch the stems. A cupful is enough.

(2) If planting is done in the fall, make a mound of soil high enough to cover the stems (usually 10–12 inches) and leave it on all winter, taking it off gradually in the spring when the danger of bitter frost is over. If planted in the spring, the soil mound is still highly effective; leave it on for about a fortnight.

(3) When the mounds are removed, mulch the plants with a porous mulch (buckwheat hulls, well-rotted cow manure, straw, or grass cuttings) about 2½ inches deep, keeping the mulch slightly away from the crown of the plant (to reduce disease hazard). Renew the mulch each spring.

(4) For established plants a yearly application of fertilizer is useful, at the same rate as that at planting time, applied in the spring when the mounds are removed.

(5) Roses do not like heat or dryness. Hence water them if you live where such conditions prevail. Never let them dry out.

(6) When spring growth starts, cut away all stems that have been frost-injured, for they will never produce leaves or flowers. As spring advances see that all dead wood is subsequently removed (if undetected earlier) and little other pruning is necessary except to remove long or leggy shoots in order to keep the bush shapely.

(7) Late in each fall cut all stems down to 6–12 inches from ground level and mound up the soil enough to cover all stems during the winter.

VARIETIES

These are so numerous that it is impossible to enumerate them here, and to make a choice it is better for the seeker to consult the catalogs of reputable rose dealers. A list of 15 such, chosen for their reliability and geographical distribution, might include:

Alameda Nursery, Denver 19, Colo.
Armstrong Nurseries, Ontario, Calif.
A R P Nursery, Tyler, Tex.
Cherry Hill Nurseries, West Newbury, Mass.
Conard-Pyle Co., West Grove, Pa.
D & D Rose Gardens, Eatontown, N.J.

Holdridge Farm Nursery, Gales Ferry, Ledyard, Conn.
Home Nursery, Lafayette, Ill.
Jackson & Perkins, Newark, N.Y.
Johnson Nurseries, Kingston, Ontario, Canada
Myers Nursery, Waterloo, Iowa
Roseway Nurseries, Beaverton, Ore.
Stark Nursery, Midland, Mich.
Thomasville Nurseries, Thomasville, Ga.
Wayside Nurseries, Mentor, Ohio

To select only 9 of the readily available varieties of these hybrid roses, all chosen from the lists in Mr. Shepherd's article in *Taylor's Encyclopedia of Gardening,* the following may be considered. Four of them are illustrated in color at p. 295.

Cardinal de Richelieu. Once-blooming. Dark rich violet; 6–8 ft.
Celestial. Once-blooming. Pale pink to light blush; 5–6 ft.
Crimson Glory. Somewhat repeating. Red; 3–4 ft.
Dr. Walter Van Fleet. Climbing to 20 ft.; pink.
George Arends. Somewhat repeating. Pink; 3–4 ft.
Golden Dawn. Somewhat repeating. Yellow; 3–4 ft.
Margo Koster. Once-blooming. Light orange, 15–18 in.
Souvenir de la Malmaison. Somewhat repeating. Flesh-pink; 4–5 ft.
White Dawn. Repeated flowering. White; 6–8 ft.

RUE FAMILY
(*Rutaceae*)

Trees or shrubs with alternate or opposite leaves that are compound in those below, but simple in *Skimmia,* for which see p. 259. The leaves are usually dotted with glands and the crushed foliage may be highly aromatic in some. They have no stipules. The stems are unarmed in the Mexican Orange and Cork-Tree, but prickly or thorny in the other groups. Petals 5 (rarely 4 as in *Evodia,* for which see p. 299). Fruit dry in all those below, except the Trifoliate Orange, which has small orange-like fruit, and the Cork-Tree.

Leaflets 3.

Leaves opposite and evergreen; not spiny. Mexican Orange

Leaves alternate, not evergreen; thorny tree. Trifoliate Orange

Leaflets 5–13.

Leaves alternate; fruit dry; stems prickly. Prickly Ash

Leaves opposite; fruit fleshy; stems not prickly. Cork-Tree

MEXICAN ORANGE (*Choisya ternata*) p. 326
A beautiful evergreen Mexican shrub, the foliage highly aromatic when crushed. **Size:** 6–8 feet high, nearly as wide. **Leaves:** Opposite, evergreen, compound, composed of 3 nearly stalkless leaflets 2–3 inches long, without marginal teeth. **Flowers:** White, fragrant, about 1¼ inches wide, arranged in 3–6-flowered slender-stalked clusters; May. **Fruit:** Small, dry, not ornamental. **Hardy:** From Zone 6 southward, but only in favorable sites, precariously so in sheltered places in the southern part of Zone 5. **Varieties:** None. **Culture:** Needs a rich soil, freedom from wind and drought, and full sun.

TRIFOLIATE ORANGE (*Poncirus trifoliata*) p. 326
A viciously thorny Chinese tree, the hardiest of all the citrus fruits although the fruit is inedible; the plant is often called the Hardy Orange. **Size:** 10–20 feet high, nearly as wide, its twigs and thorns green. Thorns stout, very sharp, nearly 3 inches long. **Leaves:** Alternate, compound, composed of 3 wedge-shaped, wavy-margined leaflets, the terminal one larger than the lateral ones, the main leafstalk winged. **Flowers:** White, fragrant (less so than orange blossoms), about 2 inches wide, blooming before the leaves expand; Apr.–May. **Fruit:** A miniature orange, its flesh acid and worthless, about 2 inches in diameter. **Hardy:** From Zone 4 southward. **Varieties:** None. **Culture:** Prefers an acid soil but will grow in any ordinary garden soil. With proper pruning it can be made into impenetrable and formidable hedges, especially in the South.

PRICKLY ASH (*Zanthoxylum americanum*) p. 326

A very prickly native shrub or small tree, its small flowers blooming before the leaves unfold. **Size:** 10–20 feet high, about three fourths as wide, its numerous prickles in pairs and about ½ inch long. **Leaves:** Alternate, compound, composed of 5–13 ovalish leaflets that are about 2 inches long, the leafstalk prickly. **Flowers:** Small, greenish, in small clusters at the leaf joints, not showy; Apr.–May. **Fruit:** Dry, small, each segment with a single black seed. **Hardy:** From Zone 3 southward. **Varieties:** None. **Culture:** Easy in any ordinary garden soil.

Related plant (not illustrated):

Hercules'-Club (*Zanthoxylum Clava-Herculis*). A tree 30–50 feet high, native from Va. southward. Leaflets 7–19. Flowers in rather large terminal clusters. Hardy from Zone 5 southward.

CORK-TREE (*Phellodendron amurense*) p. 326

An irregularly low-branching Asiatic tree, its foliage aromatic when crushed, the twigs not spiny. **Size:** 40–50 feet high, round-headed and widely spreading, the bark deeply fissured and corky. **Leaves:** Opposite, compound, composed of 5–13 ovalish leaflets 2–4 inches long, the whole leaf 10–15 inches long; yellow in the autumn but soon falling. **Flowers:** Small, greenish yellow, in small clusters 2–3 inches wide, not showy and with a turpentine odor; June. **Fruit:** Black, berrylike, shining, about ½ inch thick, in rather profuse clusters, not persistent. **Hardy:** From Zone 2 southward. **Varieties:** None. **Culture:** Easy anywhere. Not a showy tree, but its bark and branching are unusual. Sometimes escaping to the wild.

Shrubs or Trees with Compound Leaves; without Thorns or Prickles. Petals Usually 5. In this Group the Leaflets Are More Than 5, Usually Much More. (There Are Only 3 Leaflets in the Fragrant Sumac and in the Squaw-Bush.)

The 6 genera below are easily separated thus:

Flowers white.
  Shrubs.
    Petals with reddish-yellow blotch at the base. *Xanthoceras*
    Petals pure white. False Spirea
  Trees. Mountain-Ash
Flowers greenish or greenish yellow.
  Foliage evil-smelling; a tall tree. Tree-of-Heaven
  Foliage not evil-smelling (except in the Squaw-bush); shrubs. Sumac

XANTHOCERAS (*Xanthoceras sorbifolium*)
Family Sapindaceae p. 326
A handsome Chinese shrub, not easy to transplant, and not so much grown for ornament as it might be. **Size:** 10–15 feet high, about three fourths as wide. **Leaves:** Alternate, compound, composed of 9–17 narrow, toothed leaflets 1–2 inches long and pale beneath. **Flowers:** White, with a reddish blotch at the base of each petal, arranged in upright, rather showy clusters 6–10 inches long; May. **Fruit:** A greenish, bur-like pod, 2–3 inches thick, not persistent. **Hardy:** From Zone 3 southward. **Varieties:** None.

Culture: Appears to have no soil preferences, but difficult to transplant. Get only ball-and-burlapped specimens and plant in full sun.

FALSE SPIREA (*Sorbaria sorbifolia*)

Family Rosaceae p. 326

An extremely hardy Asiatic shrub, grown for its profusion of summer-blooming, spirea-like flowers. **Size:** 4–6 feet high, about as wide. **Leaves:** Alternate, compound, composed of 13–21 lance-shaped leaflets 2½–5 inches long and doubly toothed on the margin. **Flowers:** Small, profuse, white, borne in dense, erect, branching clusters 6–12 inches long and showy; June–July. **Fruit:** A small dry, hooked pod, not showy. **Hardy:** From Zone 1 southward. **Varieties:** None. **Culture:** Put in full sunshine in any ordinary garden soil. It tends to make large patches from the spreading of its underground stems.

Related plant (not illustrated):

Sorbaria aitchisoni. An Asiatic shrub, closely related to the above, but taller and with larger, later-blooming flower clusters. Hardy from Zone 4 southward.

Mountain Ash
(*Sorbus*)

Family Rosaceae

Rather thin-foliaged, medium-sized trees under which lawn grasses will grow. The mountain-ash is cultivated for its white flowers but mostly for its very showy fruit. Leaves compound (in ours), always with an odd leaflet at the tip, the leaflets toothed. Flowers small, white, but showy as they are borne in a profuse, flat-topped, branching and terminal cluster. Fruit fleshy, brightly colored.

Cultivation of the mountain-ash is simple in any ordinary garden soil, and it will stand dry sites better than most trees. Over 80 species are known, of which perhaps 25 are in cultivation in the U.S. Only the following can be included here:

Petals pointed; fruit about $1/4$ *inch thick.*
American Mountain-Ash
Petals nearly round; fruit nearly $1/2$ *inch thick.* Rowan Tree

AMERICAN MOUNTAIN-ASH (*Sorbus americana*) below
Less grown than the next species, this native tree is extremely showy when in full fruit, and its flowers, while not spectacular, are welcome additions in any garden. **Size:** 20–30 feet high, about half as wide and thin-foliaged. **Leaves:** Alternate, compound, composed of 13–15 narrow, taper-pointed leaflets with sharp marginal teeth; red in autumn. **Flowers:** Small, white, the petals pointed, the cluster flat-topped, terminal and 4–7 inches wide; May–June. **Fruit:** A profuse cluster of red, berrylike fruits about $1/4$ inch thick, persistent through Oct. and Nov. **Hardy:** From Zone 2 southward. **Culture:** See above. It prefers a somewhat acid soil and is less suited to city conditions than the next species.

Related plant (not illustrated):

Sorbus tianshanica. A small Asiatic tree, 10–15 feet high, with 9–15 narrow, toothed leaflets, white flowers, and red fruit. Hardy from Zone 5 southward.

American Mountain-Ash
(*Sorbus americana*), above.

ROWAN TREE (*Sorbus Aucuparia*) p. 326
This, the most grown of the mountain-ashes, is a Eurasian tree well suited to any garden and tolerating city conditions very well. **Size:** 30–50 feet high or even more, about three fourths as wide. **Leaves:** Alternate, compound, composed of 9–15 oblongish leaflets that have a bloom beneath and are 1–2 inches long, the sharp marginal teeth on the upper two thirds of the blade; reddish in the autumn. **Flowers:** Small, white, the petals nearly round, in a terminal, flat-topped cluster 4–6 inches wide; June. **Fruit:** Berrylike, red, about ½ inch thick, in profuse clusters, showy, persistent into Oct. **Hardy:** From Zone 2 southward. **Varieties:** Several, of which the most important are: *asplenifolia,* with deeply cut leaflets and almost fernlike foliage; *edulis,* a form with the fruit edible (in preserves); *xanthocarpa,* a form with yellow fruit. **Culture:** See above. Well suited also to city gardens.

Related plant (not illustrated):

Whitebeam (*Sorbus Aria*). A broad-headed European tree, 30–50 feet high. Leaves simple, ovalish, 2–5 inches long. Flowers white, in flat-topped clusters 2–3 inches wide. Fruit orange to scarlet. Hardy from Zone 3 southward.

TREE-OF-HEAVEN (*Ailanthus altissima*)
Family Simaroubaceae p. 326
Famous because of the novel *A Tree Grows in Brooklyn,* the Stinkweed, as it is more usually known, is *the* tree for impossible city gardens, lots, railway embankments, and other deplorable places; originally a native of China. **Size:** 20–40 feet high, about three fourths as wide, its handsome foliage suggestive of the tropics. **Leaves:** Alternate, compound, 20–30 inches long, composed of 13–25 stalked ovalish leaflets nearly 5 inches long and prominently toothed toward the base; perhaps the most foul-smelling, when crushed, of any shrub or tree. **Flowers:** Small, greenish, the male and female on separate trees, inconspicuous, but foul-smelling in the male flowers. **Fruit:** A collection of winged dry fruits that are rather handsome when rusty red in maturity (on female trees only). **Hardy:** Everywhere. **Varieties:**

None any better than the typical form. **Culture:** Practically impossible to kill, for it stands smoke, dust, heat, wind, and poor soil. In favorable places it can become a pest as it is an invasive tree.

SUMAC
(*Rhus*)
Family Anacardiaceae

Shrubs (rarely small trees) including the pestiferous poison ivy and poison sumac and the few nonpoisonous shrubs below. All those below have compound, alternate leaves, with an odd leaflet at the tip. Flowers small, rather inconspicuous, greenish yellow, in profuse clusters. Fruit (in ours) small, fleshy, red, hairy, in dense, often showy clusters.

Sumacs are mostly for informal shrubberies as some of them are apt to become weedy. Their cultivation is no problem as they will stand dry sites, neglect, and wind. Of the 15 species that are cultivated in the U.S., only those below are of much garden interest:

Leaflets 3. — Fragrant Sumac
Leaflets 9–31.
  Leafstalk not winged.
    Twigs smooth and bluish green. — Smooth Sumac
    Twigs densely brown-hairy. — Staghorn Sumac
  Leafstalk prominently winged. — Dwarf Sumac

FRAGRANT SUMAC (*Rhus aromatica*) p. 319
A sprawling native shrub with pleasantly aromatic foliage but little else of garden interest. **Size:** 18–36 inches high and often nearly prostrate. **Leaves:** Alternate, compound, composed of 3 ovalish, coarsely toothed hairy leaflets 2–3 inches long, turning yellow and red in the fall. **Flowers:** Negligible; Mar.–Apr., before the leaves expand. **Fruit:** Red, hairy, the cluster rather showy, persisting over the winter. **Hardy:** From Zone 2 southward. **Varieties:** None. **Culture:** See above. It is a good, dense low shrub for covering sandy banks.

Related plant (not illustrated):

Squaw-Bush (*Rhus trilobata*). A western relative of the above, with ill-scented foliage, and useful for dry, windy places. Hardy from Zone 3 southward.

Left: Smooth Sumac (*Rhus glabra*), below.
Center: Fragrant Sumac (*Rhus aromatica*), p. 318.
Right: Dwarf Sumac (*Rhus copallina*), p. 320.

SMOOTH SUMAC (*Rhus glabra*) above

A rather coarse shrub, its foliage without any hairs, cultivated only in informal shrubberies, and native nearly throughout North America. **Size:** 8–20 feet high, nearly as wide and often treelike. **Leaves:** Alternate, compound, composed of 11–31 oblongish, toothed leaflets 4–5 inches long, red in the fall. **Flowers:** Negligible. **Fruit:** Red, hairy, the cluster showy, persistent through the winter. **Hardy:** From Zone 2 southward. **Varieties:** *laciniata* has finely dissected leaflets. **Culture:** See above.

STAGHORN SUMAC (*Rhus typhina*) p. 326

A native shrub or small tree, its stems and twigs densely brown-hairy. **Size:** 10–30 feet high, about three fourths as wide. **Leaves:**

Alternate, compound, composed of 11–31 oblongish, toothed leaflets 4–5 inches long, brilliant red in the fall. **Flowers:** Negligible. **Fruit:** A conspicuous, very hairy cluster of small red fruits, persistent through the winter. **Hardy:** Everywhere. **Varieties:** *laciniata* has finely dissected leaflets; *dissecta* is still more finely dissected. **Culture:** See above. It is the best of all the cultivated sumacs.

DWARF SUMAC (*Rhus copallina*) p. 319
A native shrub, the main stalk of its leaf with conspicuous wings. **Size:** 7–15 feet high (often only 5 feet) and about as wide. **Leaves:** Alternate, compound, composed of 9–21 narrow leaflets that are without marginal teeth, 3–4 inches long, brilliant scarlet in the fall. **Flowers:** Negligible. **Fruit:** Crimson, the cluster showy, persistent through the winter. **Varieties:** None. **Culture:** See above. A valuable shrub for dry, sandy places.

5. Trees and Shrubs Having Flowers with the Petals United to Form a Cup-shaped, Bell-shaped, or Funnel-like Corolla

FROM here to the end of this book the shrubs and trees differ from all those previously treated in having a united corolla, which means that the petals are not separate. The division is fundamental in any scheme of classification, and should be familiar to the seeker if he really wants to identify the plant in hand. Unfortunately Nature has injected some exceptions to the general rule of having a united corolla, and these are listed below to avoid the peril of assigning any of these exceptions to categories where they do not belong.

The petals are more or less separate in *Clethra* (see p. 419), *Symplocos* and *Pterostyrax* (see pp. 420 and 421), *Ledum* (see p. 411), *Leiophyllum* (see p. 387), and apparently (but not truly) so in some species of *Azalea* (see p. 397), and there are no petals at all in any of the ash trees (see p. 323) except the Flowering Ash. It is best to look up these exceptions before using the general key to the section which follows. The distinction between flowers having separate petals and a united corolla is illustrated at the Picture Glossary.

Of all the plants below only two have their tiny tubular flowers arranged in a tight head suggesting the garden ageratum or the common boneset of our thickets. Both belong to the huge Daisy Family (the Compositae) which contains thousands of herbs but very few hardy shrubs. Our only two are:

LAVENDER COTTON (*Santolina Chamaecyparissus*)
Family Compositae p. 327

An almost herb-like low plant from southern Europe, too small for the shrubbery but useful in beds or borders. **Size:** 12–18 inches, only a little woody at the base, but persisting above ground over the winter as herbs do not. **Leaves:** Small, silvery gray, scarcely ⅓ inch long, finely dissected. **Flowers:** In small, solitary, relatively long-stalked yellow heads about ¼ inch in diameter; Sept.–Oct. **Fruit:** Negligible. **Hardy:** From Zone 5 southward, and often above this with a light, strawy winter mulch. **Varieties:** None. **Culture:** Needs a sandy, well-drained soil in full sun. It can be sheared to promote more of its finely dissected foliage and then used for edging.

GROUNDSEL BUSH (*Baccharis halimifolia*)
Family Compositae p. 327

While this, a native shrub, is scarcely a garden subject, although it can be easily cultivated, its sheen of feathery, almost snowlike fruiting twigs in Oct. is an unforgettable sight. **Size:** 6–10 feet high, much branched and nearly as wide. **Leaves:** Alternate, simple, oblongish, short-stalked, 1–3 inches long, wedge-shaped at the base, coarsely toothed, more or less resinous. **Flowers:** Minute, inconspicuous, in small heads scarcely ¼ inch wide; Aug.–Sept. **Fruit:** A white, puffy profuse mass of microscopic fruits, so densely produced that, if there are several bushes in a clump, the fruiting mass looks like hills of snow in mid-Oct.; not persistent indoors, where its shedding fluff is a nuisance. **Hardy:** From Zone 4 southward. **Varieties:** None. **Culture:** A native of salt marshes, it can also be grown in full sun in moist garden sites. Commonly called Salt Bush along the coast.

All the rest of the plants in the book can be divided into several easily recognized categories thus:

Leaves alternate. See p. 414

Leaves opposite or clustered; never alternate, except in one species of jasmine.

Leaves compound.
  Tall trees. — Ash
  Shrubs or small trees.
    Leaflets arranged finger-fashion. — Chaste Tree
    Leaflets arranged feather-fashion.
      Leaflets without marginal teeth. — Jasmine
      Leaflets always toothed. — Elder
Leaves simple. — See p. 332

ASH
(*Fraxinus*)
Family Oleaceae

Tall, handsome shade trees with opposite, compound leaves and 5 or more leaflets. In all except the Flowering Ash the flowers are inconspicuous, have no petals, and bloom before the leaves unfold. In the Flowering Ash the petals are white and bloom about the time the leaves unfold. It is the only showy tree in the group. Fruit a small nutlet embedded in a narrow winglike structure.

Most ash trees will not tolerate dry sites, but if the soil is reasonably moist they present no difficulties. Some of them, from self-sown seed, tend to become a pest. Only 4 are of much garden interest:

Petals white; flower cluster showy. — Flowering Ash
Petals none; flowers inconspicuous.
  Twigs smooth; leaflets 5–9. — White Ash
  Twigs hairy. — Green Ash

FLOWERING ASH (*Fraximus Ornus*) p. 327
The only really showy member of the group, this Eurasian round-headed tree makes a handsome lawn specimen, but it needs room. **Size:** 30–50 feet high, nearly as wide. **Leaves:** Opposite, compound, composed of 7 (sometimes 9 or 11) rather broadly oblong leaflets 2½–3½ inches long, irregularly toothed, and with rusty hairs along the midrib beneath. **Flowers:** White, fragrant, in dense terminal clusters; May, and flowering with the expan-

sion of the leaves. **Fruit:** A small nut set in a winglike structure that is about 1 inch long, blunt or notched at the tip. **Hardy:** From Zone 4 southward. **Varieties:** None. **Culture.** See above.

WHITE ASH (*Fraxinus americana*) below
A very common native tree, much planted as a shade tree, perhaps because of its ease of cultivation. **Size:** 60–120 feet high, nearly as wide and densely foliaged; its twigs smooth. **Leaves:** Opposite, compound, composed of 7 (rarely 5 or 9) ovalish leaflets 3–5 inches long, stalked, mostly without marginal teeth; yellow in the fall. **Flowers:** Negligible. **Fruit:** A small nut set in an oblong winglike structure about 1½ inches long, blunt or notched at the tip. **Hardy:** From Zone 2 southward. **Varieties:** None. **Culture:** See above. It is the easiest of all the ash trees to grow in any ordinary (and even poor) garden soil.

Related plant (not illustrated):

European Ash (*Fraxinus excelsior*). Taller than the White Ash, with 7–11 stalkless leaflets that drop in the fall while still green. Hardy from Zone 3 southward.

White Ash (*Fraxinus americana*), above.
Green Ash (*Fraxinus-pensylvanica lanceolata*), below.

GREEN ASH (*Fraxinus pensylvanica lanceolata*) above
A very common native tree, useful in dry, wind-swept places as it is indifferent as to soils. **Size:** 40–70 feet high, nearly as wide, its

foliage dense and its twigs hairy. It differs only slightly from the White Ash, mostly in having 9 leaflets instead of 7. **Hardy:** From Zone 2 southward. **Varieties:** None. **Culture:** See p. 324.

CHASTE TREE (*Vitex Agnus-castus*)

Family Verbenaceae p. 327

A very beautiful European shrub or small tree, valued for its summer-blooming, very fragrant flowers, beloved by bees. **Size:** 8–20 feet high, but often winter-killed in the North and then sending up flowering stems the following season. **Leaves:** Opposite, compound, the 5–7 narrow leaflets arranged finger-fashion, about 4 inches long, grayish-hairy, pleasantly aromatic when crushed. **Flowers:** Small, irregular (one lobe larger than the other 4), arranged in dense showy terminal spikes, lilac-blue, fragrant; July–Aug. **Fruit:** Small, plum-like, about ¼ inch thick. **Hardy:** As a permanent shrub or tree only from Zone 5 southward, but grown north of this where it winter-kills to the ground. **Varieties:** *alba* has white flowers; *macrophylla* has larger leaves and deeper-colored flowers and is the most widely cultivated form. **Culture:** Easy in any ordinary garden soil, but north of Zone 5 it should be cut back in the fall and mulched over the winter as the tops are fairly sure to winter-kill.

Related plant (not illustrated):

Vitex Negundo. An Asiatic shrub, 8–15 inches high, with 3–5 leaflets and deep lavender-blue flowers in Aug.–Sept. Its variety *incisa* has much-cut leaflets and is reasonably safe up to Zone 4.

JASMINE
(*Jasminum*)
Family Oleaceae

Extremely fragrant shrubs, often sprawling, and one is a vine, for which see p. 55. The stems are greenish and often angled. Leaves opposite, compound, the leaflets arranged feather-fashion and wholly without marginal teeth. One of the species (*Jasminum floridum*) breaks the rule by having alternate leaves, and another (*Jasminum beesianum*) has only a single leaflet.

RUE FAMILY (*Rutaceae*), 1, 6, 7, 8
ROSE FAMILY (*Rosaceae*), 2, 3
4, 5, 9 in various families

1. **Cork-Tree** (*Phellodendron amurense*) p. 313
   A low-branching Asiatic tree, the foliage aromatic. Flowers negligible. Fruit shiny black.

2. **Rowan Tree** (*Sorbus Aucuparia*) p. 317
   Tolerating city conditions. Its white flowers and persistent red fruit are both showy.

3. **False Spirea** (*Sorbaria sorbifolia*) p. 315
   A medium-sized Asiatic shrub with a profusion of summer-blooming white flowers.

4. **Staghorn Sumac** (*Rhus typhina*) p. 319
   Vigorous native shrub, related to Poison Ivy but safe, its red fruiting clusters handsome.

5. ***Xanthoceras sorbifolium*** p. 314
   A little-grown aromatic Chinese shrub with showy white and red flowers. Transplants with difficulty.

6. **Prickly Ash** (*Zanthoxylum americanum*) p. 313
   Prickly native shrub or small tree, its greenish flowers blooming in early spring, usually before leaf expansion.

7. **Mexican Orange** (*Choisya ternata*) p. 312
   Evergreen Mexican shrub with highly aromatic foliage and fragrant white flowers. Not hardy northward.

8. **Trifoliate Orange** (*Poncirus trifoliata*) p. 312
   The only hardy citrus fruit, but its orange-like fruit is inedible. Flowers white and fragrant.

9. **Tree-of-Heaven** (*Ailanthus altissima*) p. 317
   The tree that "grows in Brooklyn" and other unfavorable places. Plant only female trees — male flowers are foul-smelling.

1
2
3
4
5
6
7
8
9

1
2
3
4
5
6
7
8
9

OLIVE FAMILY (*Oleaceae*), 1, 3, 7, 8, 9
DAISY FAMILY (*Compositae*), 2, 6
HONEYSUCKLE FAMILY (*Caprifoliaceae*), 4
VERVAIN FAMILY (*Berbenaceae*), 5

1. **Flowering Ash** (*Fraxinus Ornus*) p. 323
   The only ash tree with showy, white, fragrant flowers in dense clusters, blooming with leaf expansion.

2. **Groundsel Bush** (*Baccharis halimifolia*) p. 322
   The only substantial shrub of the Daisy Family that is hardy northward, its white fruiting clusters showy.

3. **Winter Jasmine** (*Jasminum nudiflorum*) p. 328
   Valued for its very early yellow flowers borne on the naked twigs as winter wanes.

4. **Red-berried Elder** (*Sambucus racemosa*) p. 331
   A native shrub, related to the elderberry, but with red fruit. Flowers small, white, May-blooming.

5. **Chaste Tree** (*Vitex Agnus-castus*) p. 325
   Summer-blooming European shrub or small tree with compound leaves and terminal flower clusters.

6. **Lavender Cotton** (*Santolina Chamaecyparissus*) p. 322
   Not quite hardy northward, this herblike shrub can be sheared to promote more of its dissected foliage.

7. ***Jasminum beesianum*** p. 328
   A remarkable Chinese jasmine, about 3 feet high, with May-blooming pink flowers. Fruit shiny.

8. ***Osmanthus fortunei*** p. 334
   A splendid evergreen hybrid shrub with holly-like foliage and sparsely produced fragrant white flowers.

9. **Tea Olive** (*Osmanthus fragrans*) p. 333
   An extraordinarily fragrant evergreen Asiatic shrub, its white flowers winter-blooming in the South.

JASMINE cont'd:

Flowers prevailingly fragrant, so much so that some habitually call the plants Jessamine, the corolla tubular or salver-shaped. Fruit a black berry.

It is important to scan the hardiness notes below, for most of the most fragrant and desirable jasmines are not hardy in regions of severe winters. Otherwise their culture is easy as they do well in a variety of soils in the South, although they thrive best in a loamy soil. The only really hardy one is the Winter Jasmine, which may flower as early as March 1 in the latitude of New York City. Of about 200 species only the 3 below are likely to interest the northern amateur:

Flowers pink; leaflet 1. — *Jasminum beesianum*
Flowers yellow; leaflets more than 1.
  Flowering in March. — Winter Jasmine
  Flowering much later. — *Jasminum floridum*

JASMINUM BEESIANUM p. 327

A low Chinese shrub, remarkable among jasmines for having only a single leaflet and a pink flower. **Size:** Usually about 3 feet high, rarely to nearly 6 feet, about half as wide. **Leaves:** With only a single narrowly oval leaflet 1½–3 inches long, slightly hairy on both sides. **Flowers:** About 1 inch long, pink, borne in a sparse cluster; May. **Fruit:** About ⅓ inch thick, black and shining. **Hardy:** From Zone 6 southward. **Varieties:** None. **Culture:** See above.

Related plant (not illustrated):

Jasminum stephanense. A hybrid jasmine, closely related to the above, with fragrant, also pink flowers. Hardy from Zone 6 southward.

WINTER JASMINE (*Jasminum nudiflorum*) p. 327

The only really safe jasmine for northern gardens, this Chinese shrub blooms so early that its flowers are often frozen. **Size:** 3–10 feet high, its stiff, arching branches 4-angled. **Leaves:** Opposite, compound, composed of 3 oval leaflets about 1 inch long, without teeth but faintly hairy on the margin (seen only with a lens). **Flowers:** Solitary, yellow, about ¾ inch wide, borne along the bare branches of the previous season; Mar., or all winter in the South. **Fruit:** A small black berry. **Hardy:** From Zone 4 southward. **Varieties:** None. **Culture:** See above.

Related plant (not illustrated):

Jasminum mesneyi. An evergreen Chinese shrub, 6–8 feet high, closely related to the Winter Jasmine, with 3 leaflets and with yellow flowers with a darker center, sometimes double-flowered. Hardy from Zone 5 southward.

JASMINUM FLORIDUM below

A half-evergreen, rambling Chinese shrub, remarkable in having alternate leaves; its smooth branches upright or arching. **Size:** 10–20 feet high or even more. **Leaves:** Alternate, compound, composed of 3–5 oval leaflets ½–1½ inches long. **Flowers:** Golden yellow, in large terminal clusters; July–Aug. **Fruit:** A small black berry. **Hardy:** From Zone 5 southward. **Varieties:** None. **Culture:** See above.

Related plant (not illustrated):

Italian Jasmine (*Jasminum humile*). An evergreen or half-evergreen shrub from tropical Asia, with 3–7 leaflets and golden-yellow flowers blooming in Aug.–Sept. Hardy from Zone 5 southward.

Jasminum floridum, above.

ELDER
(*Sambucus*)
Family Caprifoliaceae

Rather coarse, weedy shrubs of slight garden interest, but cultivated by some who fancy elderberry wine more than better brews. They are medium-sized shrubs. Leaves opposite, compound, the leaflets with marginal teeth. Flowers small, but in profuse flat-topped or rounded clusters. Fruit fleshy, edible, small, usually very profuse and much favored by birds.

The elders are not particular as to soils, but the first species, a common native shrub, does best in moist or even wet sites. Only 3 species are worth cultivation:

Flower cluster flat-topped.
- *Fruit purplish black; leaflets 7.* — American Elder
- *Fruit black; leaflets 5.* — European Elder

Flower cluster rounded; fruit red. — Red-berried Elder

AMERICAN ELDER (*Sambucus canadensis*) p. 331
A stout, rampant, native shrub, common in swamps, and scarcely worth cultivating except for its profuse fruit used for wine and preserves. **Size:** 6–10 feet high, quite as wide, its branches brittle. **Leaves:** Opposite, compound, composed of usually 7 short-stalked leaflets 2½–6 inches long, tapering at the tip, the margins toothed. **Flowers:** Minute (about 1/10 inch wide) in a profuse flat-topped cluster about 4 inches wide; June. **Fruit:** A small purplish-black berry, very profuse. **Hardy:** Everywhere. **Varieties:** *acutiloba* has the leaflets dissected; there is also a form with yellow foliage called Golden. **Culture:** See above.

Related plant (not illustrated):

American Red Elder (*Sambucus pubens*). A native shrub, mostly in the North, 8–12 feet high, with 5–7 leaflets, white flowers in a pyramidal cluster, and scarlet, inedible berries. Hardy from Zone 2 southward.

EUROPEAN ELDER (*Sambucus nigra*) p. 331
A coarse Eurasian shrub or small tree, much resembling the American elder, but with only 5 leaflets and shiny black fruit.

Not much cultivated except in the varieties mentioned below. **Hardy:** From Zone 3 southward. **Varieties:** *albo-variegata* has white-blotched foliage; *aureo-variegata* has yellow-blotched foliage; *laciniata* has deeply dissected leaflets; there is also one yellow-fruited variety. **Culture:** See above.

RED-BERRIED ELDER (*Sambucus racemosa*) p. 327
The most commonly cultivated of the elders, this Eurasian shrub is also the best of them, being not so rampant as the American elder. **Size:** 6–10 feet high, about three fourths as wide. **Leaves:** Opposite, compound, composed of 5–7 practically stalkless ovalish leaflets 2–3½ inches long, tapering at the tip. **Flowers:** Yellowish white, very small, in a rounded cluster 1½–3 inches wide; May. **Fruit:** A small scarlet berry, very profuse. **Hardy:** From Zone 3 southward. **Varieties:** Several, much resembling those of the American elder. Two, however, are worth growing: *laciniata* has dissected leaflets; *plumosa-aurea* has golden-yellow leaflets that are cut almost to the middle. **Culture:** See above. It also does well in ordinary garden soil.

Left: American Elder (*Sambucus canadensis*), p. 330.
Right: European Elder (*Sambucus nigra*), p. 330.

Leaves Opposite Simple

Leaves without marginal teeth. See p. 356

Leaves with obvious, even if small, marginal teeth or lobes, the margins sometimes spiny.

Leaves spiny-margined (except in 2 species of Osmanthus). Tea Olive

Leaves merely toothed, not spiny-margined.

Flowers regular. See p. 341

Flowers irregular (almost regular in the Beauty-Bush).

(For definition of regular and irregular flowers, see Picture Glossary)

Flowers blue or lilac-purple.

Flowers blue.

Erect shrub, 4–6 feet high. Blue Spirea

Low under-shrub, 1–3 feet high. Russian Sage

Flowers lilac-purple. *Elsholtzia*

Flowers never blue.

Flowers yellow.

Twigs with matted hairs; foliage aromatic Jerusalem Sage

Twigs without matted hairs; foliage not aromatic. *Diervilla*

Flowers not yellow, mostly red or pink.

Fruit dry, a capsule. *Weigela*

Fruit fleshy, berrylike. Beauty-Bush

Tea Olive
(*Osmanthus*)

Family Oleaceae

Extraordinarily fragrant evergreen shrubs, often with holly-like foliage that is spiny-margined (except in 2 species). Leaves opposite, thick, leathery, and always evergreen. Flowers quite small, usually in small clusters, often wanting in young specimens, but of intoxicating fragrance. Fruit fleshy, with one stone.

The tea olives, beloved of Southern belles, are not hardy northward, and not easy to grow. They should never be moved without a tight ball of earth, closely wrapped in burlap, for their roots must not be exposed to the sun and wind. They do best in a moderately acid soil (pH 5–6), but poorly in alkaline soils. A top-dressing of acid peat is helpful.

Leaves slightly toothed or without any, not spiny-margined. Tea Olive
Leaves with obvious teeth, not spiny-margined. *Osmanthus delavayi*

Leaves spiny-margined.
Leaves 1½–2½ inches long. *Osmanthus ilicifolius*
Leaves 3–4 inches long. *Osmanthus fortunei*

TEA OLIVE (*Osmanthus fragrans*) p. 327

Perhaps the most alluring fragrance of any plant in this book is furnished by the small, essentially winter-blooming flowers of this Asiatic shrub. **Size:** 8–20 feet high, sometimes treelike in the South, about three fourths as wide. **Leaves:** Opposite, simple, leathery, ovalish or oblong, 2–4 inches long, slightly toothed on the margin or without any teeth. **Flowers:** Small, white, very fragrant, borne in small clusters along the twigs, not showy; blooming all winter in the South, but in Apr. in Zone 6. **Fruit:** Fleshy, black. **Hardy:** From Zone 6 southward. **Varieties:** None. **Culture:** See above.

OSMANTHUS DELAVAYI p. 334

A Chinese evergreen shrub, its leathery foliage handsome. **Size:** 6–10 feet high, nearly as wide, its shoots downy. **Leaves:** Opposite, simple, ½–1 inch long, short-stalked, and finely toothed on the margin. **Flowers:** Small, white, fragrant, borne in small clusters at the leaf joints, or terminal; Apr. **Fruit:** Small, roundish, blue-black. **Hardy:** From Zone 6 southward. **Varieties:** None. **Culture:** See above.

Left: *Osmanthus ilicifolius,* below.
Right: *Osmanthus delavayi,* p. 333.

OSMANTHUS ILICIFOLIUS above

A Japanese evergreen shrub with holly-like foliage (but holly has alternate leaves) and fragrant flowers. **Size:** 10–20 feet high, sometimes treelike, about three fourths as wide. **Leaves:** Opposite, simple, oblong to ovalish, 1½–2½ inches long, the margins with a few spiny teeth. **Flowers:** Small, white, fragrant, borne in small, close clusters at the leaf joints; June. **Fruit:** Berrylike, bluish black, about ½ inch long. **Hardy:** From Zone 5 southward. **Varieties:** *myrtifolius* has leaves without the spiny margin; *variegatus* has white-blotched leaves; *purpurascens* has the young foliage purple, later green and purple-tinged; *rotundifolius* has roundish leaves, with blunt teeth and is a smaller shrub. **Culture:** See above.

OSMANTHUS FORTUNEI p. 327

A stout, evergreen, hybrid shrub, the best of all for Zones 5 and 6 (not hardy northward), its foliage dense and holly-like (but the holly has alternate leaves). **Size:** 4–6 feet high and about as wide, its foliage lustrous and handsome. **Leaves:** Opposite, simple, oblongish, 3–4 inches long, very spiny on the margin. **Flowers:**

Small, white, very fragrant, in small clusters at the leaf joints, often wanting; Sept. **Fruit:** Small, bluish black, berrylike. **Varieties:** None. **Culture:** See above. It is the easiest to grow and a splendid, holly-like evergreen.

BLUE SPIREA (*Caryopteris incana*)

Family Verbenaceae p. 342

A beautiful late-flowering Asiatic shrub with a profusion of blue fringed flowers. **Size:** 4–6 feet high, about three fourths as wide. **Leaves:** Opposite, simple, ovalish, 2–3 inches long, grayish, coarsely but bluntly toothed. **Flowers:** Blue, irregular, slightly fringed, in showy clusters, mostly at the leaf joints; Sept. **Fruit:** Small, dry, separating into 4 winged nutlets. **Hardy:** From Zone 5 southward; north of this it frequently is killed to the ground, but if mulched it sends up flowering shoots the next spring. **Varieties:** One called Blue Mist is reputedly finer than the typical plant. **Culture:** Does well in a variety of soils, preferably in full sun.

Related plant (not illustrated):

Bluebeard (*Caryopteris clandonensis*). A hybrid shrub, somewhat more hardy than the Blue Spirea, but usually winter-killed north of Zone 5. A handsome free-flowering shrub.

RUSSIAN SAGE (*Perovskia atriplicifolia*)

Family Labiatae p. 342

A rather showy, aromatic Siberian under-shrub (almost herb-like), but not widely grown although it has fine late-blooming flower spikes. **Size:** 1–3 feet high, sometimes a little higher, bushy and nearly as wide. **Leaves:** Opposite, simple, nearly lance-shaped, 1½–4 inches long, coarsely toothed, sage-like in odor when crushed, covered with ashy hairs. **Flowers:** Small, irregular, blue, arranged in long terminal spikes that are 15–20 inches long; Sept. **Fruit:** Small, dry. **Hardy:** From Zone 5 southward; if grown north of this often winter-killed, but sending up flowering shoots the next year, especially if mulched. **Varieties:** None. **Culture:** Prefers open sunny places, in a sandy loam or even a gritty soil.

ELSHOLTZIA (*Elsholtzia stauntoni*). Family Labiatae p. 342
A Chinese aromatic under-shrub, chiefly valued for its late bloom. **Size:** 2–4 feet high, its twigs hairy. **Leaves:** Opposite, simple, oval-oblong, 3–5 inches long, very sticky beneath, toothed. **Flowers:** Small, lilac-purple, in a 1-sided spike 6–12 inches long, the stamens long-protruding; Sept.–Oct. **Fruit:** Small, dry. **Hardy:** From Zone 5 southward. **Varieties:** None. **Culture:** It needs a light sandy loam and open sunlight, but preferably out of north and northwest winds. Not a very decorative plant.

JERUSALEM SAGE (*Phlomis fruticosa*)
Family Labiatae p. 342
A many-branched under-shrub from southern Europe, suited to the border, and with very aromatic foliage (when crushed). **Size:** 2–4 feet high, about half as wide. **Leaves:** Opposite, simple, toothed, ovalish, wrinkled, about 4 inches long and covered with yellow matted hairs. **Flowers:** Yellow, irregular, in dense, nearly stalkless clusters at the leaf joints; June. **Fruit:** Small, dry. **Hardy:** From Zone 3 southward. **Varieties:** None. **Culture:** It demands a dry, sunny site and will not thrive where the winter slush can smother its crown and roots.

Diervilla

(*Diervilla*)

Family Caprifoliaceae

Low shrubs of chief interest because they thrive better in shady places than in the open. Leaves simple, opposite, always with marginal teeth. Flowers funnel-shaped, yellow, in small leafless clusters. Fruit a small dry capsule.

The diervillas, often called bush honeysuckle, are of secondary garden interest, but are useful in shady places, where their underground rooting stems often result in large patches. They will also grow in full sun, but not so well as in the shade.

Leaves distinctly stalked. Gravelweed
Leaves essentially stalkless. *Diervilla sessilifolia*

GRAVELWEED (*Diervilla Lonicera*) p. 342

A low native shrub, chiefly of interest as a ground cover and useful also for holding banks, preferably in the shade. **Size:** 1–2 feet high, often making large patches. **Leaves:** Opposite, simple, distinctly stalked, oblongish, 2–4 inches long, taper-pointed at the tip, the margins toothed. **Flowers:** Yellow, usually 3 in a cluster at the leaf joints, funnel-shaped, about ¾ inch long; June. **Fruit:** A small dry capsule. **Hardy:** Everywhere. **Varieties:** None. **Culture:** See above. Not a showy shrub, but useful in the shade.

DIERVILLA SESSILIFOLIA below

Closely related to the above, but with essentially stalkless leaves and more freely flowering. **Leaves:** Opposite, simple, narrowly oblong, 2½–5 inches long, nearly stalkless, toothed. **Flowers:** Yellow, funnel-shaped, ½–¾ inch long, usually in clusters of 3–7; June. **Fruit:** A small dry capsule. **Varieties:** None. **Culture:** See above.

Diervilla sessilifolia, above.

WEIGELA
(*Weigela*)

Very handsome, widely grown shrubs, particularly useful for their late May and early June showy bloom, when many spring-flowering shrubs are past flowering. Leaves opposite, simple, always toothed. Flowers generally funnel-shaped but slightly irregular, prevailingly red or pink (white or yellow in some hybrids), borne in clusters of 1–3. Fruit a woody, dry capsule.

The weigelas, which have no common name, are so popular that there are scores of horticultural forms, which makes identification a bit confusing. Being hardy nearly everywhere and of the easiest culture in any garden, they are among the most widely cultivated shrubs. The simple key below does not convey the complexity of the technical characters upon which the identification of the species of weigela is based.

Fruit smooth. — *Weigela florida*
Fruit hairy. — *Weigela japonica*

WEIGELA FLORIDA p. 342

The most widely planted sort (especially some of its hybrids), this Asiatic shrub is one of our finest garden bushes. **Size:** 8–10 feet high and quite as wide because of its many stems arising at ground level. **Leaves:** Opposite, simple, elliptic, 3–4 inches long, hairy on the veins beneath. **Flowers:** Rose-pink in the typical form (but white, pink, or carmine in some of its hybrids), funnel-shaped, about 2 inches long, profuse, very showy; June. **Fruit:** A perfectly smooth, small, dry capsule. **Hardy:** From Zone 3 southward. **Varieties:** Over 40 named forms, many of them quite similar. They are best picked out by visiting a well-labeled arboretum or nursery, for their names are in much confusion. **Culture:** See above.

Related plants (not illustrated):

Weigela middendorfiana. An Asiatic shrub, 2–4 feet high, with yellow flowers, often offered under the doubtful name of *Weigela lutea.* Hardy from Zone 4 southward.

Weigela hybrida. An invalid nurserymen's name for many

shrubs, mostly allied to *Weigela florida,* among them the superb dark maroon Eva Rathke.

Weigela japonica, below.

WEIGELA JAPONICA above

A Japanese shrub with a superficial resemblance to *Weigela florida,* but the flowers always white at first, fading to carmine. **Size:** 6–9 feet high, quite as wide. **Leaves:** Opposite, simple, oblongish, 3–4 inches long, hairy beneath, especially on the veins. **Flowers:** White at first, fading to carmine, generally in clusters of 3, about 1½ inches long, somewhat bell- or funnel-shaped; June. **Fruit:** A small, hairy, dry capsule. **Hardy:** From Zone 4 southward. **Varieties:** None. **Culture:** See above.

Beauty-Bush
(*Callicarpa*)
Family Verbenaceae

Not widely grown but handsome shrubs, planted more for their conspicuous fruit than the decidedly less attractive flowers. Leaves simple, opposite, always toothed, but the teeth bluntish. Flowers small, tubular, of various colors, ranging from white to purple, pink, or red, borne in compact clusters from the sides of the twigs. Fruit berrylike, rather showy and fairly long-persistent.

These bushes need a rich soil, sun or shade, and neither of them is hardy in the North. Otherwise their culture is easy. A related shrub, *Lantana Camara,* is usually only a summer-bedding plant in our area. It is pictured on p. 189 in Norman Taylor's *Guide to Garden Flowers.*

Flower stalk longer than the leafstalk. *Callicarpa dichotoma*
Flower stalk not as long as the leafstalk. French Mulberry

Callicarpa dichotoma, below.

CALLICARPA DICHOTOMA above

An Asiatic relative of the French Mulberry, this shrub is well worth growing for its showy, berrylike fruit. **Size:** 2–4 feet high, about three fourths as wide. **Leaves:** Opposite, simple, elliptic, 1–3 inches long, bluntly toothed toward the tip, usually toothless toward the base. **Flowers:** Pink, small, not conspicuous, the small clusters few-flowered; Aug. **Fruit:** Violet or lilac, berrylike, about ½ inch thick, very profuse and showy, persistent into Nov. **Hardy:** Certainly from Zone 5 southward and in protected sites in Zone 4. **Varieties:** None. **Culture:** See above.

FRENCH MULBERRY (*Callicarpa americana*) p. 342

Neither French nor a mulberry, this native American shrub acquired its misleading common name years ago. It is a spectacular bush when in full fruit. **Size:** 4–5 feet high, about three fourths as wide. **Leaves:** Opposite, simple, ovalish but tapering to a narrow tip, 4–6 inches long, rusty beneath. **Flowers:** Not showy, about ⅓ inch long, in dense short-stalked clusters, bluish, white, purple, lilac, or even red; May. **Fruit:** Violet, profuse, glistening, about ¼ inch thick, very showy, and persistent until Nov. **Hardy:**

From Zone 5 southward. **Varieties:** One is offered with white fruit but is not common. **Culture:** See above.

Related plant (not illustrated):

Callicarpa japonica. A Japanese shrub, 2–4 feet high. Flowers white or pink, in Aug. Fruit violet. Hardy from Zone 5 southward.

Flowers Regular

Flowers yellow, with apparently 4 strap-shaped petals. Golden Bell
Flowers never yellow. Honeysuckle Family

Golden Bell
(*Forsythia*)
Family Oleaceae

Extremely handsome spring-blooming shrubs, widely planted for ornament. Leaves simple, opposite, always with marginal teeth. Flowers prevailingly yellow, in small clusters that are borne so profusely that the bare twigs are literally smothered with bloom in some of the best varieties. Corolla regular, united below, but its 4 narrow, strap-shaped lobes look (falsely) like separate petals. Fruit a dry capsule. They need plenty of space as they are very spreading shrubs.

Nearly all the ones below bloom before the leaves expand, and some of them have arching stems, the tips of which may root, thus forming large patches. The golden bells are of the easiest culture in any ordinary garden soil and they root as easily as privet. The best three are:

Twigs hollow. *Forsythia suspensa*
Twigs not hollow.
Flowers at least 1½ inches long, very profuse. *Forsythia intermedia*
Flowers smaller, not so profuse. *Forsythia ovata*

MINT FAMILY (*Labiatae*), 3, 5, 9
OLIVE FAMILY (*Oleaceae*), 2, 7
VERVAIN FAMILY (*Verbenaceae*), 4, 8
HONEYSUCKLE FAMILY (*Caprifoliaceae*), 1, 6

1. ***Weigela florida*** p. 338
   Besides the pink flowers, this Asiatic shrub is also to be had in white or dark red forms.

2. ***Forsythia suspensa*** p. 344
   A fine Chinese golden bell, with arching branches that may root at the tip. Not so fine as No. 7.

3. ***Elsholtzia stauntoni*** p. 336
   A fall-blooming Chinese undershrub with aromatic foliage and showy flower clusters.

4. **French Mulberry** (*Callicarpa americana*) p. 340
   A native shrub, not hardy northward, grown chiefly for its decorative glistening fruit.

5. **Jerusalem Sage** (*Phlomis fruticosa*) p. 336
   An undershrub from southern Europe that does not like winter slush at its crown.

6. **Gravelweed** (*Diervilla Lonicera*) p. 337
   A native undershrub useful as a ground cover in shady places. Flowers yellow, not showy.

7. ***Forsythia intermedia*** p. 344
   Much the best of all the golden bells, this hybrid shrub blooms later than No. 2.

8. **Blue Spirea** (*Caryopteris incana*) p. 335
   A fall-flowering Asiatic shrub, often winter-killed in the North, but blooming the next season.

9. **Russian Sage** (*Perovskia atriplicifolia*) p. 335
   A Siberian undershrub with fall-blooming flowers. Suited to dry, sandy soils.

1
2
3
4
5
6
7
8
9

1
2
3
4
5
6
7
8
9

HONEYSUCKLE FAMILY (*Caprifoliaceae*), 1–9

1. ***Viburnum dilatatum*** p. 352
   A Japanese relative of our native arrow-wood, with a white, flat-topped flower cluster and persistent scarlet fruit.

2. **Witherod** (*Viburnum cassinoides*) p. 353
   A medium-sized native shrub good for informal plantings, with small white flowers and persistent fruit.

3. ***Viburnum carlesi*** p. 350
   Extremely fragrant, rather low Korean shrub, its flowers blooming in April or May.

4. **Japanese Snowball** (*Viburnum tomentosum sterile*) p. 350
   Resembling No. 8, but without the lobed leaves of that species. A shrub 7–10 feet high.

5. ***Viburnum odoratissimum*** p. 349
   An evergreen Asiatic shrub with white, fragrant May-blooming flowers. Not hardy northward.

6. **Laurustinus** (*Viburnum Tinus*) p. 348
   Handsome evergreen European shrub, its pinkish flowers unpleasantly scented. Not hardy northward.

7. ***Viburnum rhytidophyllum*** p. 348
   The best of the evergreen viburnums for the North, its foliage very handsome. Flowers in June.

8. **European Snowball** (*Viburnum Opulus sterile*) p. 349
   The lobed leaves of this shrub distinguish it from No. 4. Red fruits winter-persistent.

9. **Black Haw** (*Viburnum prunifolium*) p. 353
   A hardy native shrub or small tree, its flat-topped white flower cluster blooming in May.

FORSYTHIA SUSPENSA p. 342

A showy Chinese shrub, but not so free-flowering as the next species, its branches erect at first, but arching and even rooting at the tips in age in some forms. Twigs hollow. **Size:** 8–12 feet high, at least as wide or even more. **Leaves:** Opposite, simple, oval-oblong, 3–5 inches long, toothed, long-persistent but not colored in the fall. **Flowers:** Yellow, about 1 inch long, in clusters of 1–5, but these profuse on the bare twigs; Mar.–Apr. **Fruit:** Negligible. **Hardy:** From Zone 2 southward, but in the upper part of this range the flowers may be blasted by frost. **Varieties:** *fortunei* has chiefly erect branches and is not arching; *sieboldi* is lower and its branches are almost trailing, hence useful for covering banks. **Culture:** See above.

FORSYTHIA INTERMEDIA p. 342

By far the most showy of all the golden bells, especially in some of its varieties, this hybrid shrub bids fair to replace any forsythia here mentioned. **Size:** 7–9 feet high, at least as wide or even more,

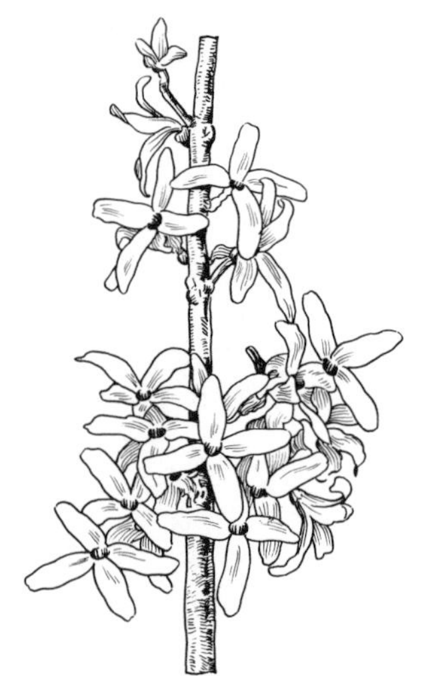

Forsythia ovata, p. 345.

its branches erect or arching, but not pendulous and rooting at the tip, the twigs not hollow. **Leaves:** Opposite, simple, oblong or ovalish, 3–5 inches long, sometimes 3-parted but usually merely toothed, persistent while still green into Nov. **Flowers:** Yellow, about 1¼ inches long, borne in small clusters on the bare twigs, but very numerous; Mar.–Apr., but later than *Forsythia suspensa.* **Fruit:** Negligible. **Hardy:** From Zone 2 southward. **Varieties:** *primulina* has pale yellow flowers; *spectabilis,* and its form known as Lynwood Gold, has larger, golden-yellow flowers that are so profuse that they completely cover the bare twigs; certainly the finest of all the golden bells. **Culture:** See above.

Related plant (not illustrated):

Forsythia viridissima. A green-twigged Chinese shrub, 7–10 feet high, the yellow flowers about 1 inch long and in sparse clusters. The only forsythia with autumnal foliage color (red or pinkish). Hardy from Zone 2 southward. It also has a dwarf variety.

FORSYTHIA OVATA p. 344

The earliest-flowering of all the golden bells, this Korean shrub has little advantage over the others. **Size:** 3–5 feet high, about as wide. **Leaves:** Opposite, simple, oval or rounder, 2–3 inches long. **Flowers:** Yellow, solitary, and not so profuse as in the others; Mar.–Apr. **Fruit:** Negligible. **Varieties:** None. **Culture:** See above.

HONEYSUCKLE FAMILY
(*Caprifoliaceae*)

A LARGE family of shrubs, trees, and a few vines, for which see p. 62. The ones below are all shrubs or trees, some of them of wide garden interest. Leaves simple, opposite, always, in those below, with marginal teeth or lobes, except in some evergreen viburnums, but the teeth are small and remote in *Abelia.* Flowers regular, white or pink, the corolla with 4–5 lobes. Fruit fleshy in *Viburnum,* but dry in *Abelia* and *Kolkwitzia.* A related woody plant, the Twinflower (*Linnaea borealis*), is so herblike that it is in *Guide to Garden Flowers,* p. 188, figure 3. Only 3 genera concern us here:

Flowers more or less wheel-shaped, flattish, scarcely if at all tubular, borne in large clusters. *Viburnum*
Flowers obviously bell-shaped or tubular, borne singly, in pairs, or in a cluster.
Flower cluster terminal. Beauty-Bush
Flowers usually solitary or in pairs. *Abelia*

VIBURNUM
(*Viburnum*)

So popular are the viburnums that of the 150 known species about a third of them are in cultivation in the U.S., but only a selection of the best of them can possibly be included here. They are mostly shrubs or small trees with opposite, simple leaves the margins of which are always toothed or lobed except in two of the evergreen species below. Flowers white or pink, small, generally flattish and wheel-shaped, rarely tubular, showy because borne in dense clusters. In some forms, known as Snowball, the flowers are sterile and double and grouped in a dense globular or flattish cluster. In some there are marginal sterile flowers larger than the central ones. Fruit fleshy, often brightly colored and persistent in some; nearly all of them favored by birds.

Because *Cornus* and *Viburnum* both have opposite leaves and small flowers in clusters, their superficial resemblance confuses many. Their chief differences are:

	LEAVES	FLOWERS
Viburnum:	Always toothed on the margin, except in two evergreen species.	With 4–5 lobes of a united corolla.
Cornus:	Margins without teeth.	With 4 separate petals.

All except the evergreen species are of the easiest culture in any ordinary garden soil, but hardiness notes should be studied as some of them are not hardy everywhere. The evergreen ones should be moved only with a ball of earth tightly roped in burlap, as their roots must not be exposed to the air. Identification of the species is not easy. In spite of rigorous selection the following 12 species are all in such common cultivation that they demand notice here. A simple key for them follows:

Leaves evergreen.
- *Leaves wrinkled above, yellowish gray beneath.* — *Viburnum rhytidophyllum*
- *Leaves not wrinkled above.*
  - *Flowers pinkish.*
    - *Fruit red at first.* — *Viburnum suspensum*
    - *Fruit black.* — Laurustinus
  - *Flowers white, fragrant.* — *Viburnum odoratissium*

Leaves not evergreen.
- *Leaves lobed.* — Cranberry Tree
- *Leaves not lobed.*
  - *Flower cluster with larger, marginal sterile flowers.* — *Viburnum tomentosum*
  - *Flower cluster without large marginal sterile flowers.*
    - *Flowers extremely fragrant, blooming before or with leaf expansion.* — *Viburnum carlesi*
    - *Flowers scarcely fragrant, blooming after the leaves expand.*
      - *Foliage rank-smelling when crushed.* — *Viburnum sieboldi*
      - *Foliage not rank-scented.*
        - *Main veins ending exactly in the leaf teeth.*
          - *Flower cluster smooth.* — *Viburnum setigerum*
          - *Flower cluster hairy.* — *Viburnum dilatatum*
        - *Main veins not ending in the leaf teeth.*
          - *Flower clusters stalked.* — Witherod
          - *Flower clusters stalkless.* — Black Haw

VIBURNUM RHYTIDOPHYLLUM p. 343

An evergreen Chinese shrub valued both for its handsome foliage and its showy flower clusters. **Size:** 7–10 feet high, about three fourths as wide, as its branches are relatively erect; twigs densely hairy when young. **Leaves:** Simple, opposite, oblongish, 5–7 inches long, dark green and wrinkled above, the under side densely grayish-felty, only slightly toothed. **Flowers:** Small, yellowish white, the showy, profuse, flat clusters nearly 8 inches wide; June. **Fruit:** Red at first, later black, not persistent. **Hardy:** From Zone 4 (where it may drop some of its leaves) to Zone 5 southward, where it is always evergreen. **Varieties:** *roseum* has flowers pink in the bud, ultimately white. **Culture:** See above. It does best in a rich soil and in partial shade.

VIBURNUM SUSPENSUM p. 351

A beautiful evergreen shrub from the Ryukyu Islands near Formosa, which, unfortunately, is not hardy in the North, but much grown in the South and Calif. **Size:** 4–6 feet high, about three fourths as wide. **Leaves:** Simple, opposite, ovalish, 3–4 inches long, toothed toward the tip. **Flowers:** Small, pinkish, borne in dense clusters that are about 1½ inches wide; June. **Fruit:** Red at first, ultimately black, not persistent. **Hardy:** From Zone 6 southward. **Varieties:** None. **Culture:** See above.

LAURUSTINUS (*Viburnum Tinus*) p. 343

A very handsome evergreen shrub from the Mediterranean region, but not hardy in the North. **Size:** 7–10 feet high, nearly as wide. **Leaves:** Opposite, simple, oblongish or broader, 2–3 inches long, without any marginal teeth, dark green. **Flowers:** Small, pinkish, in clusters that are about 3 inches wide; July–Aug., and unpleasantly scented. **Fruit:** Black, not persistent. **Hardy:** From Zone 6 southward. **Culture:** See above, but its evergreen habit demands more care. It should not be watered in late summer as this induces a lush growth that may be winter-killed. It is a good hedge plant if sheared only every other year. It has several varieties of no special merit, one of them with variegated leaves.

Related plants (not illustrated), both evergreens:

Viburnum davidi. A Chinese shrub, not over 3 feet high,

the elliptic leaves 2½–6 inches long, 3-veined. Flowers dirty white, the flat cluster 2–3 inches wide. Fruit blue. Hardy from Zone 6 southward.

Viburnum japonicum. A Japanese shrub, 4–6 feet high, the broadly oval leaves 3–6 inches long. Flowers white, fragrant, the rounded cluster 3–4 inches wide. Fruit red. Hardy from Zone 6 southward.

VIBURNUM ODORATISSIMUM p. 343

A handsome evergreen shrub stretching from the Himalayas to China, valued for its extremely fragrant flowers. **Size:** 7–10 feet high, about as wide. **Leaves:** Opposite, simple, ovalish, 4–6 inches long, shining green, practically without marginal teeth. **Flowers:** Small, white, very fragrant, the pyramidal cluster about 4 inches high and showy; May–June. **Fruit:** Red at first, ultimately black, not persistent. **Hardy:** From Zone 6 southward. **Varieties:** None. **Culture:** See above.

CRANBERRY TREE (*Viburnum Opulus*) p. 343

A Eurasian shrub very useful in cities as it will stand smoke, dust, and wind. **Size:** 8–12 feet high, nearly as wide, its foliage not evergreen. **Leaves:** Opposite, simple, maple-like, 3–5 lobed, about 3½ inches wide, hairy on the under side, beautifully red in the fall. **Flowers:** White, the cluster nearly 4 inches wide, its marginal flowers sterile, about ¾ inch wide and showy; May–June. **Fruit:** Profuse, red, persistent over the winter. **Hardy:** From Zone 2 southward. **Varieties:** *sterile* is the European Snowball, far more widely grown than the typical form as its wholly sterile flowers are in dense globelike clusters 3–5 inches thick. (For other Snowballs see *Viburnum tomentosum*); *nanum* is a dwarf form with smaller leaves.

Related plants (not illustrated):

Cranberry Bush (*Viburnum trilobum*). A native representative of the Cranberry Tree, differing, if at all, in having the leaves smooth both sides, and usually 3-lobed. Hardy from Zone 4 southward.

Dockmackie (*Viburnum acerifolium*). A native shrub, 5–6 feet high, its maple-like leaves 3-lobed and coarsely toothed. Flower cluster long-stalked, white. Fruit black-purple. Hardy from Zone 3 southward.

Poison Haw (*Viburnum molle*). A native shrub from the central U.S., its leaves unlobed, but coarsely toothed. Flowers white, in long-stalked clusters. Fruit bluish black. Hardy from Zone 3 southward.

Viburnum sargenti. An Asiatic relative of the Cranberry Tree, but with thicker, sometimes larger leaves and larger sterile marginal flowers. Hardy from Zone 5 southward.

VIBURNUM TOMENTOSUM p. 343

A widely cultivated Asiatic shrub, especially in one of its forms commonly called Japanese Snowball. **Size:** 7–10 feet high, about as wide. **Leaves:** Opposite, simple, unlobed, ovalish, 3–4 inches long, toothed on the margin, reddish in the autumn. **Flowers:** Small, white, in a flat-topped cluster nearly 4 inches wide, the marginal flowers about 1 inch wide, showy and sterile; May–June. **Fruit:** Red at first, ultimately bluish black, not persistent. **Hardy:** From Zone 3 southward. **Varieties:** *sterile,* the Japanese Snowball, is one of the most widely cultivated plants in the country, because of its immense ball-like clusters of wholly sterile white flowers.

Related plant (not illustrated):

Chinese Snowball (*Viburnum macrocephalum*). A Chinese shrub 7–12 feet high, most cultivated in the variety *sterile,* bearing a dense, ball-like cluster of white flowers, the marginal ones sterile. Hardy from Zone 5 southward.

VIBURNUM CARLESI p. 343

An extraordinarily fragrant Korean shrub, blooming before or with the expansion of the leaves. **Size:** 3–5 feet high, about three fourths as wide. **Leaves:** Opposite, simple, ovalish, 2–3½ inches wide, hairy both sides, toothed. **Flowers:** Small, white or pinkish, in dense rounded clusters 3–5 inches wide, very fragrant; Apr.–May. **Fruit:** Bluish black, much favored by birds. **Hardy:** From Zone 3 southward. **Varieties:** None. **Culture:** See above. It is perhaps the most fragrant of all the viburnums.

Related plants (not illustrated):

Viburnum fragrans. A Chinese shrub, 5–9 feet high, with elliptic leaves 2–4 inches long, pointed both ends. Flowers pink at first, ultimately white, fragrant, in loose clusters before the leaves expand. Fruit blue. Hardy from Zone 5 southward.

Viburnum carlcephalum. A hybrid shrub, closely related to *V. carlesi,* its white, very fragrant flowers opening before the leaves expand. Hardy from Zone 4 southward.

Viburnum juddi. A hybrid shrub, closely related to *V. carlesi,* with a shorter leafstalk and fragrant flowers in a rather loose cluster. Hardy from Zone 4 southward.

Viburnum burkwoodi. A hybrid, half-evergreen shrub, 6–8 feet high, closely related to *V. carlesi,* the leaves with brown veins. Flowers white, fragrant, the cluster about 3 inches wide. Hardy from Zone 4 southward.

VIBURNUM SIEBOLDI below

A stunning Japanese shrub with fine foliage, abundant flowers, and red-stalked fruits. **Size:** 8–10 feet high, nearly as wide. **Leaves:** Opposite, simple, ovalish but broader toward the tip, 4–6 inches long, roughish, red in the fall; unpleasantly scented when crushed. **Flowers:** White, small, in a cluster about 4 inches

Left: *Viburnum suspensum,* p. 348. Center: *Viburnum setigerum,* p. 352. Right: *Viburnum sieboldi,* above.

high; May–June. **Fruit:** Pink at first, ultimately bluish black, not persistent, but its showy red stalk hanging on some time after the fruit falls. **Hardy:** From Zone 3 southward. **Varieties:** None. **Culture:** See above. It is one of the best for northern gardens.

VIBURNUM SETIGERUM p. 351

A large Chinese shrub with rather striking red fruit in Oct. **Size:** 7–12 feet high, about three fourths as wide, its twigs smooth. **Leaves:** Opposite, simple, ovalish or oblong, 3½–5 inches long, toothed, the main veins ending at the leaf teeth. **Flowers:** Small, white, profuse, in clusters that are about 2 inches wide. **Fruit:** Egg-shaped, red, not persistent. **Hardy:** From Zone 5 southward. **Varieties:** *aurantiacum* has showy orange-yellow fruit. **Culture:** See above.

Related plant (not illustrated):

Viburnum wrighti. A sturdy Japanese shrub 7–9 feet high, the leaves nearly round and coarsely toothed. Flowers white, in a short-stalked cluster about 4 inches wide. Fruit red, showy. Hardy from Zone 4 southward.

VIBURNUM DILATATUM p. 343

A showy-fruited Japanese shrub which also has attractive flat-topped clusters of white flowers. **Size:** 6–10 feet high, about as wide, the young twigs softly hairy. **Leaves:** Opposite, simple, nearly round, about 4½ inches wide, hairy both sides, coarsely toothed, the veins ending at the leaf teeth. **Flowers:** Small, white, the flattish clusters nearly 5 inches wide; June. **Fruit:** Scarlet, profuse, and persistent through the winter. **Hardy:** From Zone 3 southward. **Varieties:** *xanthocarpum* has yellow fruit, but is rather uncommon in cultivation. **Culture:** See above.

Related plants (not illustrated):

Arrow-Wood (*Viburnum dentatum*). A native shrub 10–15 feet high, its ovalish leaves 2–3 inches long and coarsely toothed. Flowers white, in long-stalked flattish clusters about 3 inches wide. Fruit bluish black. Hardy from Zone 2 southward.

Viburnum pubescens. A native shrub, differing from the

Arrow-Wood only in technical characters. Neither of these related plants is superior to *Viburnum dilatatum*.

WITHEROD (*Viburnum cassinoides*) p. 343
A stout native shrub with rather late-blooming flowers; its fruit is somewhat showy in the fall. **Size:** 8–12 feet high, nearly as wide, its twigs somewhat scurfy. **Leaves:** Opposite, simple, ovalish, 3–4 inches long and finely toothed, the veins not ending at the leaf teeth. **Flowers:** Small, white, in short-stalked flattish clusters that are about 3–4 inches wide; June. **Fruit:** Ultimately bluish black, reddish when young, persistent for part of the winter. **Hardy:** From Zone 2 southward, and a useful native shrub for informal plantings. **Varieties:** None. **Culture:** See above.

Related plant (not illustrated):

Nanny-Berry (*Viburnum Lentago*). A native shrub, often treelike, 20–30 feet high, differing chiefly from the witherod in having completely stalkless flower clusters. Fruit bluish black. Hardy from Zone 2 southward.

BLACK HAW (*Viburnum prunifolium*) p. 343
A stoutish native shrub or small tree, quite widely cultivated, sometimes for its fruits, used in preserves. **Size:** 10–15 feet high, about three fourths as wide. **Leaves:** Opposite, simple, ovalish, 2–3 inches long, finely toothed, the autumnal color bright red. **Flowers:** Small, white, in a stalkless flat-topped cluster nearly 4 inches wide; May. **Fruit:** Bluish black, with a slight bloom, nearly ½ inch thick, persistent into Nov. **Hardy:** From Zone 3 southward. **Varieties:** None. **Culture:** See above.

BEAUTY-BUSH (*Kolkwitzia amabilis*) p. 358
One of the most showy of all Chinese shrubs, this deservedly popular plant is of the easiest culture. **Size:** 4–6 feet high, about three fourths as wide, its stems more or less erect. **Leaves:** Opposite, simple, ovalish, 2–3 inches long, remotely or not at all toothed on the margin, which is finely fringed with hairs (seen only with a lens). **Flowers:** Bell-shaped, about ½ inch long,

pink, but with a yellow throat, grouped in profuse terminal clusters, so numerous that the plant is spectacular; June. **Fruit:** Small, dry, bristly. **Hardy:** From Zone 3 southward. **Varieties:** None. **Culture:** Easy in any ordinary garden soil in full sun. It is an extremely attractive, relatively small shrub, introduced in 1901 and now everywhere.

Abelia
(*Abelia*)

Delightful summer-blooming shrubs, not generally hardy northward, but widely cultivated from Wilmington, Del., southward. Leaves opposite, simple, rather minutely toothed (rarely toothless), often half evergreen. Flowers profuse, mostly solitary or in pairs, rarely in leafy clusters. Corolla more or less bell-shaped or funnel-shaped, with 5 small lobes. Fruit dry and leathery, inconspicuous.

The abelias, which have no common name, are handsome shrubs which just miss being evergreen in our area, but are often so in the South. They do best in well-drained soils mixed with a fair amount of leaf mold. Plant in full sunlight. Of the 25 species the following 3 are the most popular.

Flowers pure white. — *Abelia chinensis*
Flowers blush-white or pink.
  Flowers blush-white. — *Abelia grandiflora*
  Flowers pink. — *Abelia schumanni*

ABELIA CHINENSIS p. 355

A relatively small Chinese shrub, its foliage not evergreen even in the South, its white flowers lasting from summer to autumn. **Size:** 3–5 feet high, about three fourths as wide. **Leaves:** Opposite, simple, oval, about 1 inch long, finely toothed. **Flowers:** White, funnel- or bell-shaped, about ½ inch long, grouped in dense terminal clusters; July–Sept. **Fruit:** Negligible. **Hardy:** From Zone 6 southward. **Culture:** See above. It is little known except in the South.

Abelia chinensis, p. 354.

ABELIA GRANDIFLORA p. 358

The most widely planted and best of the abelias, this hybrid is valued for its long-continued bloom and lustrous, half-evergreen foliage. **Size:** 4–6 feet high, as wide or even more, as it makes a round-headed bush in age. **Leaves:** Opposite, simple, shining green, ovalish, about 1 inch long, half evergreen in the South, less so northward, becoming bronzy in the fall. **Flowers:** Blush-white, bell-shaped, about ¾ inch long, usually borne in pairs, but the pairs very numerous, in leafy clusters; July to frost. **Fruit:** Negligible. **Hardy:** From Zone 4 southward, but precariously so in the northern part of Zone 4. **Varieties:** None superior to the typical plant (but see related plant below). **Culture:** See above.

Related plant (not illustrated):

Edward Goucher, a hybrid between *Abelia grandiflora* and *Abelia schumanni,* is a taller, more handsome shrub, sometimes nearly evergreen, and blooms from June to frost. It is often (incorrectly) offered as *Abelia zanderi* (a Chinese species little known in cultivation).

ABELIA SCHUMANNI p. 358

A Chinese shrub, not evergreen, grown mostly for its small pink flowers. **Size:** 3–6 feet high, about three fourths as wide, the young twigs dark purple. **Leaves:** Opposite, simple, ovalish, ¾–1¼ inches long, minutely toothed on the margin, or with no teeth. **Flowers:** Pink, about 1 inch long, borne singly, but quite profuse, the corolla funnel- or bell-shaped; July–Sept. **Fruit:** Negligible. **Hardy:** From Zone 6 southward. **Varieties:** None. **Culture:** See above.

Leaves without Marginal Teeth or Essentially So

Trees.
- *Flowers blooming with or before leaf expansion.* — Empress Tree
- *Flowers blooming after leaf expansion.* — *Catalpa*

Shrubs.
- *Shrubs 3–10 feet high or more.* — See p. 362
- *Plants usually less than 3 feet high.*
  - *Juice milky.* — Periwinkle
  - *Juice not milky; foliage aromatic when crushed.* — Mint Family

EMPRESS TREE (*Paulownia tomentosa*)
Family Scrophulariaceae p. 358
A striking, quick-growing Chinese tree from the wood of which the Japanese make very beautiful trays, boxes, etc. **Size:** 30–50 feet high, about three fourths as wide, its large leaves casting a dense shade. **Leaves:** Opposite, simple, wavy-margined but without teeth, oval, 5–10 inches long on mature shoots, nearly 2 feet long on young twigs or in the seedling stage; without autumn color. **Flowers:** Irregular, pale violet, but yellow-striped inside, about 2 inches long, sticky-hairy on the outside, borne in a profuse pyramidal cluster 10–15 inches long and very showy; Apr.–May, before or with leaf expansion. **Fruit:** A conical capsule about 1 inch long, many-seeded. **Hardy:** From Zone 4 southward, but the flowers are often frost-blasted in the northern part of its hardiness range. **Varieties:** None superior to the typical form. **Culture:** Difficult to transplant as it has a deep taproot which must be preserved and handled without injury. Easily raised from seed planted ½ inch deep when available (usually in Sept.). Buy only young nursery stock. In its early stages the quickest-growing of any tree — sometimes 10–15 feet in a single season. Not long-lived.

CATALPA
(*Catalpa*)
Family Bignoniaceae

Quick-growing, not very permanent trees, grown for their handsome summer-blooming flowers. Leaves long-stalked, large, opposite, simple, without marginal teeth, but sometimes obscurely lobed. Flowers irregularly cup-shaped followed by a long cylindric pod, the seeds of which have a tuft of hairs at each end.

The catalpas, often called Indian Bean, are rampant growers in any ordinary garden soil, but their wood is weak and they should not be planted for permanent effects. Their showy flowers are produced in large branched clusters.

Leaves with a short point; flowers about 1¾ inches wide.
Indian Bean

Leaves with a long point; flowers about 2½ inches wide.
Hardy Catalpa

INDIAN BEAN (*Catalpa bignonioides*) p. 358
A native of the southeastern states, this showy, summer-blooming tree is widely planted in spite of its comparatively short life. **Size:** 30–40 feet high, about as wide; very quick-growing in youth, round-headed in age. **Leaves:** Opposite, simple, oval, without marginal teeth, 6–8 inches long, bad-smelling when crushed, long-stalked, and short-pointed at the tip. **Flowers:** White, but yellow-striped inside, about 1½ inches long, irregular, borne in large, showy, upright clusters, 6–9 inches high; June–July. **Fruit:** A cylindric pod, pencil-thick, 9–14 inches long, persistent into the autumn. **Hardy:** From Zone 3 southward. **Varieties:** *nana,* the Umbrella Catalpa, is a dwarf tree with a dense, umbrella-shaped crown, useful for formal planting. It is commonly, but incorrectly, sold as *Catalpa bungei.*

HONEYSUCKLE FAMILY (*Caprifoliaceae*), 2, 6, 8
MINT FAMILY (*Labiatae*), 4, 7
TRUMPET-CREEPER FAMILY (*Bignoniaceae*), 1, 9
3, 5 in various families

1. **Indian Bean** (*Catalpa bignonioides*) p. 358
The flowers of this tree are not quite so showy as those of No. 9, and the tree is not as hardy.

2. **Beauty-Bush** (*Kolkwitzia amabilis*) p. 353
A superlatively showy, medium-sized Chinese shrub, its bell-shaped flowers blooming in June.

3. **Empress Tree** (*Paulownia tomentosa*) p. 356
Only the upper leaves are shown. The mature leaves are much larger. Flowers bloom usually before the leaves expand.

4. **Winter Savory** (*Satureia montana*) p. 361
A low half-evergreen shrub, related to the annual summer savory. Not quite hardy northward.

5. **Periwinkle** (*Vinca minor*) p. 360
One of the best evergreen ground covers, prostrate and vine-like; good for shade or open places.

6. ***Abelia schumanni*** p. 355
A lower relative of No. 8, rarely over 3 feet high, its flowers pink and summer-blooming.

7. **Rosemary** (*Rosmarinus officinalis*) p. 361
The aromatic foliage retains its odor even when dry. Of easy culture in open, sandy places.

8. ***Abelia grandiflora*** p. 355
A valuable half-evergreen hybrid shrub, 4–6 feet high, its white, or pink flowers long-blooming.

9. **Hardy Catalpa** (*Catalpa speciosa*) p. 360
A very sturdy but not long-lived midwestern tree with more showy flowers than No. 1. Very hardy.

2
3
4
5
6
7
8
9

1
2
3
4
5
6
7
8
9

OLIVE FAMILY (*Oleaceae*), 2, 5, 9
HONEYSUCKLE FAMILY (*Caprifoliaceae*), 4, 6
1, 3, 7, 8 in various families

1. **Oleander** (*Nerium Oleander*) p. 362
   A subtropical shrub with handsome pink or white flowers, but dangerously poisonous juice and fruit.

2. **Fontanesia** (*Fontanesia fortunei*) p. 369
   A favorite in midwestern gardens, this Chinese shrub is 6–12 feet high and blooms in May or June.

3. **Button-Bush** (*Cephalanthus occidentalis*) p. 366
   A native shrub favoring moist places, its showy, ball-like white flower clusters blooming in summer.

4. ***Lonicera fragrantissima*** p. 363
   Half evergreen in the North, this Chinese shrub is ravishingly fragrant when in bloom around the end of winter.

5. **Wax Privet** (*Ligustrum japonicum*) p. 365
   The best of all the evergreen privets, this Asiatic shrub is not safely hardy north of Wilmington, Del.

6. ***Lonicera nitida*** p. 364
   A tiny gem from China, its very small, crowded evergreen leaves a delight. Not hardy northward.

7. ***Gardenia jasminoides fortuniana*** p. 363
   A moderately hardy form of the florists' gardenia, but not certainly hardy north of Norfolk, Va.

8. ***Buddleia davidi*** p. 367
   An outstanding Chinese shrub, its showy flowers blooming from July to frost. Other forms with pink, red, or white flowers are available.

9. **Mock Privet** (*Phillyrea decora*) p. 364
   An early-blooming, evergreen Asiatic shrub with small white flowers. Not hardy northward.

HARDY CATALPA (*Catalpa speciosa*) p. 358
A tough tree from the Midwest, showy in flower and suited to dry, windy sites. **Size:** 40–60 feet high, nearly as wide, the bark reddish brown. **Leaves:** Opposite, simple, oval, long-stalked, long-pointed at the tip, 8–12 inches long, densely hairy on the under side, not malodorous when crushed. **Flowers:** White, irregular, yellow-striped inside, about 2 inches long, borne in a pyramidal cluster about 7 inches high; June. **Fruit:** A cylindric pod, pencil-thick, 12–20 inches long. **Hardy:** From Zone 3 southward. **Varieties:** None. **Culture:** See above. An extremely hardy if somewhat rank tree.

PERIWINKLE (*Vinca minor*). Family Apocynaceae p. 358
A splendid evergreen ground cover, often called Ground Myrtle, from the Old World, growing especially well in the shade, but also standing open sunshine. **Size:** Prostrate, vinelike, never more than 6 inches high, its stems wiry and with a milky juice. **Leaves:** Opposite, simple, lance-shaped, dark lustrous green, about 2 inches long. **Flowers:** Light blue, solitary at the leaf joints, about ¾ inch wide; Apr. and sporadically through the summer. **Fruit:** Small, dry, cylindric, inconspicuous. **Hardy:** From Zone 4 southward, but not satisfactory in the far South. **Varieties:** *alba* has white flowers; *atropurpurea* has pink flowers; Bowles variety is a stronger-growing form with light blue flowers; there are several others with variegated foliage. **Culture:** Will grow anywhere, in all garden soils, but does best in the shade.

MINT FAMILY
(*Labiatae*)

A HUGE family of mostly herbaceous plants, but also with a few low shrubs all having opposite, simple leaves, without marginal teeth in those below, the foliage very aromatic when crushed, and in some without crushing. Flowers always irregular, usually small and often crowded in small tight clusters. Fruit a collection of small nutlets. The stems are usually 4-sided or at least angled. Some that are technically shrubs, but low and herblike, are the lavender and thyme. For these see *Guide to Garden Flowers,* pp. 189 and 201.

For other plants in this family, which have leaves that are toothed on the margin, see Russian Sage, *Elsholtzia,* and Jerusalem Sage (pp. 335 and 336). The 2 below may be separated thus:

Flowers blue. Rosemary
Flowers white or purplish. Winter Savory

ROSEMARY (*Rosmarinus officinalis*) p. 358
A delightfully fragrant evergreen sub-shrub from the Mediterranean region, its foliage retaining its aromatic scent even when dry. **Size:** 1–3 feet high, rarely more, and in maturity making large patches 6–8 feet wide. **Leaves:** Opposite, simple, narrowly lancelike, about 1 inch long, without marginal teeth, but the margins rolled, very profuse. **Flowers:** Irregular, pale blue, in clusters at the leaf joints, as fragrant as the foliage; Apr. (or earlier in the South), sporadically thereafter. **Fruit:** Small, dry, nutlike, and inconspicuous. **Hardy:** From Zone 4 (in protected sites) southward. **Varieties:** There is a white-flowered form, little known here, more common in England. **Culture:** Needs a relatively dry sandy or gritty soil, preferably in full sun but sheltered from north winds, especially in Zone 4. A rampant grower in the South.

WINTER SAVORY (*Satureia montana*) p. 358
A low, aromatic, half-evergreen, herblike shrub from southern Europe and north Africa, grown mostly as a seasoning plant. **Size:** 8–15 inches high, its slightly hairy stems erect. **Leaves:** Opposite, simple, rigid, lancelike, about 1 inch long, without marginal teeth, the tip almost prickly. **Flowers:** Irregular, white or purplish, small, crowded in loose clusters that are leafy and terminal; Aug.–Sept. **Fruit:** Small, nutlike, inconspicuous. **Hardy:** From Zone 5 southward, and in protected sites in Zone 4. **Varieties:** *subspicata* has the flowers in terminal spikes. **Culture:** Easily grown in any ordinary garden soil, preferably in full sun.

Shrubs 3–10 Feet High or More

Leaves normally falling in autumn, not evergreen. See p. 365
Leaves evergreen or half evergreen.
  Juice milky, dangerously poisonous. Oleander
  Juice not milky.
    Flowers very fragrant, summer-blooming. Gardenia
    Flowers, if fragrant, blooming in winter or early spring. *Lonicera fragrantissima*
    Flowers small, inconspicuous.
      Flowers on the side of twigs. Mock Privet
      Flowers in terminal clusters.
        Leaves scarcely ½ inch long. *Lonicera nitida*
        Leaves more than 1 inch long. Wax Privet

OLEANDER (*Nerium Oleander*)
Family Apocynaceae p. 359
A beautiful evergreen shrub or small tree, its juice and fruits dangerously poisonous if eaten, but harmless to the touch. **Size:** 8–25 feet high, rarely reaching the highest figure except in the Deep South; juice milky. **Leaves:** Opposite or in clusters, simple, without marginal teeth, narrowly oblong, 6–8 inches long, dark green above, paler and with a prominent midrib beneath. **Flowers:** White, red, pink, or purple, about 2½ inches wide, often double in some varieties, the corolla slightly twisted; midsummer in the North, from Apr. to Oct. in the Deep South. **Fruit:** A long cylindric pod. **Hardy:** From Zone 6 southward, precariously hardy in protected sites in Zone 5 (definitely risky). **Varieties:** There are many, and a choice might include the following:

Brilliant. Single, rose-red.
Cardinal. Brilliant red, and the hardiest.
Jannock. Single, bright red.
Mrs. Roeding. Double, salmon-pink.
Mrs. Swanson. Double, soft pink.
Sister Agnes. Single, pure white.

Culture: Will grow in any ordinary garden soil, in full sun. In the South it stands dust, smoke, wind, and salt spray and is much used to line walks and drives; set at least 10 feet apart as it has a tendency to spread. Treelike only if all but one stem is pruned off at ground level.

GARDENIA (*Gardenia jasminoides fortuniana*)
Family Rubiaceae p. 359
A spicily fragrant, very aromatic evergreen Chinese shrub, long called the Cape Jasmine because it was erroneously credited to South Africa. **Size:** 2–5 feet high, about as wide, and densely furnished with shining evergreen foliage. **Leaves:** Opposite or in threes, simple, without marginal teeth, broadly lance-shaped, 3–4 inches long, thick and leathery. **Flowers:** White, very fragrant, 2–3½ inches wide, solitary, slightly twisted, the deeply cut corolla appearing (falsely) as of separate petals; midsummer in the North. **Fruit:** Leathery or fleshy, stalkless, inconspicuous. **Hardy:** From Zone 6 southward, precariously so in protected sites in Zone 5 and definitely risky there. **Varieties:** This shrub is itself a variety of *Gardenia jasminoides,* which is the familiar florists' gardenia — a greenhouse plant. Our hardy variety, the only one safe for outdoor planting, has one variety known as Belmont which has somewhat larger flowers; the so-called *Gardenia radicans* is a form of *Gardenia jasminoides* with lower stature and smaller flowers. **Culture:** Not easy. It needs a good rich, decidedly acid soil (pH 5.5–6), preferably a moist site (not wet), full sunshine (although it will grow in partial shade), and a 2-inch mulch of pine needles, sawdust, or other light mulching material. Keep it out of windy sites. Often offered as *Gardenia veitchi.*

LONICERA FRAGRANTISSIMA
Family Caprifoliaceae p. 359
An extremely fragrant Chinese shrub, evergreen in the South and half evergreen northward, with rather small but numerous flowers. **Size:** 8–10 feet high, about as wide, and round-headed in maturity, the twigs smooth. **Leaves:** Opposite, simple, without marginal teeth, oval, thick, 1–2 inches long. **Flowers:** Creamy white, slightly irregular, about ½ inch long, very fragrant, almost

stalkless, and numerous; Jan.–Mar., or even Apr. in the North. **Fruit:** A small red berry, not persistent and much eaten by birds. **Hardy:** From Zone 4 southward, possibly in protected sites in Zone 3. **Varieties:** None. **Culture:** Easy in any ordinary garden soil. With shearing it makes a fairly good informal hedge

Related plant (not illustrated):

Lonicera standishi. A partly evergreen Chinese shrub, 6–8 feet high, with bristly young twigs, fragrant white flowers (Apr.), and red fruit. Hardy from Zone 3 southward.

MOCK PRIVET (*Phillyrea decora*)

Family Oleaceae p. 359

An upright evergreen Asiatic shrub, not widely grown, but with attractive foliage and clusters of small white flowers. **Size:** 6–9 feet high, about three fourths as wide. **Leaves:** Opposite, simple, leathery, oblongish or narrower, 2½–5 inches long, yellowish green on the under side, without marginal teeth or with a few remote ones. **Flowers:** Small, white, in a pointed, dense cluster; Apr.–May. **Fruit:** Berrylike, black, football-shaped; persistent into Nov. **Hardy:** From Zone 5 southward. **Varieties:** None. **Culture:** Easy in any ordinary garden soil.

Related plant (not illustrated):

Phillyrea angustifolia. A dense European evergreen shrub, 8–10 feet high, the narrow leaves about 2½ inches long. Flowers small, white, fragrant; May–June. Fruit black. Hardy from Zone 5 southward.

LONICERA NITIDA. Family Caprifoliaceae p. 359

A dainty evergreen Chinese shrub, distinguished by its very numerous small leaves and its wide-spreading slender branches. **Size:** 2–4 feet high (rarely more) and nearly or quite as wide. **Leaves:** Opposite, simple, evergreen, ovalish or oblong, scarcely ½ inch long, very numerous. **Flowers:** Minute, white, fragrant, often wanting; May. **Fruit:** Berrylike, small, purplish. **Hardy:** From Zone 5 southward, and perhaps in mild parts of Zone 4. **Varieties:** None. **Culture:** Easy in any ordinary garden soil, but choose a place sheltered from north and west winds. A delightful little evergreen.

Related plant (not illustrated)

Lonicera pileata. A closely related Chinese evergreen shrub, scarcely over 1 foot high, its leaves about 1 inch long. Flowers white, small. Hardy from Zone 5 southward.

WAX PRIVET (*Ligustrum japonicum*)

Family Oleaceae p. 359

A very vigorous evergreen shrub from Japan and Korea, much used as a hedge in the South but not hardy northward. **Size:** 7–10 feet high, about as wide, and densely foliaged. **Leaves:** Opposite, simple, oblong-oval, 2–4 inches long, without marginal teeth, thick and leathery. **Flowers:** Small, white, the clusters 4–6 inches long; July–Sept. **Fruit:** Small, berrylike, black, winter-persistent. **Hardy:** From Zone 5 southward. **Varieties:** Several with variegated leaves; more useful is the variety *rotundifolium,* which is a lower shrub with many roundish leaves that are darker green (sometimes offered as *Ligustrum coriaceum*). **Culture:** Easy in any ordinary garden soil, in full sun.

Related plant (not illustrated):

Ligustrum lucidum. Closely related to the above, this Chinese plant may be treelike, 15–30 feet high, its leaves 4–6 inches long, and the flower cluster 8–9 inches high. Hardy from Zone 5 southward.

Leaves Normally Falling in Autumn, Not Evergreen, the Margins Not Toothed or Only Slightly So

Leaves with stipules (see Picture Glossary).

Flowers in a ball-like cluster. Button-Bush

Flowers in terminal spikes. Butterfly-Bush

Leaves without stipules (see Picture Glossary). See p. 368

BUTTON-BUSH (*Cephalanthus occidentalis*)
Family Rubiaceae p. 359
While an inhabitant of swamps over much of North America, this shrub will grow in ordinary garden soil and is grown mostly for its late flowers which bloom when few shrubs are in flower. **Size:** 5–12 feet high, nearly as wide, the branches spreading. **Leaves:** Simple, opposite or in clusters, ovalish, 3–5 inches long, stalked, and without marginal teeth. **Flowers:** Small, tubular, fragrant, white, crowded in dense, ball-like, stalked clusters that are about 1½ inches thick; midsummer and later. **Fruit:** Small, dry, nutlike. **Hardy:** Everywhere. **Varieties:** None. **Culture:** Easy in wet or swampy places, but tolerant of most garden soils. Not suited to dry sites.

BUTTERFLY-BUSH
(*Buddleia*)
Family Loganiaceae

Spectacularly beautiful summer-blooming shrubs, unfortunately not quite hardy everywhere, so that the hardiness notes below should be studied before planting. Leaves simple, opposite (except in one species), without marginal teeth or with very remote and small ones; usually scurfy or hairy, especially on the under side. Flowers small, in dense clusters in one but in long spirelike spikes in the other. Fruit a small dry capsule. They are often called Summer Lilac.

The butterfly-bush needs full sun and a good rich garden soil. In many places in the North the plants may be winter-killed to the ground, but will bloom next season from new shoots. A winter mulch of strawy manure may avoid some winter-killing. The following are the ones most likely to be grown:

Leaves alternate; flower cluster dense. *Buddleia alternifolia*
Leaves opposite; flower cluster spirelike. *Buddleia davidi*

BUDDLEIA ALTERNIFOLIA p. 367
The only cultivated butterfly-bush with alternate leaves, this beautiful Chinese summer-flowering shrub is not hardy north-

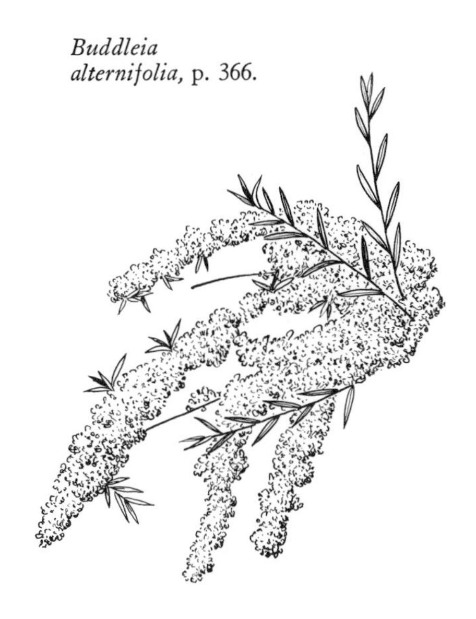

Buddleia alternifolia, p. 366.

ward, but much grown in the South. **Size:** 6–10 feet high, and as wide or even wider from its wide-spreading, arching branches. **Leaves:** Simple, opposite, lance-shaped, 1–4 inches long, without marginal teeth, grayish and scurfy beneath. **Flowers:** Lilac-purple, small, tubular, crowded in dense, leafy clusters and very showy; June–July. **Fruit:** A small capsule, usually surrounded by the withered flower. **Hardy:** From Zone 5 southward. **Varieties:** None. **Culture:** See above. It cannot be cut to the ground as is often necessary with the next species.

Related plant (not illustrated):

Buddleia globosa. A Peruvian shrub, 10–15 feet high, its foliage nearly evergreen. Flowers yellow in long-stalked, dense heads. Hardy only in protected sites in Zone 6 and southward.

BUDDLEIA DAVIDI p. 359

The finest of the cultivated butterfly-bushes and blooming from mid-July to frost, this Chinese shrub is an outstanding garden plant. **Size:** 4–10 feet high if left alone, but as cut back to the ground, usually less. **Leaves:** Simple, opposite, lance-oval, 6–9 inches long, very finely toothed on the margin, green above, white-scurfy beneath. **Flowers:** Fragrant, lilac, but orange-spotted at the throat, small, tubular, crowded in a nodding spike 5–12 inches long and very showy; July to frost. **Fruit:** A small capsule. **Hardy:** From Zone 5 southward, without cutting the stems to the ground for the winter. If this is necessary, as in Zones 4 and 3, the new spring shoots will flower that summer. Not hardy north of Zone 3 and sometimes precariously so even there. **Varieties:** Over 30; a reasonable selection might include: Charming, pink; Dubonnet, dark purple; Empire Blue, bluish; Fascination, orchid-pink; Isle de France, dark purple; Royal Red, red; White Bouquet, white. Most of these varieties are superior to the typical form. **Culture:** Easy in any ordinary garden soil, but see above, especially the hardiness notes.

Leaves without Stipules

(For definition and illustration of stipules see the Picture Glossary.)

Flowers regular.
  (*For difference between regular and irregular flowers see the Picture Glossary.*)
    Fruit dry.
      Fruit winged. — *Fontanesia*
      Fruit not winged. — Lilac
    Fruit fleshy.
      Flower cluster drooping. — Fringe-Tree
      Flower cluster not drooping, usually terminal and erect. — Privet
Flowers irregular (only slightly so in Clerodendron and in some bush honeysuckles).
    Flower cluster 6–8 inches wide. — *Clerodendron*

Flower cluster less than 6–8 inches wide.
Flowers in clusters; fruit a pulpy berry. *Symphoricarpos*
Flowers mostly in pairs (in a small cluster in Lonicera heckrotti). Bush Honeysuckle

FONTANESIA (*Fontanesia fortunei*)
Family Oleaceae p. 359
An upright Chinese shrub, not particularly showy, but grown for ornament as the flowers are borne in a leafy cluster. It is more popular in gardens in the Midwestern states than eastward. **Size:** 6–12 feet high, about half as wide as its branches are rather upright. **Leaves:** Simple, opposite, narrowly lance-shaped, 2–4 inches long, without marginal teeth and shining green. **Flowers:** Small, greenish white, in leafy clusters, not showy; May–June. **Fruit:** A small, flat, ovalish, winged nutlet about ¾ inch long. **Hardy:** From Zone 4 southward. **Varieties:** None. **Culture:** Easy in any ordinary garden soil, in full sun.

LILAC
(*Syringa*)
Family Oleaceae

The Latin name of the lilacs is *Syringa* and confuses many as Syringa is also the (incorrect) common name of what is better called Mock-Orange (*Philadelphus*), for which see p. 157. The lilacs are sweet-smelling shrubs, rarely low trees, with simple, opposite leaves that have no marginal teeth, but are sometimes lobed in the Persian Lilac. The flowers, prevailingly fragrant, are small, salver-shaped, and with 4 short, usually spreading lobes; showy because they are borne in profuse large clusters. Many of the hybrids are double-flowered. Fruit a dry capsule, of no decorative value.

While there are only about 30 species of lilac, the French have produced hundreds of hybrids that are finer, but often not so fragrant as some of the older species. A list of some of the more desirable of these horticultural varieties follows the list of species.

Lilacs are not difficult to grow and thrive in a great variety of garden soils. They are somewhat slow to start, but a top-dressing of well-rotted manure, spaded in, will increase growth and the size of the flowering truss, but this should not be done more than once in two or three years as, if the soil is too rich, it may promote leafy growth at the expense of flowers. Remove all withered blossoms and the capsules that follow them, to improve bloom the next year. Lilacs are apt to send up many suckers. Most of these should be cut off at ground level, unless you want the shrubs to spread beyond the space you have chosen for them. Plants slow to flower may often be in too rich a soil, and should not be manured until flowering becomes regular. In old plants it is well to remove, every three years, the old stems, allowing newer growth to take over as it usually bears flowers more profusely than overmature wood.

Of the 15 species in the 4th edition of *Taylor's Encyclopedia of Gardening,* the following 7 are of chief interest to the amateur. The characters in the key apply only to the *species* below, not to the horticultural forms, which are very variable.

Flowers whitish yellow. Amur Lilac
Flowers lilac, lilac-purple or lilac-violet.
  Flower clusters from terminal buds.
    Flower lobes upright. Hungarian Lilac
    Flower lobes spreading. *Syringa villosa*
  Flower clusters from lateral buds.
    Leaves hairy on under side, not over 1½ inches long. *Syringa microphylla*
    Leaves smooth on under side, over 2 inches long.
      Leaves ovalish. Common Lilac
      Leaves more or less oblong.
        Flower clusters 2–3 inches long. Persian Lilac
        Flower clusters 4–6 inches long. Rouen Lilac

AMUR LILAC (*Syringa amurensis*) p. 374
An Asiatic shrub, but often a small tree or treelike, valued for its late bloom, which comes about a month after the Common Lilac. **Size:** 5–15 feet high, about three fourths as wide. **Leaves:** Simple, smooth, opposite, without marginal teeth, broadly oval and nearly 5 inches long. **Flowers:** Whitish yellow, the cluster nearly 6 inches long; June. **Fruit:** Negligible. **Hardy:** From Zone 3 southward. **Varieties:** The variety *japonica* is sometimes a tree 30 feet high and has hairy leaves. **Culture:** See above.

Related plant (not illustrated):

Syringa pekinensis. A showy Chinese shrub, 10–15 feet high, with whitish-yellow, unpleasantly scented flowers in June. Hardy from Zone 4 southward.

HUNGARIAN LILAC (*Syringa Josikaea*) p. 372
A Hungarian shrub, not particularly showy, but blooming nearly a month later than the Common Lilac. **Size:** 7–12 feet high, about three fourths as wide. **Leaves:** Simple, opposite, without marginal teeth, glossy green above, ovalish, 2–5 inches long, whitish beneath. **Flowers:** Fragrant, deep lilac to violet, in rather sparse clusters; June. **Fruit:** Negligible. **Hardy:** From Zone 3 southward. **Varieties:** None. **Culture:** See above.

Related plant (not illustrated):

Syringa reflexa. A Chinese shrub, 7–12 feet high, its nodding flower cluster pink and fragrant in June. Hardy from Zone 3 southward.

SYRINGA VILLOSA p. 373
An upright, strong-growing Chinese shrub, flowering later than the Common Lilac and not quite so fragrant. **Size:** 7–10 feet high, about half as wide as its branches are more or less erect. **Leaves:** Simple, opposite, without marginal teeth, ovalish but pointed both ends, 3–6 inches long, whitish beneath. **Flowers:** Fragrant, lilac-pink or paler, borne in terminal hairy-stalked clusters 5–9 inches long and showy; May–June. **Fruit:** Negligible. **Hardy:** From Zone 3 southward. **Varieties:** None. **Culture:** See above.

Related plant (not illustrated):

Syringa sweginzowi. A Chinese shrub 5–9 feet high with leaves 2–4½ inches long. Flowers reddish or pale lilac, fragrant, in June. Hardy from Zone 3 southward.

SYRINGA MICROPHYLLA p. 373

A small-leaved Chinese shrub having also small flowers and sometimes blooming again in the fall. **Size:** 3–5 feet high, about as wide or wider as its slender branches are more spreading than in most lilacs. **Leaves:** Simple, opposite, without marginal teeth, roundish, ½–1½ inches long, hairy on the under side. **Flowers:** Fragrant, lilac, small, and in lateral clusters 1–3 inches long, hence not showy; May–June, and sometimes, sparingly, in the fall. **Fruit:** Negligible. **Hardy:** From Zone 3 southward. **Varieties:**

Left: Rouen Lilac (*Syringa chinensis*), p. 376.
Right: Hungarian Lilac (*Syringa Josikaea*), p. 371.

One offered as *superba* (a doubtfully authentic name) is reputed to be finer than the typical plant. **Culture:** See above.

Related plant (not illustrated):

Syringa velutina. A small Chinese and Korean shrub, 3–8 feet high, with oblongish leaves 2–3 inches long, densely hairy beneath. Flowers lilac in hairy clusters 4–8 inches long; May. Hardy from Zone 3 southward.

COMMON LILAC (*Syringa vulgaris*) p. 374

The lilac of history and the poets, this supremely fragrant shrub of southeastern Europe (especially Hungary) has been grown for centuries by Europeans and since colonial times by us. **Size:** 10–20 feet high, widely spreading, sometimes treelike and always with numerous suckers (which should be removed; see above). **Leaves:** Simple, opposite, without marginal teeth, heart-shaped to oval, 2–6 inches long. **Flowers:** Fragrant, lilac (except in horticultural varieties), borne in showy clusters 6–8 inches long; May. **Fruit:** Negligible. **Hardy:** From Zone 2 southward. **Varieties:** Hundreds; see below. **Culture:** See above.

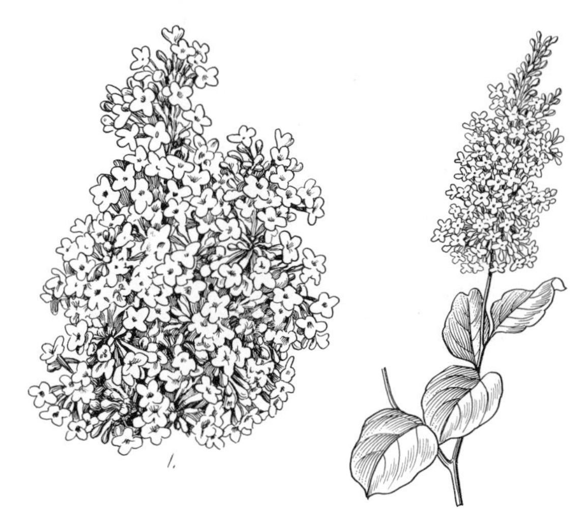

Left: *Syringa microphylla,* p. 372. Right: *Syringa villosa,* p. 371.

OLIVE FAMILY (*Oleaceae*), 1, 4, 5, 6, 7, 9
HONEYSUCKLE FAMILY (*Caprifoliaceae*), 2, 3
VERVAIN FAMILY (*Verbenaceae*), 8

1. ***Ligustrum sinense*** p. 380
An almost evergreen Chinese privet, much planted in the South but not certainly hardy north of Norfolk, Va.

2. **Tartarian Bush Honeysuckle** (*Lonicera tatarica*) p. 384
A commonly planted Asiatic shrub, good for shrub borders or a screen but not showy in bloom.

3. **Snowberry** (*Symphoricarpos albus laevigatus*) p. 381
The persistent white fruits of this city-tolerant shrub make it a welcome plant for unfavorable sites.

4. **Persian Lilac** (*Syringa persica*) p. 376
A low Asiatic shrub, its fragrant flowers borne in a flat-topped, very profuse cluster. Quite hardy.

5. **Fringe-Tree** (*Chionanthus virginicus*) p. 377
A native shrub or small tree with lacy white flowers in hanging clusters. The male flowers are most showy.

6. **California Privet** (*Ligustrum ovalifolium*) p. 380
This Japanese shrub is the most widely grown of any privet, since it can be sheared and is reasonably hardy.

7. **Amur Lilac** (*Syringa amurensis*) p. 371
A tree-like Asiatic shrub, valued for its bloom, which comes about a month after the Common Lilac.

8. ***Clerodendron trichotomum*** p. 380
This Asiatic shrub is much more showy in fruit than in flower. Easily grown in ordinary garden soil.

9. **Common Lilac** (*Syringa vulgaris*) p. 373
The most widely grown of all lilacs and very fragrant. Many fine hybrids have been developed, and some lack the fragrance of the original species. For lists see the text.

1
2
3
4
5
6
7
8
9

1
2
3
4
5
6
7
8
9

HEATH FAMILY (*Ericaceae*), 1–4, 7, 9
5, 6, 8 in various families

1. **Rhodora** (*Rhodora canadensis*) p. 397
A beautiful native azalea-like shrub that blooms before or with the expansion of the leaves.

2. **Cross-leaved Heath** (*Erica Tetralix*) p. 389
A small evergreen European shrub with tiny leaves and many small flowers. Culture not easy.

3. **Cornish Heath** (*Erica vagans*) p. 392
Not a foot high, this European shrub is a profuse bloomer. Flowers in late summer. Culture not easy.

4. **Sand Myrtle** (*Leiophyllum buxifolium*) p. 387
A native evergreen shrub. Difficult to grow outside of its moist, highly acid sites.

5. **Matrimony-Vine** (*Lycium chinense*) p. 394
An Asiatic shrub good for covering dry banks, its twigs prickly. Flowers in June and later.

6. **Storax** (*Styrax japonica*) p. 395
An attractive Asiatic, medium-sized tree, its waxy, fragrant white flowers blooming in summer.

7. ***Phyllodoce empetriformis*** p. 388
A Pacific Coast evergreen, not over 1 foot high. Urn-shaped flowers bloom in early spring.

8. **Persimmon** (*Diospyros virginiana*) p. 394
A native tree, its male and female flowers on separate trees. Fruit wanting on male trees.

9. **Mollis Azalea Hugo Koster** p. 399
For details see text.

PERSIAN LILAC (*Syringa persica*) p. 374

One of the smallest of the lilacs, this shrub is native from Persia to northwestern China, and is a profuse bloomer. **Size:** 5–6 feet high, about as wide, of compact habit. **Leaves:** Simple, opposite, without marginal teeth, but sometimes lobed, lance-shaped, about 2½ inches long. **Flowers:** Fragrant, small, pale lilac, in short, broad clusters that are about 3 inches wide and very profuse; May. **Fruit:** Negligible. **Hardy:** From Zone 3 southward. **Varieties:** *laciniata* (considered a species by some) has deeply lobed leaves. **Culture:** See above.

ROUEN LILAC (*Syringa chinensis*) p. 372

A hybrid lilac originating in the botanical garden at Rouen during our Revolution was the result of a chance crossing of the Common Lilac and the Persian Lilac. It is widely cultivated and is about half the size of the Common Lilac. Its lilac-purple flower clusters are looser and half as long as that species; May. **Hardy:** From Zone 4 southward. **Varieties:** *alba* has white flowers; *saugeana* has lilac-red flowers. **Culture:** See above.

Hybrid Lilacs

The horticultural forms of the Common Lilac comprise hundreds of varieties, most of them hybrids of French origin. To list even a quarter of them here is impossible, but a selection of 25 of them will enable the amateur to choose the color of his choice, but not always the fragrance. Sometimes fragrance has been lost in the process of hybridizing. Those known to be fragrant are indicated by a dagger (†), and the list is separated into those with single or double flowers.

DOUBLE-FLOWERED

White: †Miss Ellen Willmott, Edith Cavell
Lilac or rosy lilac: †Waldeck Rousseau, President Fallières
Purple-red: †De Saussure, Paul Thirion, Mrs. Edward Harding

Blue: †President Grévy, Olivier de Serres
Pink: Mme. Antoine Buchner, Katherine Havemeyer, Montaigne

SINGLE-FLOWERED

White: Vestal, Jan van Tol, Mont Blanc
Pink: †Lamartine, Lucie Baltet, Macrostachya, †Leon Gambetta
Lilac: Jacques Callot
Purple-red: †Congo, †Diderot, †Ludwig Spaeth
Blue: †President Lincoln, Decaisne, Maurice Barrés

FRINGE-TREE (*Chionanthus virginicus*)
Family Oleaceae p. 374
One of the most beautiful of native shrubs (sometimes treelike), found wild from N.J. to Fla. and Tex. and profuse in its production of lacy white flowers. **Size:** 10–25 feet high, nearly as wide. **Leaves:** Simple, opposite, without marginal teeth, oblongish, 6–8 inches long; late-expanding. **Flowers:** The male and female often on separate plants, the male more showy, comprising a hanging cluster 5–7 inches long of white flowers with 4 narrow petals only slightly united at the base; May–June. **Fruit:** On female plants only, fleshy, egg-shaped, about 1 inch long, blue, in grape-like clusters, not persistent. **Hardy:** From Zone 3 southward. **Varieties:** None. **Culture:** Easy in most ordinary garden soils, in full sun. An extremely decorative shrub.

PRIVET
(*Ligustrum*)
Family Oleaceae

A few of the privets are true evergreens and will hence be found at pp. 364 and 365. The ones below are not truly evergreen, although the California Privet will hold its leaves for a good deal of the winter in the milder sections of the country. Leaves opposite, simple, without marginal teeth, usually unpleasantly scented when crushed; sometimes half evergreen or at least persistent. Flowers small, white, in terminal clusters unpleasantly scented. Fruit fleshy, black or blackish.

It is almost impossible to kill privet, which will make rampant growth in any garden soil. For mild regions the best one for hedges is the California Privet, which just misses being a real evergreen. Both the California Privet and the common one are widely grown for hedges. The former is more nearly evergreen, but not quite so hardy as the Common Privet, which is the safest to use north of Zone 4. For hedges choose young plants (15–30 inches high) and plant about 18 inches apart. Cut back the twigs to about half their unpruned length. As the plants develop, shear them so that the width of the hedge is slightly broader at the base than at the top, thus allowing light to reach the base of the hedge, which will promote bushy growth and prevent the plants becoming "leggy." This is an important feature at the start of a hedge, and should be maintained during all subsequent clipping.

Of the 50 known species the following 4 are most in cultivation:

Leaves minutely hairy on the margin. Amur Privet
Leaves not minutely hairy on the margin.
  Leaves completely hairless.
    Twigs and flowering stalk minutely hairy; individual flowers stalked. Common Privet
    Twigs and flowering stalk smooth; individual flowers essentially stalkless. California Privet
  Leaves hairy on the midrib beneath. *Ligustrum sinense*

AMUR PRIVET (*Ligustrum amurense*) p. 379
One of the hardiest of the privets, from northern China, much resembling the California kind, but not quite so useful as a hedge plant. **Size:** 10–15 feet high, about half as wide, as its branches are erect; densely foliaged. **Leaves:** Simple, opposite, without marginal teeth, half evergreen, 1½–2½ inches long, hairy on the midrib beneath, and minutely so on the margin. **Flowers:** Small, white, in clusters about 3 inches long; June–July. **Fruit:** Dull, black, berrylike, pea-size. **Hardy:** From Zone 3 southward. **Vari-**

eties: None. **Culture:** See above. A very useful privet for the North, often sheared as a fairly good hedge plant.

Left: Common Privet (*Ligustrum vulgare*), below.
Right: Amur Privet (*Ligustrum amurense*), p. 378.

COMMON PRIVET (*Ligustrum vulgare*) above

A ubiquitous European shrub, so widely used for hedges that its monotony palls on some gardeners. It is, however, so reliable that it overcomes what one expert called its "fungus-like rapidity of growth and its stinking odor." **Size:** 6–15 feet high and, if not clipped into a hedge, wide-spreading. **Leaves:** Simple, opposite, without marginal teeth, oblong-oval, 1¾–2½ inches long, completely smooth, scarcely half evergreen. **Flowers:** Hardly ever produced on clipped hedges, otherwise white, small, malodorous, in clusters not over 2 inches long; July. **Fruit:** Black, shining, berrylike, pea-size. **Hardy:** From Zone 3 southward. **Varieties:** Many, mostly variations in the color of the leaves (golden, white, variegated, etc.) and not superior to the typical plant. **Culture:** See above.

CALIFORNIA PRIVET (*Ligustrum ovalifolium*) p. 374
By far the most popular hedge plant in mild regions, this Japanese shrub is far more popular in the East than in Calif. **Size:** 5–25 feet high if not clipped, rather widely spreading. **Leaves:** Half evergreen, often persistent through part of the winter in mild regions, about 2 inches long, glossy green, completely smooth. **Flowers:** Small, white, malodorous, the cluster 3–4 inches long; July. **Fruit:** Pea-size, black, berrylike. **Hardy:** From Zone 4 southward but often planted in Zone 3 where severe winters usually winter-kill its twigs or may kill the plant outright. **Varieties:** Many, mostly variations in leaf color (golden, white, variegated, etc.), none of which is as good a hedge plant as the typical form. **Culture:** See above. It is one of the most widely used hedge plants in the U.S.

Related plant (not illustrated):

Ibota Privet (*Ligustrum obtusifolium,* often called *Ligustrum Ibota*). A spreading Japanese shrub 6–10 feet high, its leaves hairy beneath. Flowers white, in July. A lower variety of it is Regel's Privet. Hardy from Zone 3 southward.

LIGUSTRUM SINENSE p. 374
A much-planted privet in the South, this Chinese shrub is almost evergreen there, but only half evergreen in the northern part of its hardiness range. **Size:** 7–12 feet high, nearly as wide, as its branches are spreading. **Leaves:** Simple, opposite, without marginal teeth, oblongish, 1½–3 inches long, hairy on the midrib beneath. **Flowers:** Small, white, the branching cluster 3–5 inches long, so profuse that this is the showiest of the privets here included. **Fruit:** Black, berrylike, pea-size. **Hardy:** From Zone 6 southward. **Varieties:** None superior to the typical form. **Culture:** See above.

CLERODENDRON (*Clerodendron trichotomum*)
Family Verbenaceae p. 374
An unusual Asiatic shrub, perhaps more showy in fruit than in

flower; not as much cultivated as it should be. **Size:** 7–20 feet high, but usually about 6–8 feet as cultivated and about three fourths as wide, its branches more or less horizontal. **Leaves:** Simple, opposite, ovalish or elliptic, 3–7 inches long, tapering at the tip, remotely or not at all toothed. **Flowers:** Slightly fragrant, white, in long-stalked clusters, and beneath each flower a reddish-brown calyx; Aug. **Fruit:** Blue, fleshy, partly enclosed by the reddish starlike calyx, and quite showy, about 1/4 inch in diameter, persistent until late Oct. **Hardy:** From Zone 4 southward. **Varieties:** None. **Culture:** Easy in any ordinary garden soil, in full sun. It has a tendency to have too many stems and for best flowers and fruit it is best to remove all but 2 or 3 main stems.

SYMPHORICARPOS
(*Symphoricarpos*)
Family Caprifoliaceae

Useful shrubs for inhospitable places as they will stand city grime, heat, dust, smoke, and wind; grown mostly for their fruits as the flowers are not showy. Leaves simple, opposite, without marginal teeth (rarely lobed on vigorous shoots), rather short-stalked. Flowers slightly irregular (see Picture Glossary), small, in rather small clusters. Fruit a generally showy berry with 2 seeds, often quite persistent.

These shrubs will grow in all garden soils, either in partly shady places or in the open. They are fine for city back yards.

Of the 16 known species the 2 below are most useful.

Fruit white.	Snowberry
Fruit red or purplish red.	Indian Currant

SNOWBERRY (*Symphoricarpos albus laevigatus*) p. 374
An extremely hardy North American shrub, widely grown in city gardens for its showy white fruits that persist for half the winter. **Size:** 5–7 feet high, about three fourths as wide, as its slender branches are nearly erect. **Leaves:** Simple, opposite, with-

out marginal teeth, ovalish or oblong, 1–2 inches long, blunt at the tip. **Flowers:** Small, not showy, but much liked by bees, pinkish, in small clusters; midsummer. **Fruit:** A berry, usually borne in pairs, pure white, about ½ inch thick, profuse and persistent into midwinter. **Hardy:** Everywhere. **Varieties:** This is itself a variety and about twice the height of the typical form. It is the best of the snowberries. **Culture:** See above.

INDIAN CURRANT (*Symphoricarpos orbiculatus*) below
A native North American shrub, widely grown for its fruit as the flowers are of little interest; often called Coralberry. **Size:** 5–7 feet high and about three fourths as wide, the branches more or less erect. **Leaves:** Simple, opposite, without marginal teeth, ovalish or elliptic, 1½–2½ inches long, turning red in autumn. **Flowers:** Small, white, in small clusters; midsummer. **Fruit:** A fleshy reddish purple berry, not white-dotted, borne in great profusion along the twigs and persistent into Jan. **Hardy:** From Zone 3 southward. **Varieties:** *variegatus* has yellow-blotched leaves. **Culture:** See above. It is an excellent shrub for city gardens or in a shrub border.

Indian Currant (*Symphoricarpos orbiculatus*), above.

BUSH HONEYSUCKLE
(*Lonicera*)
Family Caprifoliaceae

Rather coarse, somewhat rampant shrubs, useful as screen planting but of slight decorative value, except for the colored fruit of some species. It is a very large genus, comprising some handsome vines with highly irregular flowers (see p. 62) and evergreen shrubs (see p. 364). All those below lose their leaves in winter. Leaves simple, opposite, without marginal teeth. Flowers small, not showy, irregular in some, but essentially regular in others, mostly borne in pairs, rarely in clusters. Fruit a berry, often brightly colored, but small.

The bush honeysuckles are medium-sized shrubs of easy culture in any ordinary garden soil, preferably in full sun. They do better in reasonably moist sites than in dry ones. Of the 180 species only 5 of the bush honeysuckles are worth much garden attention and usually only as screen planting as their decorative value is slight. Of the 5 below the Tartarian Bush Honeysuckle is the best and the most widely grown, especially in some of its varieties. The species are not easy to identify. All those below have hollow twigs, without solid pith, except *Lonicera maximowiczi* (which is listed under Tartarian Bush Honeysuckle) and *Lonicera heckrotti*.

Flowers in small spikelike clusters. *Lonicera heckrotti*
Flowers in pairs.
  Leafstalks shorter than the flower stalks.
    Flowers red or pinkish, not fading to yellow.
      Leaves smooth both sides.
        Tartarian Bush Honeysuckle
      Leaves hairy on the under side. *Lonicera korolkowi*
    Flowers pinkish, fading to yellow. *Lonicera bella*
  Leafstalks as long as or longer than the flower stalks. *Lonicera maacki*

LONICERA HECKROTTI p. 386

A possibly hybrid shrub, almost vinelike in habit, grown for its spiky clusters of flowers which are quite different from any of the bush honeysuckles here treated. **Size:** 3–5 feet high, as wide or wider due to its sprawling habit. **Leaves:** Simple, opposite, without marginal teeth, the upper pair united, oblong or oval, 1–2½ inches long, nearly stalkless, and whitish beneath. **Flowers:** Purple outside, yellowish inside, about 1½ inches long, in terminal, spikelike clusters, the corolla decidedly irregular (see Picture Glossary); June to frost. **Fruit:** A juicy berry. **Hardy:** From Zone 3 southward. **Varieties:** None. **Culture:** See above.

TARTARIAN BUSH HONEYSUCKLE

(*Lonicera tatarica*) p. 374

An extremely hardy Asiatic shrub and by far the best of all the bush honeysuckles. **Size:** 8–10 feet high, about three fourths as wide, its branches more or less erect. **Leaves:** Simple, opposite, without marginal teeth, oblongish, 1–2½ inches long, pointed at the tip, pale beneath, the leafstalks shorter than the flower stalks. **Flowers:** Irregular, about ¾ inch long, pinkish (rarely white), borne in pairs; May–June. **Fruit:** A small red berry, rather profuse, not persistent. **Hardy:** From Zone 2 southward. **Varieties:** Several, of which the best are *alba,* with white flowers; *grandiflora rosea,* with rosy-pink flowers; *rosea,* with flowers rosy pink outside but paler inside; *rubra,* with red flowers. **Culture:** See above.

Related plant (not illustrated):

Lonicera maximowiczi. A Korean shrub, 5–8 feet high, its purplish twigs with a solid pith. Flowers violet-red. Fruit red, more or less persistent. Hardy from Zone 4 southward.

LONICERA KOROLKOWI p. 385

An attractive shrub from Turkestan valued for its bluish foliage, red fruit, and graceful arching habit. **Size:** 8–10 feet high, as wide or wider, with spreading branches. **Leaves:** Simple, opposite, without marginal teeth, oval, about 1 inch long, hairy, bluish green, pointed at the tip, and striking because of their color. **Flowers:** Pale rose-colored, about ⅔ inch long, in pairs; May–

June. **Fruit:** Red, not persistent. **Hardy:** From Zone 3 southward. **Varieties:** *floribunda* has more profuse flowers; Zabel's Bush Honeysuckle (variety *zabeli*) differs from the typical form only in having leaves somewhat broader at the base. **Culture:** See above.

Related plant (not illustrated):

European Fly Honeysuckle (*Lonicera Xylosteum*). A Eurasian shrub 8–10 feet high, with ovalish leaves hairy beneath, yellow flowers often tinged with red, and red fruit. Hardy from Zone 3 southward.

Left: *Lonicera bella,* below.
Right: *Lonicera korolkowi,* p. 384.

LONICERA BELLA above

A hybrid shrub, unusual among the bush honeysuckles in having white flowers that fade to yellow. **Size:** About 10 feet high, nearly as wide, due to its spreading branches. **Leaves:** Simple, opposite, without marginal teeth, ovalish, 1–2 inches long, sometimes hairy beneath. **Flowers:** White, fading to yellow, in pairs; May–June. **Fruit:** Red, not persistent. **Hardy:** From Zone 3 southward. **Varieties:** *albida* has pure white flowers. **Culture:** See above.

LONICERA MAACKI below

A very vigorous Asiatic shrub with wide-spreading branches, useful in screen planting. **Size:** 10–15 feet high, as wide or even wider. **Leaves:** Simple, opposite, without marginal teeth, oval, 1½–3 inches long, with a long slender tip, the leafstalk as long as or longer than the flower stalk. **Flowers:** White, fading to yellow, about ⅔ inch long, in pairs; May–June. **Fruit:** Bright red, not persistent. **Hardy:** From Zone 3 southward. **Varieties:** *podocarpa* has broader leaves and even more spreading branches, but it is not significantly different from the type. **Culture:** See above.

Related plant (not illustrated):

Lonicera ruprechtiana. An Asiatic shrub, 8–10 feet high, its lance-oval leaves 2–4 inches long. Flowers white, fading to yellow. Fruit purple. Hardy from Zone 3 southward.

Left: *Lonicera maacki,* above.
Right: *Lonicera heckrotti,* p. 383.

Leaves Alternate

NOTE: An exception is the Swamp Laurel (see p. 412)

Leaves with normally expanded leaf blades, mostly 2 inches long or more. See p. 393
Leaves usually scalelike or needlelike, never more than 1 inch long. Heath Family

HEATH FAMILY
(*Ericaceae*)

Here are grouped only those shrubs with needlelike or scalelike leaves that are less than 1 inch long. They are usually so small and some of them hug the twigs so closely that the plant appears (falsely) as though leafless. Other sections of the family, with normally expanded leaves, or with leaves that have marginal teeth, will be found at p. 395, where a general description of the Heath Family will be found.

Flowers small, urn-shaped, perfectly regular (the Sand Myrtle has 5 separate petals and is a striking exception to the rule) and dry fruits. Some of them are so low as to be herblike and are hence treated in the *Guide to Garden Flowers*. The spike heath, Irish heath, and the heather will be found at p. 156 of that book.

The 3 genera here considered may be separated thus:

Flowers with separate petals. Sand Myrtle
Flower a united urn-shaped corolla.
Flowers dropping off when withered. *Phyllodoce*
Flowers persisting long after death. Heaths

SAND MYRTLE (*Leiophyllum buxifolium*) p. 375
A low native evergreen shrub found in highly acid soils from N.J. to S. Car. and suited only to garden sites having sandy loams. **Size:** Not usually over 18 inches high, but of compact habit and about half as wide. **Leaves:** Simple, alternate, smooth, ovalish,

not over ½ inch long, and very numerous. **Flowers:** Small, white, in terminal clusters that are about 1½ inches wide, the 5 petals separate; May. **Fruit:** A tiny dry capsule which is inconspicuous. **Hardy:** From Zone 4 southward. **Varieties:** None readily available. **Culture:** Exacting. It needs full sun, a sandy, light loam that is acid, a pH of at least 4.5–5. Start with young plants from pots, and it is better to plant several in a clump rather than singly. Makes a good ground cover in suitable sites.

PHYLLODOCE (*Phyllodoce empetriformis*) p. 375
A dense, tufted evergreen shrub from the Pacific Coast, not at all easy to grow and best suited to the rock garden. **Size:** 6–9 inches high, about as wide and of dense habit. **Leaves:** Very numerous, alternate, simple, linelike, not over ½ inch long, usually less, the margin rolled. **Flowers:** Reddish purple, about ¼ inch long, urn-shaped, in small nodding clusters; Apr.–May. **Fruit:** A tiny dry capsule. **Varieties:** None. **Culture:** Difficult. It needs a cool, moist, shady place in the rock garden. Use a definitely acid peat (pH 4.5–5) mixed half and half with good garden loam. Emphatically not a plant for the hasty or careless grower, but most attractive in bloom. Not suited to the warm coastal plain.

HEATHS
(*Erica*)

An enormous group of shrubs (over 500 species), mostly South African, but a few from northern Europe and the Mediterranean region, which are the only ones that can be grown within the range of this book. Leaves evergreen, very small, never over 1 inch long, usually less, needlelike and very numerous. Flowers small, urn-shaped, sometimes solitary, more usually in small clusters, persisting, without falling, long after death. Fruit an inconspicuous dry capsule.

The heaths, besides being precariously hardy, are not easy to grow and impossible without close attention to the following details. If the soil is ordinary garden loam, or has clay or silt in it, dig out all soil for a depth of 12–15 inches and fill in with the following mixture: 6 parts pure sand, 3 parts chopped acid peat. If your soil is naturally sandy or gritty, omit the sand but mix in the acid peat. Water only with rain water, and buy young specimens in pots. The soil mixture should be a pH of 4–5. Do not cultivate the soil, and hand-pull any weeds. The plants demand full sun or half shade and a thin mulch of pine needles will help. They should be planted in groups. Over 20 species are in cultivation in the U.S., most of them on the Pacific Coast, but only the following are recommended for the average grower in the East. None is suited to the prairie states.

Leaves minutely hairy on the margin. — Cross-leaved Heath
Leaves not minutely hairy on the margin.
  Low shrubs, not over 18 inches high, often sprawling.
    Twigs finely hairy. — Twisted Heath
    Twigs not hairy. — Cornish Heath
  Erect shrub, 6–10 feet high. — *Erica mediterranea*

CROSS-LEAVED HEATH (*Erica Tetralix*) p. 375
A small evergreen shrub from western and northern Europe, its foliage grayish green and minutely hairy. **Size:** Not over 2 feet high and of sprawling habit. **Leaves:** Simple, usually in fours, not over 1/8 inch long, minutely hairy on the margin and on the surface (lens needed to see this), evergreen. **Flowers:** Rose-red, urn-shaped, about 1/4 inch long, crowded in small, dense terminal clusters; June to frost. **Fruit:** Negligible. **Hardy:** From Zone 3 southward. **Varieties:** *mollis* has grayish foliage and white flowers. **Culture:** See above.

Related plant (not illustrated):

Fringed Heath (*Erica ciliaris*). A European shrub, scarcely 12 inches high, its tiny leaves hairy on the margin. Flowers red, about 1/2 inch long, in clusters up to 5 inches long. Hardy only precariously in Zone 5, safely from Zone 6 southward.

HEATH FAMILY (*Ericaceae*), 1–8

1
2
3
4
5
6
7
8

1
2
3
4
5
6
7
8
9

HEATH FAMILY (*Ericaceae*), 1–9

1. ***Rhododendron fortunei*** p. 408
   A showy evergreen Chinese shrub, its flowers fragrant. It grows 8–12 feet high.

2. **Glenn Dale Azalea Martha Hitchcock** p. 401
   For details see text.

3. **Mountain Rose-Bay** (*Rhododendron catawbiense*) p. 408
   A magnificent native evergreen shrub, 12–18 feet high. Not suited to hot, dry, windy places.

4. **Glenn Dale Azalea Helen Fox** p. 401
   For details see text.

5. **Flame Azalea** (*Azalea calendulacea*) p. 402
   A spectacular shrub from the mountains from Penn. to Ga., its flowers yellow, orange, or scarlet.

6. **Carolina Rhododendron** (*Rhododendron carolincanum*) p. 409
   A low evergreen shrub from N. Car. mountains, with small rusty leaves and small flowers.

7. **Great Laurel** (*Rhododendron maximum*) p. 409
   A native evergreen shrub, its attractive flowers half hidden by the foliage; hence not as showy as some hybrids.

8. **White Swamp Azalea** (*Azalea viscosa*) p. 403
   A native shrub, not evergreen, found in bogs and swamps and partial to such sites.

9. **Pinkster-Flower** (*Azalea nudiflora*) p. 403
   A native shrub, blooming before or with the expansion of the leaves.

TWISTED HEATH (*Erica cinerea*) p. 393
A densely branched European shrub valued for its long-continued bloom. **Size:** 12–18 inches high, apt to sprawl and needing space; its twigs hairy. **Leaves:** Generally in threes, scarcely ½ inch long. **Flowers:** Rosy purple, about ¼ inch long, in dense terminal clusters that may be 4 inches long; June–Sept. **Fruit:** Negligible. **Hardy:** From Zone 3 southward and naturalized on Nantucket. **Varieties:** At least a dozen, of which a reasonable selection might include: *alba,* with white flowers; *atropurpurea,* with deep purple flowers; *fulgida,* with red flowers (this is usually offered, incorrectly, as variety *coccinea*), *pallida,* with light rose-colored flowers. **Culture:** See above.

CORNISH HEATH (*Erica vagans*) p. 375
A very showy European evergreen shrub of small stature but profuse bloom. **Size:** 8–12 inches high, spreading but not prostrate, its twigs not hairy. **Leaves:** In fours or fives, not over ½ inch long. **Flowers:** Rose-red, about ⅕ inch long, crowded in dense, leafy clusters that may be 6 inches long; Aug.–Oct. **Fruit:** Negligible. **Hardy:** From Zone 4 southward. **Varieties:** *alba* has white flowers; *rubra* has deeper-colored flowers. **Culture:** See above.

Related plant (not illustrated):

Spring Heath (*Erica carnea*). A low European shrub 8–10 inches high. Leaves in fours. Flowers red, about ¼ inch long, in one-sided clusters. Hardy from Zone 3 southward.

ERICA MEDITERRANEA p. 393
Much the tallest of all the heaths here considered but not very vigorous. **Size:** 6–10 feet high, its branches upright. **Leaves:** In fours or fives, needlelike, about ½ inch long, very numerous. **Flowers:** Deep red, about ¼ inch long, in terminal clusters that may be 4–5 inches long; Apr. **Fruit:** Negligible. **Hardy:** From Zone 6 southward, precariously so in protected sites in Zone 5. **Varieties:** *alba* has white flowers. **Culture:** See above.

Related plant (not illustrated):

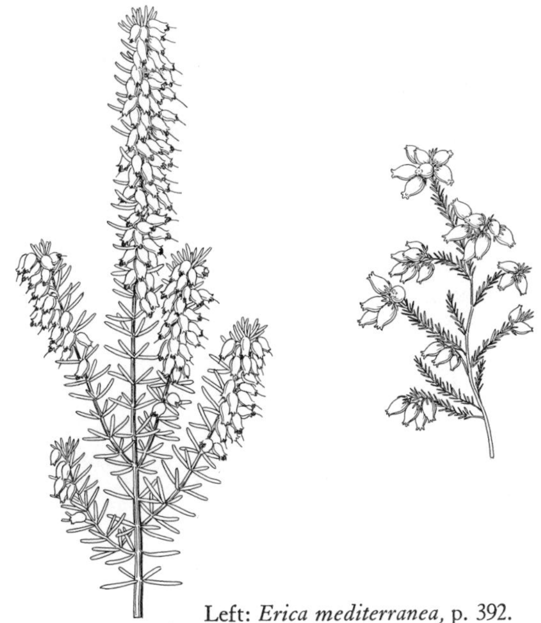

Left: *Erica mediterranea,* p. 392.
Right: Twisted Heath (*Erica cinerea*), p. 392.

Erica darleyensis. A hybrid heath, somewhat taller than the spring heath, but lower than *Erica mediterranea* (which are its parents), winter-blooming, and doubtfully hardy above Zone 7.

Leaves with Normally Expanded Leaf-Blades, Mostly 2 Inches Long or More

Leaves with marginal teeth. p. 414

Leaves without marginal teeth or with remote, very small teeth.

Tall trees. Persimmon

Shrubs or small trees.
Shrub usually prickly; fruit scarlet. Matrimony-Vine
Plants never prickly.
Small tree or shrub with hanging white flowers. Storax
Flower clusters not usually hanging. Heath Family

PERSIMMON (*Diospyros virginiana*)
Family Ebenaceae p. 375

A native American tree of secondary decorative value, but prized by some for its fruit, which, however, is inferior to the Japanese persimmon. **Size:** In the open 30–50 feet high, but in the forest 70–100 feet, about three fourths as wide, the bark thick and deeply cut into rectangular plates. **Leaves:** Simple, alternate, without marginal teeth, ovalish, pointed at the tip, 4–7 inches long, yellow in the fall. **Flowers:** Male and female on separate trees, neither conspicuous, yellowish white; May–June. **Fruit:** Produced only on female trees (with a male tree in the vicinity), a fleshy orange berry, about 1½ inches thick, persistent into Dec. The fruit is worthless until increasing cold weather removes most of its astringency. **Hardy:** From Zone 3 southward. **Varieties:** None superior to the typical form for decorative purposes. For fruit, Early Golden and Ruby are preferred. **Culture:** The plant has a deep taproot and moving it is not easy. It is better to start with quite young specimens, *i.e.,* before the taproot is too long. It will grow in any ordinary garden soil.

MATRIMONY-VINE (*Lycium chinense*)
Family Solanaceae p. 375

A rather rambling or sprawling prickly Asiatic shrub, now replacing the old Matrimony-Vine (*Lycium halimifolium*) because it is more showy. Both are equally good for covering dry banks or walls. **Size:** 6–12 feet long, always sprawling and sometimes vinelike. **Leaves:** Simple, alternate, without marginal teeth, variable as to shape, 1–4 inches long. **Flowers:** Small, purplish, solitary or in small clusters; June and later. **Fruit:** An orange to scarlet egg-shaped berry, about ¾ inch long, not persistent. **Hardy:** From Zone 3 southward. **Varieties:** None. **Culture:** Easy anywhere, even in poor soils.

Related plant (not illustrated):

Washington's Bower (*Lycium halimifolium*). A Eurasian relative of the above, still widely planted, but with smaller flowers and fruit than *Lycium chinense.* Hardy from Zone 3 southward.

STORAX (*Styrax japonica*). Family Styracaceae p. 375
A very attractive Asiatic lawn shrub or small tree, valued for its profusion of white flowers. **Size:** 15–20 feet high, about three quarters as wide, sometimes higher, treelike and up to 30 feet high. **Leaves:** Simple, alternate, without marginal teeth, or very remotely toothed, broadly oblong, pointed at the tip, 2–3 inches long. **Flowers:** Waxy, white, fragrant, bell-shaped, about ½ inch long, in few-flowered hanging clusters that are numerous and showy, borne on the under side of leafy twigs; June–July. **Fruit:** Rather dryish, berrylike, egg-shaped, about ½ inch long. **Hardy:** From Zone 3 southward. **Varieties:** None. **Culture:** Easy in any light, well-drained soil, and it does reasonably well in most ordinary garden soils.

Related plant (not illustrated):

Styrax Obassia. A more showy plant from Japan, its leaves nearly round, 3½–8 inches long. Flowers white, fragrant, in clusters 6–8 inches long, but half hidden by the foliage. Hardy from Zone 3 southward.

HEATH FAMILY
(*Ericaceae*)

On p. 387 there is a brief description of this highly important family of flowering shrubs and trees, but it is restricted to those with very small needlelike or scalelike leaves. Those below have normally expanded leafblades, mostly over 2 inches long, and the leaves are simple, alternate, and without marginal teeth (an exception is the Opposite-leaved Swamp Laurel). Flowers irregular in *Azalea* and a few related groups, but bell-shaped, urn-shaped or funnel-shaped, and perfectly regular in all the others except *Ledum* and *Clethra* which have separate petals. Fruit dry in those below.

The Heath Family, comprising over 70 genera and perhaps 1500 species, contains some of our most magnificent broad-leaved evergreens such as rhododendron and mountain laurel, as well as the azaleas, most but not all of which drop their leaves in the winter. The cultivation of them is not simple as nearly all of them require an acid soil and no exposure of their roots to the sun and wind—hence are always to be moved with a ball of earth, tightly roped up in burlap. This is also important because practically all the Heath Family rely on some microscopic soil organisms for proper function of their fine root hairs.

Most of them are quite unsuited to hot, dry, open, windy places and are hence useless in the prairie states. Beautiful, low, trailing, herblike plants, both actually woody, in this family are the Bearberry and the Trailing Arbutus which are pictured at p. 156 of *The Guide to Garden Flowers.* Those below may be separated thus:

Flowers irregular, never urn-shaped.
- *Flowers somewhat irregular, often 2-lipped; leaves not evergreen (except in some azaleas); never scurfy beneath.*
  - *Flowers distinctly 2-lipped.* — *Rhodora*
  - *Flowers only slightly 2-lipped.* — *Azalea*
- *Flowers only slightly irregular, not 2-lipped; leaves evergreen and usually scurfy beneath.* — *Rhododendron*

Flowers regular, more or less urn-shaped, bell-shaped, or funnel-shaped (except in Ledum which has separate petals).
- *Leaves evergreen.*
  - *Petals separate.* — Labrador Tea
  - *Petals not separate.*

Flowers without pouchlike structures. Bog Rosemary

Flowers with pouchlike structures. *Kalmia*

Leaves not evergreen. Stagger-Bush

NOTE: *This key to the Heath Family is admittedly inferior to one based on the technical characters of the Ericaceae, which are too difficult to include here. Most specialists, for instance, insist that Rhodora, Azalea, and Rhododendron all belong to the one comprehensive genus Rhododendron. While this is technically correct, gardeners generally prefer to think of the rhodora, azalea, and rhododendron as separate entities and they have been so treated here for the convenience of the garden public. If fruit plants were included, the blueberry and huckleberry would come here as both belong to the heath family.*

RHODORA (*Rhodora canadensis*) p. 375
A beautiful native bog shrub, immortalized in Emerson's poem, well suited to the North, but languishing in the South and completely unsuited to the prairie states. **Size:** 2–3 feet high, about half as wide, its twigs bare in winter. **Leaves:** Simple, alternate, without marginal teeth, 1½–2 inches long, gray-hairy beneath. **Flowers:** Showy, rose-purple, solitary or in small, few-flowered clusters, the corolla 2 inches long, irregular and 2-lipped; Mar.–Apr., before the leaves expand. **Fruit:** A small dry capsule. **Hardy:** From Zone 4 (only away from the coast) northward. **Varieties:** None. **Culture:** Generally similar to *Azalea,* to which it is closely related, but thriving best in a moist acid site.

AZALEA
(*Azalea*)

Very beautiful flowering shrubs with alternate simple leaves, without marginal teeth, and dropping in the autumn (except in some evergreen or partly evergreen forms), never scurfy on the under side. It is a very large genus and in addition to the species there are nearly 2500 named horticultural forms, which it is impossible to include here. For all but a fraction of these, as noted below, the seeker is referred to *The Azalea Book,* by Frederick P. Lee, issued by the American Horticultural Society (Washington, D.C., 1958). Flowers somewhat irregular, slightly 2-lipped, usually in clusters, these often few-flowered, and in some the flower is solitary. Fruit a more or less oblong capsule, the seeds very numerous, sometimes as many as 500 in a single pod.

The garden azaleas are relatively shallow-rooted and do best in a sandy loam which is naturally acid or can be made so by the the addition of acid peat. When ready for planting the soil should test pH 4–5, and there must be no standing water at their roots. While the plants will stand full sun, they flower better if there is some filtered light, *i.e.,* passing shade. Generally they do not tolerate deep shade. Never cultivate around them, and keep them mulched with 3 inches of pine needles or oak leaves. While some of them will stand more sun and wind than most rhododendrons, they must never be allowed to dry out. If watering is necessary use rain water as many village water supplies are too alkaline. All of them should be planted only with a tight ball of earth around the roots — never planted "bare-root."

The complexity of the group is so great that it is next to impossible to make a key to either the species or the horticultural named forms. Such a key would not help the amateur and would certainly provoke the professionals. Thus the usual treatment in this book will be modified in the direction of clarity. *Azalea* is an admittedly difficult group, thoroughly understood by only a handful of experts. These, quite rightly, insist that all azaleas belong to the genus *Rhododendron,* and should be included with them. But the garden use of the genus name *Azalea* for all azaleas is so universal that it is retained here.

Modern azaleas have been derived from a bewildering number of species, some native in America, but most from China, Japan, and Korea. For hundreds of years breeders of azaleas in the Orient, in England, Holland, and Belgium have crossed and recrossed these wild species, and often backcrossed many of the hybrids. Later came a group of American breeders who have developed many fine hybrids. Hundreds of horticultural forms are now available and the confusion of names may well discourage the amateur.

The confusion would be baffling if it were not possible to group these beautiful shrubs into various classes based on their habit, color, culture, or other important features. Even of these *classes* of azaleas only the most important can be included here, with a few varieties listed under each group. Such a selection will give any amateur a good start on the azaleas trail. To all who want to go beyond this, Mr. Lee's book is earnestly recommended.

1. MOLLIS AZALEAS (not evergreen) p. 375

   A group of azaleas derived by crossing some Asiatic species and possibly an American one. All bloom in May or June and the flowers, mostly in clusters of 7–13, are prevailingly in the range of yellow, orange, or reddish, rarely white. The individual flower is showy and usually about 2½ inches wide. Generally hardy up to Zone 4. All are erect shrubs of medium height. A choice might include:

   Comte de Papadopoli. Flowers nearly 3 inches wide, orange-red, blotched yellowish orange.

   Comte de Gomer. Flowers violet red, about 2½ inches wide, with orange yellow blotch. (Sometimes offered as Consul Ceresola.)

   Hugo Koster. Flowers reddish orange with an orange blotch.

   Miss Louisa Hunnewell. Flowers nearly 3 inches wide, orange-yellow.

   Christopher Wren. Flowers yellow, with a yellowish-orange blotch.

2. GHENT AZALEAS (not evergreen) p. 390

   An immense group of azaleas, originating in Ghent, Belgium, mostly derived from crossing our native Flame Azalea, the Pinkster-Flower, our Swamp Azalea, and a couple of Asiatic species. They are medium-sized erect shrubs, more hardy than the Mollis azaleas, and usually safe up to Zone 3. Flowers average from 1½–2¼ inches wide, are often fragrant, and have a wide color range, all spring-blooming. A few are double-flowered. A selection might include:

Altaclarensis. Late flowering, single-flowered, white, but orange-blotched.

Charlemagne. Flowers single, orange, but yellow-blotched, late-flowering.

Daviesi. Flowers single, nearly 2¼ inches wide, white to pale yellow, and late-blooming.

Narcissiflora. Double-flowered, yellow and fragrant.

Raphael de Smet. Double-flowered, white, but edges flushed with orange-red.

3. Kurume Azaleas (evergreen) p. 390

A large group of hybrid azaleas, originating in Kurume on the island of Kyushu, Japan, about 1800, but practically unknown here until they were brought to the San Francisco Exposition in 1915. They are, as usually grown here, low, very compact and twiggy shrubs, so covered with small, early-blooming flowers that the evergreen foliage is hidden. Color range is from white to deep maroon. They are only precariously safe in Zone 4, but perfectly so in Zone 5 and southward. They are frequently forced by florists. Of over a hundred varieties, many with confusing and not always accurate Japanese names, the following 6 might provide a good start:

Avalanche. Flowers pure white.
Christmas Cheer. Flowers red.
Coral Bells. Flowers coral-pink.
Pink Pearl. Flowers salmon-pink.
Orange Beauty. Flowers orange-pink.
Cheerfulness. Flowers vermilion-red.

4. Southern Indian Azaleas (evergreen) p. 390

A very large group of evergreen azaleas mostly developed in Belgium and England, little known here before 1840 when Magnolia Gardens at Charleston, S. Car., began importing them to make one of the most magnificent azalea gardens in this country. They are mostly single-flowered, often flaked or striped, and usually 2–3½ inches wide. Very few are hardy in the North, but the selection below includes those that are hardy in protected sites in Zone 5 and of course southward. Some of them are quite high. Often offered by nurserymen as *"Azalea indica"* or as "Indicas."

Cavendish. A low spreading shrub, the flowers red, with a darker blotch, red-and-white-striped.

Fielder's White. A medium-sized shrub, with flowers about 2¾ inches wide, white and frilled.

Flag of Truce. A medium-sized shrub, the flowers about 2 inches wide, semidouble, white and frilled.

George Lindley Taber. A medium-sized shrub, the flowers about 3 inches wide, white but magenta-tinted and with a darker blotch.

William Bull. A medium-sized spreading shrub, the flowers about 1¾ inches wide, double and orange-red.

5. KAEMPFERI AZALEAS (evergreen) p. 390

A group of hybrid azaleas, originated in Holland, but now widely grown here for their showy flowers and because they are hardier than most evergreen azaleas. They are medium-sized or tall shrubs (5–9 feet), the flowers 1½–2½ inches wide with a considerable color range. They are hardy up to Zone 4 in protected sites. A selection could include:

Carmen. A tall shrub, the single flowers about 2½ inches wide, crimson, with a brown blotch.

Cleopatra. A tall shrub, the single flowers deep red.

Gretchen. A medium-sized shrub, the single flowers about 2 inches wide, mallow-purple, with a dark blotch.

Mary. A medium-sized shrub, the single flowers about 2 inches wide and rose-pink.

Swan White. A medium-sized shrub, the single flowers white, about 1¾ inches wide, the throat faintly yellowish.

6. GLENN DALE AZALEAS (evergreen) p. 391

An immense group (perhaps 350 different sorts) of much hybridized azaleas all produced at Glenn Dale, Md., by the work of Mr. B. Y. Morrison, former head of the Plant Introduction Service of the U.S. Department of Agriculture. Their outstanding values are the large flowers and their hardiness. They are hardy up to Zone 3 (in protected places) and everywhere from Zone 4 southward. Their color range is very wide and the flowers are from 2½–4 inches wide. Perhaps no breeding of azaleas in this country exceeds this in importance. The number of forms of these Glenn Dale azaleas is so great that the seeker is advised to visit a nursery when they are in flower before making a selection.

These six classes of azaleas represent those most likely to interest the amateur, but they include less than a third of the groups now in cultivation among specialists. In addition there are a few wild species of azalea native in the eastern states which any azalea enthusiast will surely want to grow. Brief notes on the most important might include the following, of which none are evergreen.

Azalea vaseyi, below.

AZALEA VASEYI above

A very handsome shrub from N. Car., but hardy up to Zone 4. It is 6–12 feet high. Flowers rose-colored but spotted with orange or brownish orange; Apr.–May. Considered by some as allied to *Rhodora.*

FLAME AZALEA (*Azalea calendulacea*) p. 391

The most spectacular of our wild azaleas, native in the mountains from Pa. to Ga., and hardy from Zone 4 southward. It is a spreading shrub, 6–12 feet high. Flowers yellow, orange, or scarlet; May–June.

PINKSTER-FLOWER (*Azalea nudiflora*) p. 391
A native shrub in the woods of the eastern U.S., hardy up to Zone 4. Rarely over 5 feet high and rather open and spreading in habit. Flowers pink or whitish pink, about 1½ inches long, opening before the leaves expand; May.

WHITE SWAMP AZALEA (*Azalea viscosa*) p. 391
A shrub 6–10 feet high, native in bogs or swamps from Me. to Tenn. and Ohio, and preferring such sites in cultivation. It is 6–10 feet high and the flowers open long after the leaves expand. The flower is about 2 inches long, white or pinkish, very fragrant and opens in June.

RHODODENDRON
(*Rhododendron*)

Without much question these are the most gorgeous flowering shrubs in the world, comprising scores of species and many more named horticultural varieties. They differ from *Azalea* only in technical characters, except that all rhododendrons are true broad-leaved evergreens (in ours) and the under side of the leaves is often scurfy, which is not true in *Azalea*. Leaves alternate, rarely in groups, simple, and without marginal teeth. The leaves are very sensitive to cold, quickly drooping when freezing days arrive, but becoming normally erect with warmer weather. Flowers large, in often large and very showy, usually terminal clusters, the corolla only slightly irregular and not 2-lipped as in many azaleas. Fruit a dry capsule with numerous seeds.

Rhododendrons will not stand too much or too-long-continued heat, hence are practically useless in the South. Neither will they stand wind and drought so they are equally hazardous in the prairie states and many of them will not stand intense cold. They naturally inhabit cool, moist sites in the mountains, where as one expert wrote, "they have long periods of gray weather and abundant moisture in the air." They do fairly well along the Atlantic seaboard from New Jersey northward, but the finest rhododendrons in America today are grown in the Pacific Northwest.

In planting they must always be dug with a tight ball of earth with no exposure of their roots to the sun and wind. Their soil requirements are practically the same as for the azaleas (which see above), but they require, or do better in, partially shady places. Once established they should not need watering, but if this is necessary after planting, use rain water, not tap water. Keep them heavily mulched with oak leaves or pine needles, at least 6 inches thick, and they must never be cultivated.

Nearly 75 species are in cultivation in this country, many of them suited only to Oregon, Washington and northern California. A selection suited to the range of this book is noted below, leaving for the end a list of the best of the hybrids for the amateur grower. The key is useful only for the species below—not for the hybrids. It should be understood that, because of the climatic restrictions of rhododendrons, the hardiness notes state only the zones in which they can be successfully grown, having regard both for heat and cold. As in *Azalea,* the *species* of *Rhododendron* listed are far less important to the gardener than the hundreds of hybrid named forms of the plant, a selection of which is given after the description of the species. Many of these fine hybrids arose from crossing various Asiatic species, especially from the Himalayas and China, which have the greatest concentration of rhododendrons of any region in the world. A few American species were also used, notably the Mountain Rose-Bay.

Flowers in small clusters or solitary.
Rhododendron racemosum

Flowers never solitary, mostly in showy clusters.

Leaves not much or not at all scurfy on the lower side.

Leaves without hairs on the under side, or with a few hairs confined to the midrib.
Mountain Rose-Bay

Leaves definitely hairy beneath, sometimes densely so.

Hairs on under side of leaf, scattered or few.
Rhododendron fortunei
Hairs on under side of leaf dense. Great Laurel
Leaves definitely scurfy on the under side.
Carolina Rhododendron

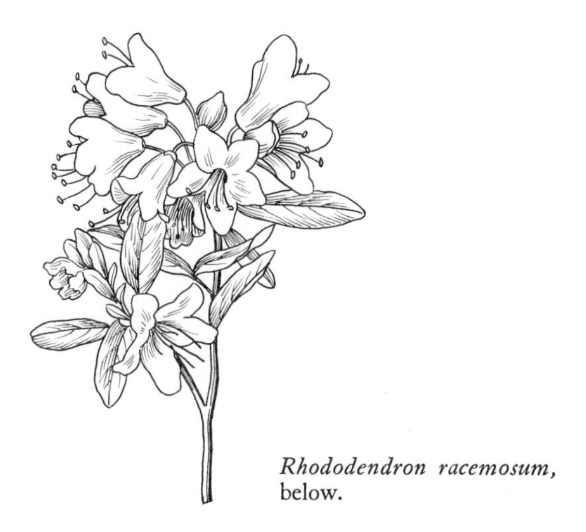

Rhododendron racemosum, below.

RHODODENDRON RACEMOSUM above
A Chinese evergreen shrub, perhaps better grouped with the azaleas, but considered as a true rhododendron by most experts. **Size:** 3–5 feet high, about as wide, its handsome foliage bluish green beneath. **Leaves:** Elliptic or broader, ¾–2 inches long, smooth above, bluish green and scurfy beneath. **Flowers:** Pink, in clusters of 2–5, or solitary at the leaf joints, the corolla about 1 inch wide; Apr.–May. **Fruit:** Negligible. **Hardy:** In Zones 4 and 5, but scarcely north of this. **Varieties:** None. **Culture:** See above.

Related plant (not illustrated):

Rhododendron praecox. A hybrid evergreen shrub 3–5 feet high, its elliptic leaves rusty beneath. Flowers sparse, about 2 inches wide, rose-purple. Hardy in Zones 4 and 5.

HEATH FAMILY (*Ericaceae*), 1–9

1. **Hybrid Rhododendron *Everestianum*** p. 410
For details see text.

2. **Stagger-Bush** (*Lyonia mariana*) p. 414
Suited mostly for informal plantings, this native shrub is exceeded in attractiveness by its evergreen relatives.

3. **Hybrid Rhododendron *Roseum elegans*** p. 410
For details see text.

4. **Labrador Tea** (*Ledum groenlandicum*) p. 411
This northern shrub is suited only to regions of long winters and no summer heat.

5. **Mountain Laurel** (*Kalmia latifolia*) p. 412
A native evergreen shrub, easier to grow than the more showy azaleas and rhododendrons.

6. ***Pieris japonica*** p. 416
A deservedly popular Japanese evergreen shrub and one of the best of the Heath Family for amateur growers.

7. **Bog Rosemary** (*Andromeda glaucophylla*) p. 411
A far-northern native shrub, suited only to cool regions and a moist, acid site.

8. **Strawberry Tree** (*Arbutus Unedo*) p. 414
A European evergreen tree, cultivated with some difficulty and not hardy northward.

9. **Hybrid Rhododendron Mrs. C. S. Sargent** p. 410
For details see text.

1
2
3
4
5
6
7
8
9

1
2
3
4
5
6
7
8
9

HEATH FAMILY (*Ericaceae*), 1, 2, 5, 6, 7
3, 4, 8, 9 in various families

1. ***Enkianthus campanulatus*** p. 420
   An extremely handsome Japanese shrub blooming in May. Autumn foliage bright red. Culture difficult.

2. **Fetter-Bush** (*Leucothoe catesbaei*) p. 417
   A native evergreen shrub from the mountains in the southeastern states, but quite hardy northward.

3. ***Symplocos paniculata*** p. 421
   A little-grown Chinese tree or large shrub blooming in May–June, the flowers not very lasting.

4. **Tisswood** (*Halesia monticola*) p. 422
   A tall relative of No. 9, even more showy, but not quite so hardy. Trees 20–40 feet high.

5. **Sweet Pepperbush** (*Clethra alnifolia*) p. 419
   A useful native shrub, 3–9 feet high, its flowers the most fragrant of summer-blooming shrubs.

6. **Sourwood** (*Oxydendrum arboreum*) p. 419
   A slow-growing native tree, its summer-blooming flowers very fragrant. Fall foliage scarlet.

7. **Salal** (*Gaultheria Shallon*) p. 418
   A Pacific Coast evergreen shrub with flowers blooming in May–June. Culture difficult in the East.

8. **Epaulette-Tree** (*Pterostyrax hispida*) p. 420
   A little-grown Asiatic tree, 20–45 feet high, its hanging flower clusters very fragrant. June.

9. **Silver-Bell Tree** (*Halesia carolina*) p. 422
   About half the height of No. 4, but much more widely grown. Flowers in May, unscented.

MOUNTAIN ROSE-BAY (*Rhododendron catawbiense*) p. 391
A magnificent American evergreen shrub, native in the mountains from Va. to Ga., widely grown for its showy bloom and one of the parents of many hybrids. **Size:** 12–18 feet high, its stunning foliage dense. **Leaves:** Elliptic, 3–5 inches long, shining green above, paler beneath, without any hairs on the under side, or with a few on the midrib. **Flowers:** Bell-shaped, only slightly irregular, lilac-purple, about 2 inches wide, very profuse and showy, often spotted with olive-green; May–June. **Fruit:** Negligible. **Hardy:** From protected places in Zone 3 southward, but only in the uplands. Not suited to the hot coastal plain. **Varieties:** Many, some of which are listed below, among them the commonly cultivated Everestianum. **Culture:** See above.

Related plant (not illustrated):

Rhododendron morelianum. A hybrid shrub closely related to the above but with lilac-violet flowers. Hardy from Zone 3 southward, but not on the hot coastal plain.

RHODODENDRON FORTUNEI p. 391
A very showy Chinese evergreen shrub, its flowers unusual in being clustered in racemes. **Size:** 8–12 feet high, nearly as wide, its twigs smooth. **Leaves:** Oblongish, 6–10 inches long, the under side with a few scattered hairs, pale green above, bluish green beneath. **Flowers:** Pinkish, fragrant, 3–4½ inches wide, in a showy cluster; May–June. **Fruit:** Negligible. **Hardy:** In Zone 5 and protected places in Zone 4. **Varieties:** Several, some of which are in the list below.

Related plants (not illustrated):

Rhododendron williamsianum. A low evergreen Chinese shrub, not over 5 feet high, with nearly round leaves 1–3 inches long. Flowers solitary or in twos, pale pink, about 2 inches wide. Hardy in Zone 5 and protected parts of Zone 4.

Rhododendron decorum. A Chinese evergreen shrub 6–10 feet high, its oblong leaves 4–6 inches long and bluish green beneath. Flowers funnel-shaped, pink or white. Hardy in Zone 5 and protected parts of Zone 4.

Rhododendron fargesi. A Chinese evergreen shrub or small tree, 8–20 feet high, its elliptic leaves 2–4 inches long and grayish beneath. Flowers white or rose and red-spotted. Hardy in Zone 5 and protected parts of Zone 4.

GREAT LAUREL (*Rhododendron maximum*) p. 391

The only wild rhododendron in the northeastern states and much planted for its foliage, as its handsome flowers are often half hidden among the leaves, unlike the Himalayan and Chinese hybrids in which the flower cluster is not hidden by the foliage. **Size:** 10–25 feet high, about three fourths as wide. **Leaves:** Lance-elliptic, 7–10 inches long, densely hairy beneath, very leathery. **Flowers:** Bell-shaped, only slightly irregular, rose-pink but green-spotted, about 1½ inches wide, the showy clusters half hidden by the foliage; June–July. **Fruit:** Negligible. **Hardy:** From Zone 3 southward, but not tolerating the heat of the coastal plain from Zone 5 southward. **Varieties:** Several, some of which are noted below. **Culture:** See above, but it is definitely more at home in the shade. A splendid evergreen to line a shady drive.

CAROLINA RHODODENDRON
(*Rhododendron carolinianum*) p. 391

A handsome evergreen shrub from the mountains of N. Car. and of naturally compact habit. **Size:** 4–6 feet high and about as wide, its foliage dense. **Leaves:** Elliptic, 2–3 inches long, rusty and scaly beneath. **Flowers:** About 1½ inches wide, rose-purple, not scurfy on the outside and unspotted; May. **Fruit:** Negligible. **Hardy:** From Zone 3 southward, but not recommended for the hot coastal plain. **Varieties**: *album* has white flowers. **Culture:** See above. Small tubbed specimens make good plants for a penthouse garden.

Related plant (not illustrated):

Rhododendron minus. A straggling evergreen native shrub, 4–5 feet high, its leaves 2–4 inches long and scurfy beneath. Flowers pink and green-spotted, about 1¼ inches wide. Hardy from Zone 4 southward, but not suited to the hot coastal plain.

To rhododendron enthusiasts this will seem like a very meager list of the *species* of *Rhododendron*. It was restricted by the range of this book, by the availability of the plants, and most of all by the fact that few of the many fine species from the Himalayas or China do well anywhere in America except in the Pacific Northwest.

HYBRID RHODODENDRONS p. 406

To the amateur the hybrid horticultural forms of rhododendrons are far more important than the species, for their flowers are more showy, stand well above the foliage, and the plants are hardier. All those below are safe up to the northern edge of Zone 4 and in favorable sites in Zone 3, if near the coast. None is really happy on the hot coastal plain in the South, but do well in the Piedmont in Zones 5 and 6. The culture of them is the same as that for the species, as noted above.

The number of hybrid rhododendrons is far greater than the number of species. The hybrids have been derived from crossing Asiatic and some American species and there are over 500 named forms. Only a few can be noted here, chosen for their availability and color range. All are tall or medium-sized shrubs and need plenty of space in maturity. An alphabetical list of 20 of them follows:

Album elegans; white
Album grandiflorum; white, but lavender-tinged
Astrosanguineum; red with purple markings
Blue Peter; bluish purple
Boule de Neige; white
Caractacus; red
Charles Dickens; red
Cunningham's White; white
Essex Scarlet; scarlet
Everestianum; frilled and purple
Goldsworth Yellow, yellow
Henrietta Sargent; rose-pink
Kettledrum; deep red
Lady Armstrong; pale rose
Lee's Dark Purple; dark purple
Loder's White; white
Mrs. C. S. Sargent; pinkish-red
Pink Pearl; pink
Purple Splendor; dark purple
Roseum elegans; rose-pink, but purple-tinged

LABRADOR TEA (*Ledum groenlandicum*) p. 406
An evergreen bog shrub of the Far North, suited only to low, wet, acid sites, but handsome when in bloom. **Size:** 18–30 inches high, about three quarters as wide. **Leaves:** Alternate, simple, without marginal teeth but the margins rolled under, 1½–3 inches long, rusty-hairy beneath. **Flowers:** White, about ¾ inch wide, with 5 separate petals and crowded in dense terminal clusters; May–June. **Fruit:** A small dry capsule. **Hardy:** From Zone 4 northward to the Arctic Circle, and unsuited to warm regions. **Varieties:** None. **Culture:** Fit only for the bog garden or similar acid soils that are constantly wet.

Related plant (not illustrated):

Wild Rosemary (*Ledum palustre*). A close relative of the above with narrower and shorter leaves. Flowers white. Hardy only from Zone 3 northward.

BOG ROSEMARY (*Andromeda glaucophylla*) p. 406
An attractive evergreen bog shrub, often incorrectly offered as *Andromeda Polifolia* (which is scarcely grown as a garden plant). The Bog Rosemary is confined to northern bogs and is unsuited to warm regions. **Size:** 8–12 inches high, often making large patches from the spreading of its rootstocks. **Leaves:** Simple, alternate, without marginal teeth but the margins rolled under, oblongish, about 1½ inches long, white-felty beneath. **Flowers:** Urn-shaped, about ¼ inch long, pink or whitish, crowded in nodding terminal clusters; May–June. **Fruit:** A small dry capsule. **Hardy:** From Zone 3 northward. **Varieties:** None better than the typical form. **Culture:** Fit only for the bog garden in the North. It needs a wet, acid soil and is quite unsuited to ordinary garden soils or to the warm coastal plain. *Andromeda* is a rather confusing name to most gardeners. As here restricted it includes only the Bog Rosemary. But many nurseries still list as *Andromeda* plants that properly belong to the genera *Pieris* and *Leucothoe,* which are treated with those plants of the Heath Family that have marginal teeth on the leaves. See pp. 415–418.

KALMIA
(*Kalmia*)

Evergreen shrubs (the Mountain Laurel treelike in the mountains of N. Car.) with alternate, opposite, or clustered leaves without marginal teeth. Flowers showy, in terminal or lateral clusters, prevailingly white or pinkish, the flower pouchlike. Fruit a dry capsule.

The kalmias are quite diverse in habit and growth requirements. The Mountain Laurel needs essentially the same conditions as the azaleas (which see), except that it will stand a bit more exposure to the sun and wind. It should be mulched with oak or beech leaves or, even better, with pine needles. The Sheep Laurel can be grown exactly as are azaleas. The Swamp Laurel is useful only in highly acid, wet places (bog gardens) and is hence unsuited to ordinary garden soils. The juice of all kalmias is poisonous.

Flowers in terminal clusters.
Leaves alternate, 2–4 inches long. Mountain Laurel
Leaves opposite, ½–1½ inches long. Swamp Laurel
Flower clusters scattered along the sides of the stem.
Sheep Laurel

MOUNTAIN LAUREL (*Kalmia latifolia*) p. 406
A splendid native broad-leaved evergreen shrub, sometimes treelike, with handsome foliage and profuse flowers. **Size:** As ordinarily cultivated 4–10 feet high and as wide, but in the mountains of N. Car. sometimes treelike and 20–30 feet high, the bark cinnamon-brown. **Leaves:** Alternate, simple, without marginal teeth, narrowly oval, 2–4 inches long, tough and leathery. **Flowers:** In very showy terminal clusters, pink or whitish, the corolla pouchlike and about ¾ inch wide; May–June. **Fruit:** A small roundish capsule. **Hardy:** From Zone 3 southward. **Varieties:** None. **Culture:** See above. It is a superb bush for naturalistic plantings or as a lawn specimen, preferably in partly shady places.

SWAMP LAUREL (*Kalmia polifolia*) p. 413
An evergreen bog shrub suited only to the North and in wet, acid sites. **Size:** 1–2 feet high, about half as wide, its twigs 2-edged.

Leaves: Opposite, simple, without marginal teeth, but the edges rolled under, ovalish, ½–1½ inches long, white-felty beneath. **Flowers:** Small, pouchlike, rose-colored, ½–¾ inch wide, in terminal clusters; May–June. **Fruit:** A small dry capsule. **Hardy:** From Zone 4 northward; unsuited to the warm coastal plain. **Varieties:** *microphylla* has smaller leaves, but is little known as a garden plant. **Culture:** See above. Unless one has a cool bog this shrub is to be avoided. Sometimes offered as *Kalmia glauca.*

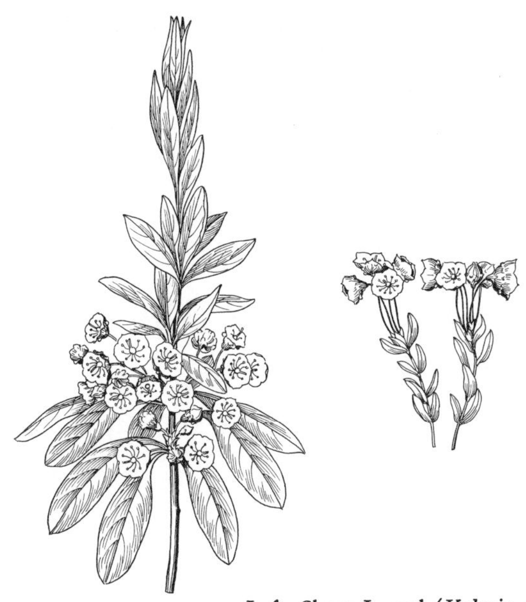

Left: Sheep Laurel (*Kalmia angustifolia*), below.
Right: Swamp Laurel (*Kalmia polifolia*), p. 412.

SHEEP LAUREL (*Kalmia angustifolia*) above

A native evergreen shrub useful for massed plantings but not very decorative as an individual specimen. **Size:** 2–3 feet high, nearly as wide, due to its thin, open habit. **Leaves:** Simple, opposite or in clusters, evergreen, oblongish, 1–2 inches long, probably poisonous to animals, hence the name Lambkill is often applied to it. **Flowers:** Numerous, small, pouchlike, lavender-rose, scattered along the sides of the stem; June. **Fruit:** A small dry capsule. **Hardy:** From Zone 1 southward. **Varieties:** *rubra* has dark purple flowers. **Culture:** See above. Attractive only in massed plantings, as the individual bush is rather thin-foliaged.

STAGGER-BUSH (*Lyonia mariana*) p. 406
A native shrub suited mostly to informal plantings, as the close relatives, *Pieris* and *Leucothoe,* are both evergreen while the Stagger-Bush loses its leaves in winter. **Size:** 4–6 feet high, about three fourths as wide. **Leaves:** Simple, alternate, without marginal teeth, ovalish or elliptic, 1–2½ inches long, the juice presumptively poisonous. **Flowers:** Urn-shaped, about ½ inch long, white, in nodding clusters at the leaf joints, the clusters a little leafy; May–June. **Fruit:** A small dry capsule. **Hardy:** From Zone 3 southward. **Varieties:** None. **Culture:** Needs a sandy loam, decidedly acid, and preferably a partially shady place, although it will grow in the open.

Leaves Alternate, Simple, the Margins with Obvious Teeth, but the Teeth Blunted or Small in Some Species

Leaves not evergreen. See p. 419
Leaves evergreen.
  Tree, 15–30 feet high. Strawberry Tree
  Shrubs, never more than 9 feet high.
    Fruit dry, a capsule.
      Flowers in nodding terminal clusters. *Pieris*
      Flowers in lateral clusters, often at the leaf joints. *Leucothoe*
    Fruit fleshy, berrylike. Salal

STRAWBERRY TREE (*Arbutus Unedo*)
Family Ericaceae p. 406
A fall-flowering, somewhat unusual evergreen tree from southern Europe, unfortunately not hardy over much of the range of this

book and best suited to Calif. **Height:** 15–30 feet, nearly as wide, its outer bark brown, often splitting and disclosing the bright red inner bark. **Leaves:** Alternate, simple, toothed, elliptic or oblong, about 3½ inches long, bright shining green. **Flowers:** White or pink, urn-shaped, about ¼ inch long, in a drooping terminal cluster about 2 inches long; Oct.–Dec. **Fruit:** Orange-red, strawberry-like, about ¾ inch thick, the flesh edible but insipid, persistent into Dec. **Hardy:** In nonhumid parts of Zone 6, and hazardously so in protected parts of Zone 5. **Varieties:** None. **Culture:** Difficult. It needs a well-drained sandy loam, decidedly acid, and full sunlight, but protection from strong winds. As it is hard to transplant, it is best started from a small specimen (1–2 feet high) with the roots in a ball of earth tightly wrapped in burlap.

PIERIS
(*Pieris*)
Family Ericaceae

Valuable evergreen shrubs of medium height with alternate simple leaves, the margins toothed. Flowers in terminal clusters which are rather showy, although the individual flowers are small, urn-shaped, and white. The flower buds are conspicuous throughout the winter before they bloom in early spring. Fruit a small dry capsule.

Pieris contains 2 widely planted shrubs good for the shrub border and often used in foundation planting. They need full sun or partial shade, a reasonably acid soil (no limestone), and should always be moved with a ball of earth tightly roped in burlap. They do better in places that are not too windy, preferably cool and reasonably moist. Keep a mulch of oak or beech leaves on them continuously, and if both are unavailable use sawdust or an acid peat. Both species are often offered under the incorrect name of *Andromeda*.

Twigs with stiff hairs; flower cluster upright. Mountain Fetter-Bush

Twigs smooth; flower cluster nodding. *Pieris japonica*

MOUNTAIN FETTER-BUSH (*Pieris floribunda*) below
An evergreen shrub from the mountains in the southeastern states more hardy than the Japanese shrub that follows, but not quite so showy. **Size:** 3–4 feet high, three fourths as wide, the twigs covered with stiff hairs. **Leaves:** Simple, alternate, sparsely toothed, ovalish to elliptic, about 3 inches long, evergreen but often bronzy in the winter. **Flowers:** White, urn-shaped but constricted at the tip, about ¾ inch long, each flower nodding in the erect terminal cluster that may be 3–4 inches high; Apr. **Fruit:** Negligible. **Hardy:** From Zone 3 southward. **Varieties:** None. **Culture:** See above.

Related plant (not illustrated):

Pieris taiwanensis. A Formosan evergreen shrub, 4–6 feet high, its lance-shaped leaves 2–5 inches long. Flowers small, white, but in dense, showy clusters. Hardy from Zone 6 southward.

Mountain Fetter-Bush
(*Pieris floribunda*), above.

PIERIS JAPONICA p. 406
The finest of all the species of *Pieris,* this Japanese evergreen shrub is very widely planted in shrub borders and much used in foundation planting. **Size:** 3–8 feet high, about three fourths as wide, the twigs smooth. **Leaves:** Alternate, simple, evergreen, but bronzy in the winter, oblongish, 1½–3 inches long, the marginal teeth rounded. **Flowers:** White, urn-shaped, about ½ inch long, in nodding terminal clusters 3–5 inches long, profuse and showy; Apr.–May. **Fruit:** Negligible. **Hardy:** From Zone 4 southward. **Varieties:** *variegata* has smaller, white-margined leaves. **Culture:** See above. It does better in partly shaded sites.

Leucothoe
(*Leucothoe*)
Family Ericaceae

Widely used evergreen shrubs, closely related to *Pieris* and differing (in ours) in having the flowers in lateral, drooping clusters, often at the leaf joints, never terminal. Leaves simple, alternate, the marginal teeth small, often remote, and easily missed without close scrutiny. Flowers white, small, urn-shaped, in dense, close drooping clusters borne along the sides of the twigs. Fruit a dry capsule. Both the species are often offered under the incorrect name of *Andromeda*.

The culture of *Leucothoe* is the same as for *Pieris* (see above). The first species is widely used, especially in foundation planting, for its evergreen foliage and early spring bloom.

Leaves with a long-tapering tip, very finely toothed. Fetter-Bush

Leaves with an abruptly pointed tip, the fine teeth remote. *Leucothoe axillaris*

FETTER-BUSH (*Leucothoe catesbaei*) p. 407
A valuable native evergreen shrub found wild in the mountains of the southeastern states. **Size:** 2–6 feet high, usually less, the branches arching, hence as wide or wider. **Leaves:** Alternate, simple, evergreen, but bronzy in winter, oval to lance-shaped, 3–7 inches long, with a long-tapering tip, the margins closely but finely toothed. **Flowers:** White, waxy, about ¼ inch long, urn-shaped, borne in dense, showy, drooping clusters, 2–3 inches long, along the under side of the twigs; May. **Fruit:** Negligible. **Hardy:** From Zone 3 southward. **Varieties:** *nana* is lower and useful along the edges of foundation planting. **Culture:** The same as for *Pieris,* which see above. The plant is called by some *Leucothoe editorum*.

LEUCOTHOE AXILLARIS p. 418
A native evergreen shrub from Va. to Fla. much resembling the Fetter-Bush, and differing from it in having abruptly pointed

Leucothoe axillaris, p. 417.

leaves (not tapering) and with the fine marginal teeth remote. **Hardy:** From Zone 4 southward. **Varieties:** None. **Culture:** The same as for *Pieris,* which see above.

SALAL (*Gaultheria Shallon*). Family Ericaceae p. 407

A Pacific Coast evergreen shrub to be grown in the East with caution and only by following the directions below. **Size:** 12–18 inches high, widely spreading from the expansion of its underground stems. **Leaves:** Alternate, simple, evergreen, round-oval, 3–5 inches long, finely and sharply toothed. **Flowers:** Pink or white, urn-shaped, about ½ inch long, in a loose terminal cluster 2–5 inches long; May–June. **Fruit:** Berrylike, purple at first, ultimately black, about ½ inch in diameter, not persistent. **Hardy:** From Zone 4 southward, but only as restricted below. **Varieties:** None. **Culture:** Difficult in the East. It needs protection from the sun and wind, a cool, moist site, and a moderately acid soil that is sandy or gritty. An attractive ground cover best confined to the rock garden in the East and wholly unsuited to the prairie states.

Leaves Not Evergreen

Flowering in midsummer.
Tree, 20–60 feet; flower cluster drooping. Sourwood
Shrubs or small trees; flower cluster erect. Sweet Pepperbush

Flowering from April to early June.
Flowers yellow or orange. *Enkianthus*
Flowers white.
Flowers very fragrant.
Flower cluster 6–12 inches long. Epaulette-Tree
Flower cluster 2–4 inches long. *Symplocos*
Flowers not noticeably fragrant. *Halesia*

SOURWOOD (*Oxydendrum arboreum*)
Family Ericaceae p. 407

A slow-growing native tree, found wild from Pa. to Fla. and La., cultivated for its showy midsummer flowers. **Size:** 30–50 feet high in the wild, much less as usually cultivated. **Leaves:** Alternate, simple, bitter-tasting, finely toothed, oblongish, 5–8 inches long, brilliantly scarlet in the fall. **Flowers:** Small, white, fragrant, cylindric, about ½ inch long, in very showy drooping clusters 8–10 inches long; July–Aug. **Fruit:** A small, dry, hairy capsule. **Hardy:** From Zone 3 southward. **Varieties:** None. **Culture:** Not at all difficult in any ordinary garden soil, in full sun. An extremely handsome small tree, but slow-growing.

SWEET PEPPERBUSH (*Clethra alnifolia*)
Family Ericaceae p. 407

Perhaps the most fragrant of all our native summer-blooming shrubs, common as a wild plant from Me. to Fla. and Tex. **Size:** 3–9 feet high, its branches more or less erect, hence about three fourths as wide. **Leaves:** Alternate, simple, toothed, oblongish but pointed at the tip, 2½–5 inches long, orange-yellow in the fall. **Flowers:** Small, white, very fragrant, about ½ inch long, the 5 petals separate, crowded in dense, spirelike, hairy terminal clusters;

Aug.–Sept. **Fruit:** A small dry capsule. **Hardy:** From Zone 2 southward. **Varieties:** *rosea* has pink flowers. **Culture:** While naturally growing in moist or wet decidedly acid places, it does reasonably well in most garden soils, always in full sun. Preferably to be grown in a reasonably moist site.

Related plant (not illustrated):

Clethra barbinervis. A Japanese shrub or small tree, 10–25 feet high, its oblongish leaves 3–6 inches long. Flowers fragrant, white, the clusters more or less horizontal. Hardy from Zone 4 southward.

ENKIANTHUS (*Enkianthus campanulatus*)

Family Ericaceae p. 407

A very handsome Japanese shrub blooming before the leaves expand, its autumn foliage bright red. **Size:** 15–25 feet high, about as broad. **Leaves:** Simple, alternate, but apt to be crowded at the ends of the twigs, elliptic, 1–3 inches long, the small marginal teeth minutely bristly. **Flowers:** Yellowish orange, but red-veined, urn-shaped or bell-shaped, about ½ inch long, in loose, nodding clusters; May. **Fruit:** A small, dry, ultimately rusty capsule. **Hardy:** From Zone 3 southward. **Varieties:** A reputedly red-flowered form is offered by some nurseries, but its true status is unknown. **Culture:** Needs an acid soil (pH 4.5–6.0) that is a sandy loam. It moves with difficulty and should never have its roots exposed to the sun and wind. Prefers a partially shaded place.

EPAULETTE-TREE (*Pterostyrax hispida*)

Family Styracaceae p. 407

An Asiatic tree with handsome fragrant flowers, not so well known as it should be, valued for its showy, hanging clusters of bloom. **Size:** 20–45 feet high, about as wide, its branches spreading and forming an open crown. **Leaves:** Simple, alternate, very finely toothed, oblongish, 5–7 inches long. **Flowers:** White, fragrant, with 5 separate petals, nearly stalkless, arranged in hanging clusters 7–10 inches long that terminate lateral twigs, very showy; June. **Fruit:** Dry, bristly and 10-ribbed, about ½ inch long. **Hardy:** From Zone 3 southward. **Varieties:** None. **Culture:** Will grow in any ordinary garden soil, preferably in the sun and in a reasonably moist site.

SYMPLOCOS (*Symplocos paniculata*)
Family Symplocaceae p. 407
An Asiatic shrub or small tree, not too well known, but attractive in fruit, which, however, is not persistent. **Size:** 20–35 feet high, about three fourths as wide, its young twigs hairy. **Leaves:** Alternate, simple, short-stalked, oblongish, 2–3 inches long, the margins finely toothed. **Flowers:** White, fragrant, small, the petals scarcely united, in clusters that are 2–3 inches long, often terminal but also at the sides of the twigs; May–June, rather fleeting. **Fruit:** Bright blue, berrylike, about ½ inch long in dense showy clusters, which, however, are not persistent. **Hardy:** From Zone 3 southward. **Varieties:** None. **Culture:** Easy in any ordinary garden soil, preferably in full sun.

HALESIA
(*Halesia*)
Family Styracaceae

Rather showy medium-sized trees or shrubs, valued for their early bloom. Leaves alternate, simple, not evergreen, the marginal teeth obvious. Flowers in small, drooping clusters, the flower bell-shaped, white, not noticeably fragrant. Fruit more or less fleshy, but apparently without much flesh and dryish; winged.

Halesia, which contains 3 or 4 species, is rather widely planted for its profuse white bloom. They thrive in a variety of soils, but do best in partial shade (at least part of each day), as their natural habitat is in the undercanopy of the forest. The site should be cool and reasonably moist — not windy.

Shrub or small tree; bark scales small and close.
Silver-bell Tree
Tall tree, 20–40 feet high, the bark scales loose and large.
Tisswood

SILVER-BELL TREE (*Halesia carolina*) p. 407
A beautiful native tree or large shrub, wild from W.Va. to Fla. and Tex., much planted for its bell-like flowers. **Size:** 10–20 feet high, about three fourths as wide, its bark scales small and close. **Leaves:** Simple, alternate, ovalish, finely but bluntly toothed, 2–4 inches long, yellow in the fall. **Flowers:** Bell-shaped, white, about ¾ inch long, not fragrant, usually in clusters of 2–5; May. **Fruit:** Oblongish, somewhat fleshy, 4-winged, 1–1½ inches long. **Hardy:** From Zone 3 southward. **Varieties:** None. **Culture:** See above. Sometimes offered as *Halesia tetraptera.*

TISSWOOD (*Halesia monticola*) p. 407
Larger and more showy than the above, this native tree is found wild from N. Car. to Tenn. and Ga., mostly in the mountains, but it is not quite so hardy as the Silver-Bell Tree. **Size:** 20–40 feet high, its bark scales large and loose. **Leaves:** Alternate, simple, rather remotely toothed, elliptic, 5–9 inches long, tapering at the tip, yellow in the fall. **Flowers:** Bell-like, about 1 inch long, in clusters of 2–5, not fragrant; May. **Hardy:** From Zone 4 southward. **Varieties:** None, except a reputedly pink-flowered form that is unavailable from ordinary sources. **Culture:** See above.

Bibliography
Picture Glossary
Systematic Tabulation
Index

Bibliography

Bailey, L. H. *The Cultivated Evergreens.* 434 pages. New York: Macmillan, 1929.

—— *Hortus II.* 778 pages. New York: Macmillan, 1941.

—— *Manual of Cultivated Plants.* 1116 pages. New York: Macmillan, 1949.

Graves, G. *Trees, Shrubs and Vines for the Northeastern States.* 267 pages. New York: Oxford University Press, 1945.

Haworth-Booth, M. *Flowering Shrub Garden.* 174 pages. New York: Charles Scribner's, 1939.

Lamson, M. D. *Gardening with Shrubs and Small Flowering Trees.* 295 pages. New York: Barrows, 1946.

Lemmon, R. S. *The Best Loved Trees of America.* 254 pages. Garden City: Doubleday, 1952.

Levison, J. J. *The Home Book of Trees and Shrubs.* 525 pages. New York: Alfred A. Knopf, 1949.

Rehder, A. *Manual of Cultivated Trees and Shrubs.* 996 pages. New York: Macmillan, 1940.

Taylor, N. *Taylor's Encyclopedia of Gardening.* 1329 pages. Boston: Houghton Mifflin, 1961.

Van Melle, P. *Shrubs and Trees for the Small Place.* 298 pages. New York: Charles Scribner's, 1943.

Wilson, E. H. *Aristocrats of the Trees.* 249 pages. Boston: Stratford, 1930.

Wyman, D. *Shrubs and Vines for American Gardens.* 442 pages. New York: Macmillan, 1949.

—— *Trees for American Gardens.* 376 pages. New York: Macmillan, 1951.

—— *The Arnold Arboretum Garden Book.* 354 pages. Princeton: Van Nostrand, 1954.

PICTURE GLOSSARY

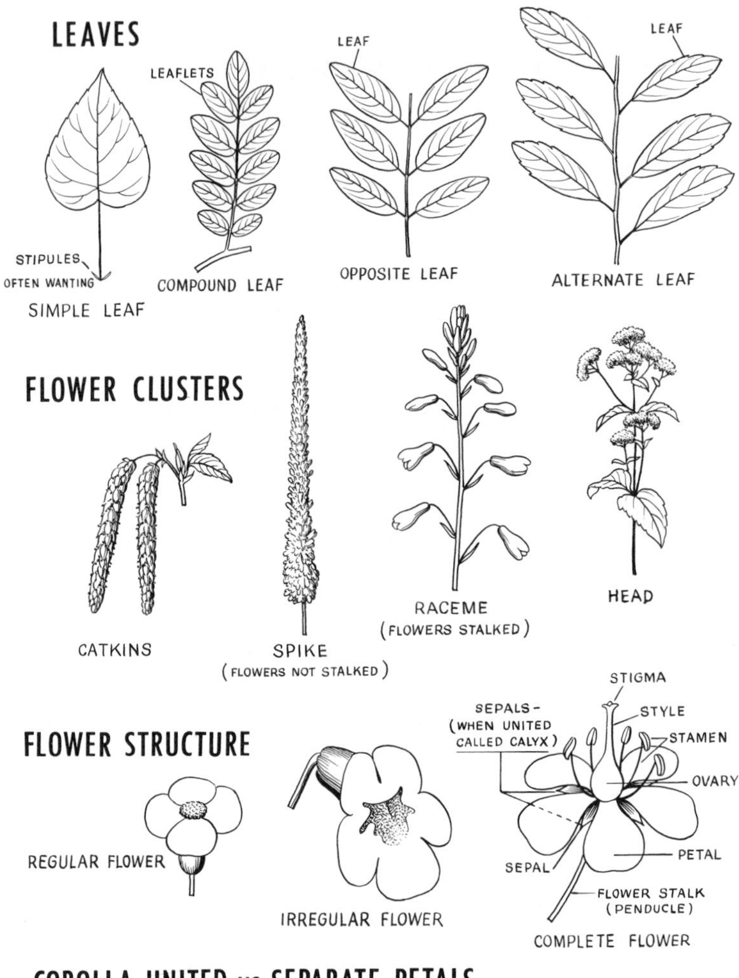

COROLLA UNITED vs SEPARATE PETALS

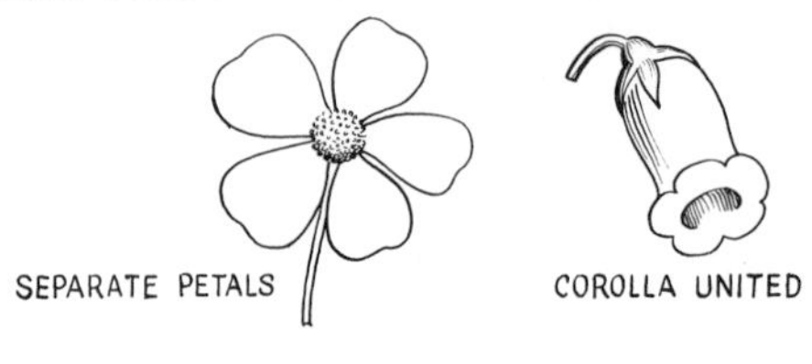

Systematic Tabulation

THE ARRANGEMENT of the plants in this book does considerable violence to the accepted procedure in technical manuals. It was adopted because knowedge of plant families and the sequence of them is a quite unnecessary discipline to ask the average gardener to conquer.

But plants are technically arranged in families or their arrangement might be chaotic. Throughout the book each plant is ascribed to its proper family, but that gives us no idea of what other plants are in that family, nor any notion of their relative position in the assumed evolutionary sequence of plant development.

To atone for that deficiency the following tabulation is presented. It is arranged strictly in the usually accepted sequence of plant families. This attempts to show the assumed evolutionary development of the plant kingdom. That it really does so is questioned by several experts, but they do not agree among themselves. In view of their conflicting ideas, the scheme below is the traditional one and is still standard in most technical books.

The sequence is based upon the supposed development of plant families from the simplest to the most complex. Under each family are listed the groups in that family which are treated in the text, and the numbers following each generic name indicate the pages where the genus is described.

The list is divided into four main divisions:

1. Coniferous trees and shrubs, without flowers, as that word is understood by most gardeners.
2. Trees and shrubs which have flowers, but these flowers lack petals; mostly shade trees.
3. Plants having flowers with separate petals.

4. Plants having flowers that comprise a united corolla, *i.e.,* with no separate petals.

In technical books there would be many subdivisions of these four main categories, but such complexities are omitted here.

The families and their genera follow:

1. Coniferous shrubs and trees

Ginkgoaceae: Ginkgo Family
Ginkgo, 2

Taxaceae: Yew Family

Cephalotaxus, 5	Taxus, 7	Torreya, 6
Podocarpus, 5		

Pinaceae: Pine Family

Abies, 32	Juniperus, 9	Pseudotsuga, 46
Araucaria, 32	Larix, 3	Sciadopitys, 21
Cedrus, 21	Libocedrus, 20	Sequoia, 46
Chamaecyparis, 16	Metasequoia, 4	Taxodium, 3
Cryptomeria, 45	Picea, 40	Thuja, 18
Cunninghamia, 36	Pinus, 25	Thujopsis, 18
Cupressus, 15	Pseudolarix, 2	Tsuga, 37

2. Trees and shrubs the flowers of which lack petals

Salicaceae: Willow Family

Populus, 89	Salix, 93	

Myricaceae: Bayberry Family

Comptonia, 88	Myrica, 85	

Juglandaceae: Walnut Family

Carya, 82	Juglans, 79	Pterocarya, 79

Betulaceae: Birch Family

Alnus, 111	Carpinus, 113	Ostrya, 112
Betula, 108	Corylus, 111	

Fagaceae: Beech Family

Castanea, 106	Fagus, 104	Quercus, 97

Ulmaceae: Elm Family

Celtis, 120	Ulmus, 116	Zelkova, 120

Moraceae: Mulberry Family
Broussonetia, 115 Maclura, 115 Morus, 114

3. PLANTS HAVING FLOWERS WITH SEPARATE PETALS

Aristolochiaceae: Birthwort Family
Aristolochia, 66
Polygonaceae: Buckwheat Family
Polygonum, 66
Ranunculaceae: Buttercup Family
Clematis, 55 Xanthorhiza, 293
Lardizabalaceae: No common name
Akebia, 73
Berberidaceae: Barberry Family
Berberis, 225 Mahonia, 296 Nandina, 296
Trochodendraceae: No common name
Cercidiphyllum, 173
Magnoliaceae: Magnolia Family
Liriodendron, 189 Magnolia, 271
Calycanthaceae: Sweet-Shrub Family
Calycanthus, 165 Chimonanthus, 156
Annonaceae: Custard-Apple Family
Asimina, 259
Lauraceae: Laurel Family
Laurus, 260 Lindera, 265 Sassafras, 189
Saxifragaceae: Saxifrage Family
Deutzia, 178 Itea, 238 Ribes, 233
Hydrangea, 180 Philadelphus, 157 Schizophragma, 61
Hamamelidaceae: Witch-Hazel Family
Corylopsis, 232 Fothergilla, 225 Loropetalum, 277
Hamamelis, 222 Liquidambar, 187 Parrotia, 214
Platanaceae: Plane Tree Family
Platanus, 187
Rosaceae: Rose Family
Amelanchier, 255 Chaenomeles, 252 Exochorda, 284
Amygdalus, 204 Cotoneaster, 285 Holodiscus, 244
Aronia, 254 Crataegus, 208 Kerria, 235

Rosaceae: Rose Family (continued)

Laurocerasus, 282
Malus, 195
Photinia, 251
Physocarpus, 243
Potentilla, 293
Prunus, 202
Pyracantha, 280
Rhodotypos, 185
Rosa, 301
Sorbaria, 315
Sorbus, 315
Spiraea, 244
Stephanandra, 243
Stransvaesia, 280

Leguminosae: Pea Family

Albizzia, 125
Amorpha, 142
Caragana, 137
Cercis, 126
Cladrastis, 141
Colutea, 138
Cytisus, 131
Genista, 129
Gleditsia, 127
Gymnocladus, 127
Laburnum, 133
Lespedeza, 137
Poinciana, 127
Pueraria, 73
Robinia, 138
Sophora, 142
Spartium, 128
Ulex, 128
Wistaria, 75

Rutaceae: Rue Family

Choisya, 312
Evodia, 299
Phellodendron, 313
Poncirus, 312
Ptelea, 292
Skimmia, 259
Zanthoxylum, 313

Simaroubaceae: No common name

Ailanthus, 317

Meliaceae: Mahogany Family

Melia, 291

Buxaceae: Box Family

Buxus, 147
Sarcococca, 260

Anacardiaceae: Sumac Family

Cotinus, 264
Rhus, 318

Aquifoliaceae: Holly Family

Ilex, 216

Celastraceae: Staff-Tree Family

Celastrus, 67
Euonymus, 174

Staphyleaceae: Bladder-Nut Family

Staphylea, 143

Aceraceae: Maple Family

Acer, 168

Hippocastanaceae: Horse-Chestnut Family

Aesculus, 144

Sapindaceae: Soapberry Family
Koelreuteria, 299 Xanthoceras, 314

Rhamnaceae: Buckthorn Family
Ceanothus, 240 Rhamnus, 220

Vitaceae: Grape Family
Ampelopsis, 69 Parthenocissus, 72 Vitis, 69

Tiliaceae: Linden Family
Tilia, 190

Malvaceae: Mallow Family
Hibiscus, 239

Dilleniaceae: Silver Vine Family
Actinidia, 68

Theaceae: Tea Family
Camellia, 235 Gordonia, 238 Stewartia, 193

Hypericaceae: St. John's-Wort Family
Hypericum, 153

Tamaricaceae: Tamarisk Family
Tamarix, 269

Cistaceae: Rockrose Family
Cistus, 156

Flacourtiaceae: Indian Plum Family
Idesia, 214

Thymelaeaceae: Mezereon Family
Daphne, 262 Dirca, 265

Elaeagnaceae: Oleaster Family
Elaeagnus, 266 Hippophae, 266

Lythraceae: Loosestrife Family
Lagerstroemia, 166

Punicaceae: Pomegranate Family
Punica, 166

Araliaceae: Ginseng Family
Acanthopanax, 292 Fatshedera, 257 Hedera, 50
Aralia, 300 Fatsia, 257

Cornaceae: Dogwood Family
Aucuba, 174 Davidia, 214 Nyssa, 259
Cornus, 161

4. Plants the flowers of which have a united corolla

Ericaceae: Heath Family

Andromeda, 411
Kalmia, 412
Oxydendrum, 419
Arbutus, 414
Ledum, 411
Phyllodoce, 388
Azalea, 397
Leiophyllum, 387
Pieris, 415
Clethra, 419
Leucothoe, 417
Rhododendron, 403
Enkianthus, 420
Loiseleuria, 481
Rhodora, 397
Erica, 388
Lyonia, 414
Vaccinium, 53
Gaultheria, 52, 418

Ebenaceae: Ebony Family

Diospyros, 394

Symplocaceae: No common name

Symplocos, 421

Styracaceae: Storax Family

Halesia, 421
Pterostyrax, 420
Styrax, 395

Oleaceae: Olive Family

Chionanthus, 377
Fraxinus, 323
Osmanthus, 332
Fontanesia, 369
Jasminum, 325
Phillyrea, 364
Forsythia, 341
Ligustrum, 377
Syringa, 369

Loganiaceae: No common name

Buddleia, 366
Gelsemium, 65

Apocynaceae: Dogbane Family

Nerium, 362
Trachelospermum, 65
Vinca, 360

Asclepiadaceae: Milkweed Family

Periploca, 64

Verbenaceae: Vervain Family

Callicarpa, 339
Clerodendron, 380
Vitex, 325
Caryopteris, 335

Labiatae: Mint Family

Elsholtzia, 336
Phlomis, 336
Satureia, 361
Perovskia, 335
Rosmarinus, 361

Solanaceae: Potato Family

Lycium, 394

Bignoniaceae: Trumpet-Creeper Family

Bignonia, 54
Campsis, 54
Catalpa, 357

Index

Floral Timetable

The lists below are arranged to show when the shrubs, trees and vines first begin to bloom. It includes only those whose flowers are sufficiently striking to be of interest to the gardener. The latitude of New York City has been chosen as fairly typical of the country between 39° and 41° North Latitude. Proximity to cool water, elevations above 1000 feet and unseasonable heat or cold will always retard or hasten bloom.

Also plants growing north of 41° or south of 39° will bloom a few days later or earlier. The lists are hence a *guide* to blooming time rather than an inflexible criterion of it.

The figures following the plant names refer to page numbers.

Winter Blooming – November to February

Camellia japonica, 236
Corylopsis pauciflora, 232
Corylopsis spicata, 232
Hamamelis, 222
Lonicera fragrantissima, 363
Tea olive, 332

March

TREES

Box-elder, 169
Cornelian cherry, 162
Red maple, 173

SHRUBS

Chaenomeles japonica, 253
Chimonanthus praecox, 156
Daphne odora, 263
Flowering quince, 252
Forsythia ovata, 345
Forsythia suspensa, 344
Leatherwood, 265
Loropetalum chinense, 277
Magnolia stellata, 273
Mezereon, 264
Rhodora, 397
Spicebush, 265
Spurge laurel, 263
Winter jasmine, 328

April

TREES

Cherry plum, 204
Double-flowered peach, 204
Empress tree, 356
Japanese flowering cherry, 207
Magnolia, 271
Norway maple, 171
Redbud, 126
Rosebud cherry, 207
Sassafras, 189
Sugar maple, 172

SHRUBS OR VINES

Azalea vaseyi, 402
Bridal wreath, 245
Carolina jasmine, 65
Daphne Genkwa, 264
Flowering almond, 205, 206
Forsythia intermedia, 342
Furze, 128
Garland-flower, 262
Mahonia beali, 297
Mountain fetter-bush, 416
Nanking cherry, 205
Oleander, 362
Oregon grape, 297
Osmanthus delavayi, 333
Pearl bush, 284
Phyllodoce empetriformis, 388

Pieris japonica, 416
Rhododendron racemosum, 405
Rosemary, 361
Shadbush, 255
Spiraea thunbergi, 248
Viburnum, 346

May

TREES

Acer Ginnala, 170
American mountain-ash, 316
Bird cherry, 204
Chinaberry, 291
Clammy locust, 141
Crabapple, 195
Cranberry tree, 349
Crataegus, 208
Dove tree, 214
English hawthorn, 212
Flowering ash, 323
Flowering dogwood, 162
Fringe-tree, 377
Honey locust, 127
Horse-chestnut, 144
Large-leaved cucumber tree, 275
Malus theifera, 201
Pea-tree, 137
Rose acacia, 139
Silver-bell tree, 422
Sweet bay, 276
Sweet buckeye, 146
Sycamore maple, 171
Washington thorn, 212
Yulan, 273

SHRUBS

Bastard indigo, 142
Berberis, 225
Black haw, 353
Blue dogwood, 163
Bog rosemary, 411
Broom, 131
Carolina rhododendron, 409
Chinese snowball, 350
Cotoneaster, 285
Crataegus intricata, 209
Deutzia gracilis, 178
Enkianthus campanulatus, 420
Everlasting thorn, 281
Father Hugo's rose, 303
Fetter-bush, 417
Flame azalea, 402
Flowering currant, 233
French mulberry, 340
Genista, 129
Golden currant, 234
Hybrid azaleas, 397
Hydrangea quercifolia, 184
Japanese barberry, 228
Jasminum beesianum, 328
Jetbead, 185
Kerria japonica, 235
Lilac, 369
Lonicera, 383
Magnolia liliflora, 274
Malus sargenti, 196
Mountain currant, 234
Mountain laurel, 412
Mountain rose-bay, 408
Persian lilac, 376
Pinkster-flower, 403
Pyracantha, 280
Red-berried elder, 331
Rhododendron fortunei, 408
Rouen lilac, 376
Sand myrtle, 387
Spiraea, 244
Swamp laurel, 412
Syringa microphylla, 372
Tartarian dogwood, 165
Viburnum, 346
Warminster broom, 133
Wintergreen barberry, 228
Witch-alder, 225

ERECT OR PROSTRATE VINES

Chinese wisteria, 76
Clematis montana, 58
Cotoneaster, 285
Cross-vine, 54

cont'd on back endpapers

FLORAL TIMETABLE
cont'd from front endpapers

Cytisus, 131
Italian honeysuckle, 64
Japanese wisteria, 75
Trumpet honeysuckle, 64

June

TREES

Amur lilac, 371
Black locust, 139
Catalpa, 357
Cornus Kousa, 162
Epaulette-tree, 420
Evergreen magnolia, 275
Japanese maple, 170
Kentucky coffee-tree, 127
Laburnum watereri, 136
Linden, 190
Magnolia sieboldi, 275
Pomegranate, 166
Red buckeye, 146
Rowan tree, 317
Scotch laburnum, 136
Storax, 395
Tulip-tree, 189
Yellow-wood, 141

SHRUBS

American elder, 330
Beauty-bush, 339
Berberis, 225
Buddleia, 366
California barberry, 297
Cherry laurel, 284
Cotoneaster, 285
Damask rose, 305
Deutzia, 178
False spirea, 315
Gravelweed, 337
Gray dogwood, 164
Great laurel, 409
Harrison's yellow rose, 304
Hungarian lilac, 371
Hybrid rhododendrons, 410
Hydrangea macrophylla, 184
Lonicera heckrotti, 383
Magellan barberry, 227
Nandin, 296
Osmanthus ilicifolius, 334
Philadelphus, 157
Provence rose, 304
Salt cedar, 270
Scotch rose, 303
Sheep laurel, 413
Spiraea cantoniensis, 248
Stephenandra, 243
Strawberry-shrub, 165
Sweetbrier, 307
Syringa, 369
Twisted heath, 392
Viburnum, 346
Weigela, 338
White swamp azalea, 403
Witherod, 353
Woadwaxen, 130

VINES

Blue-blossom, 241
Clematis jackmani, 58
Common white jasmine, 55
Dutchman's-pipe, 66
Golden clematis, 57
Italian jasmine, 329
Japanese honeysuckle, 63
Matrimony vine, 394
Vine bower, 56

July

TREES

Chaste tree, 325
Evodia danielli, 299
Golden-rain tree, 299
Pagoda tree, 142